Of Climate and Gender

Anja Zürn

Of Climate and Gender

An Ecofeminist Hegemony Analysis of the Indian and EU Climate Policy Discourses and Identities

Springer VS

Anja Zürn
Würzburg, Germany

Dissertation, University of Würzburg

ISBN 978-3-658-46773-9 ISBN 978-3-658-46774-6 (eBook)
https://doi.org/10.1007/978-3-658-46774-6

This Springer VS imprint is published by the registered company Springer Fachmedien Wiesbaden GmbH, part of Springer Nature.
The registered company address is: Abraham-Lincoln-Str. 46, 65189 Wiesbaden, Germany

To all the remarkable feminist scholars, thinkers and activists upon whose shoulders the climate movement stands today

Acknowledgements

Writing a book begins with an idea and the courage to pursue it, but it takes many more resources and, most importantly, the support of countless individuals to bring it to life. To all those who contributed in various ways—whether through encouragement, collaboration, or invaluable insights—I extend my deepest gratitude.

And to all the procrastinators out there who, like me, often find themselves reading the acknowledgments before getting into the actual content—welcome! I'm glad you're here, and I hope this book captures your interest as much as it did mine while writing it.

First and foremost, I owe a deep debt of gratitude to my dissertation supervisor, Prof. Gisela Müller-Brandeck-Bocquet. Thank you for giving me the opportunity to realise this project and for sharing your curiosity, openness, and expertise, which allowed me to merge my two greatest scientific passions—climate and gender. My interest in India goes back to my time studying with Prof. Müller-Brandeck-Bocquet, and I undoubtedly have her to thank that I became so taken with the EU! I am also incredibly grateful to Assistant Prof. Ulrike Zeigermann for her invaluable second supervision, which provided critical insights and fruitful discussions. My thanks extend to Prof. Elke Wagner, who served as the third examiner, and Prof. Christiane Gross, who chaired the defence committee—your contributions were deeply appreciated. Additionally, I wish to thank Dr. Christoph Mohamad-Klotzbach for his encouragement and diligently taking minutes during the defence.

The long journey of writing this book was filled with many highs, along with a fair share of dry spells and challenging lows. I was fortunate to have the support of many remarkable people throughout this process. I must first acknowledge the wonderful former team at the Jean Monnet Chair of European Studies and International Relations at the University of Würzburg. Thank you for the countless hours spent in theoretical discussions, reflecting on my findings, and offering much-needed moral support. Special thanks go to Dr. Manuela Scheuermann and Timo Lowinger for their critical advice and encouragement. Manuela's unwavering support kept me going, while Timo's persistent nudging finally convinced me to take the plunge and commit to this dissertation with a sophisticated theoretical approach, which he never tired of discussing and reflecting on.

I am also grateful to Dr. Carolin Rüger for the insightful discussions on my EU-related findings and to Johannes Greubel for reviewing some of the chapters and for his highly appreciated feedback. Additionally, thanks to my other former colleague Marco Zeiss for his never-ending encouragement and Catharina Crasser, my then research assistant for her help in research, data preparation and organisational relief, which is so important when you are giving numerous seminars and taking on special tasks in addition to your doctoral thesis. Doris Weth also provided crucial support, offering not only invaluable advice on administrative matters but also endless reassurance. My gratitude also extends once more to Timo Lowinger and to my former colleague Manuel Pietzko, for our writing group, which provided both motivation and a space to discuss the findings and tackle the inevitable challenges along the way.

One person from the former team deserves special mention: Dr. Philipp Gieg. Thank you for being a great office-mate, a steadfast friend, and for your unwavering support as we grew even closer over time. Your patience, encouragement, and well-timed distractions were crucial in helping me push through the project and to the finish line. And of course, without your help, the layout would be far less appealing.

Several others played key roles in the development of this book. I want to thank Mandy Gratz and Lydia Kray, whose contributions at a joint panel discussion back in 2017 inspired me to dive deeper into ecofeminism—your input laid the foundation for a rich and productive theoretical journey. I also learned a great deal from my mentor, Prof. Doris Fuchs, and I am grateful for the opportunity to have discussed my project during your PhD day. Throughout the years of writing, the support of the Skype group formed after joining the women's mentoring programme at the University of Würzburg was invaluable. Thank you, Dr. Katrin Anneser, Dr. Franziska Fröhlich and Assistant Prof. Dr. Antonia Mayr for your encouragement, motivation, and sisterhood—we did it together, even though I was the last to finish! However, I hope you miss the penguin gifs as much as I do.

Catherine Bennewitz, your meticulous copy-editing brought the linguistic precision this project needed, and I am extremely thankful for your patient and accurate work over the years.

I also extend my gratitude to everyone I met at various conferences who engaged in thoughtful discussions and provided critical feedback. I must also thank the interviewees who generously shared their insights during my research trip to India in autumn 2019, offering valuable perspectives on the gendered structures within the Indian climate discourse. This trip was made possible by the BMBF as part of the Indo-German-Partnership project "The Foundations of the Indo-European Strategic Partnership", for which I am very thankful, as I am also for the M.S. Merian – R. Tagore International Centre of Advanced Studies 'Metamorphoses of the Political' (ICAS) for its support, which included conference participation and fostering the Indo-German network. The copy-editing of this book was primarily made possible through the support of the Faculty of Human Sciences at the University of Würzburg, whom I would like to thank. The other part of funding for the copy-editing was provided by the former Jean Monnet Chair—I would like to extend my thanks to Dr. Manuela Scheuermann for this opportunity.

Lastly, I want to acknowledge my incredible friends and family as well as my new colleagues of the Würzburger Stabsstelle Klima und Nachhaltigkeit. They all kept cheering me on, pulling me out of the occasional low points, and always striving to lift my spirits. My heartfelt thanks go out to Julia Kazmaier, Esther Schießer, Marc Suhr, Theresa Wiedenhöfer, Fabienne Uhl, Hannah Lipp, Hanna Herzog, Reena Ngurang, Dr. Riccardo Altieri, Isabella Albert, Daniel Gaittet, Anna-Lena and Simon Franke, Christian Gäbele, Flora Tietgen, Martin Zürn (especially for the extrinsic motivation you provided), Moni Stenzenberger, Irmgard Bronner, Sonja Manteufel, Michael Zürn, Michi Brändle, and my parents.

Thank you all!

Würzburg
in the summer of 2024

Anja Zürn

Contents

About the Author

Anja Zürn is a Climate Mitigation Manager and Political Scientist. She holds a Ph.D. from the University of Würzburg where she was a Researcher and Member of Faculty at the Jean Monnet Chair of European Studies and International Relations, Institute of Political Science and Sociology. In her research, Anja Zürn focuses on global climate change policy and poststructuralist as well as feminist theories.

Abbreviations

AIWC	All Indian Women's Conference
ALDE	Alliance of Liberals and Democrats for Europe
AR	Assessment Report
BASIC	Brazil, South Africa, India and China
BJP	Bharatiya Janata Party
BRICS	Brazil, Russia, India, China, South Africa
CAA	Citizenship Amendment Act
CAN	Europe Climate Action Network Europe
CBAM	Carbon Border Adjustment Mechanism
CBDR-RC	Common but Differentiated Responsibilities and Respective Capabilities
CCS	Carbon Capture and Storage
CDM	Clean Development Mechanism
CDRI	Coalition for Disaster Resilient Infrastructure
CE	Circular Economy
CEAP	Circular Economy Action Plan
CJ	Climate Justice
COP	Conference of the Parties
CSE	Centre for Science and Environment
DG	Climate Directorate-General Climate
DG	Directorate-General
DG REGIO	Directorate General for Regional and Urban Policy
DRR	Disaster Risk Reduction
EC	European Commission

ECCC	Executive Committee on Climate Change
ECCP	European Climate Change Programme
ECF	European Climate Foundation
ECL	European Climate Law
ECP	European Climate Pact
ECR	European Conservatives and Reformists
EEA	European Environment Agency
EFDD	Europe of Freedom and Direct Democracy
EGD	European Green Deal
EIGE	European Institute for Gender Equality
EL	European Left
ENF	Europe of Nations and Freedom
ENGOs	Environmental Non-Governmental Organisations
ENV	Environment Council configuration
ENVI	European Parliament Committee on the Environment, Public Health and Food Safety
EP	European Parliament
EPP	European's People Party
ETS	Emissions trading System
ETUC	European Trade Union Confederation
EU	European Union
EUCO	European Council
FFF	Fridays for Future
FICCI	Federation of Indian Chambers of Commerce and Industry
FotEE	Friends of the Earth Europe
FPE	Feminist Political Ecology
G20	Group of 20
GDP	Gross Domestic Product
GE	Green Economy
GGG	Green Growth Group
GHG	Greenhouse Gas
GN	Global North
GoI	Government of India
GPI	Greenpeace India
GS	Global South
HAC	High Ambition Coalition
IBSA	India, Brazil, South Africa
INC	Indian National Congress
INCCA	Indian Network for Comprehensive Climate Change Assessment

IPCC	Intergovernmental Panel on Climate Change
ISA	International Solar Alliance
JI	Joint Implementation
JTF	Just Transition Fund
KP	Kyoto Protocol
L&D	Loss and Damage
LDCs	Least Developed Countries
LULUCF	Land Use and Land Use Change and Forestry
MDGs	Millennium Development Goals
MEF	Major Economies Forum on Energy and Climate
MEP	Member of the European Parliament
MoEF	Ministry of Environment and Forests
MoEFCC	Ministry of Environment, Forest and Climate Change
MP	Member of Parliament
MRV	Measurement, Reporting and Verification
MS	Member State
NAPCC	National Action Plan on Climate Change
NDCs	Nationally Determined Contributions
NGOs	Non-Governmental Organisations
PA	Paris Agreement
PMCCC	Prime Minister's Council on Climate Change
S&D	Progressive Alliance of Socialists and Democrats
SAPCCs	State Action Plans on Climate Change
SBI	Subsidiary Body for Implementation
SD	Sustainable Development
SDGs	Sustainable Development Goals
TERI	The Energy and Resources Institute
TFEU	Treaty on the Functioning of the European Union
UNEP	United Nations Environment Programme
UNFCCC	United Nations Framework Convention on Climate Change
UNGA	United Nations General Assembly
VP	Vice President
WIM	Warsaw International Mechanism for Loss and Damage
WMO	World Meteorological Organisation
WRI	World Resources Institute
WWF	World Wide Fund for Nature

List of Figures

List of Tables

1 Introduction

"Leaving gender out of attempts to conceptualize, examine, and correct environmental problems means that our understanding is deeply inadequate and our solutions at the least partial." (Sturgeon 2017, *21)*

In the midst of a tumultuous era marked by an unprecedented confluence of crises, humanity finds itself grappling with the intricate interplay of resurfacing war in Europe, the aftermath of the global Covid pandemic and an alarming surge in right-wing ideologies. These multiple crises are compounded by the ominous shadow cast by the almost inconceivable threat of climate catastrophe—an existential crisis that looms large as the paramount challenge of our time.

Faced with the complex puzzle of simultaneous adversities, the world has established a reliance on summits and rituals to try to combat climate change. Whether in the annual gatherings of the Conferences of the Parties (COPs) or in the sophisticated political dialogues unfolding at national and regional levels, one single question pervades every discussion: How to deal with climate change? What steps and measures do we need to take to contain the climate catastrophe and to adapt where the harm is irreversible? This question is answered differently in different regions of the world—depending on the context—and yet the responses are similar in one crucial respect: It seems that technical solutions and economic incentives are perceived as almost the only adequate answers to the climate crisis. But why do we perceive them as more important than, for example, sufficiency or saving greenhouse gases through stricter regulation of fossil fuels? To understand this, we need to tap into the climate discourses of key players, shaped within and through different power structures, which in turn might also be influenced or stabilised by the shared ideas and norms of our societies, such as gender structures.

A. Zürn, *Of Climate and Gender*, https://doi.org/10.1007/978-3-658-46774-6_1

1.1 Object of Study and Research Interest

This project analyses how the discursively produced identities of two of the most important state climate policy actors, namely India and the European Union (EU), are constructed, whether and how they are gendered by implicit power structures and what this means for the respective climate policies and, beyond that, for global efforts to stem the climate crisis.

As "the way in which we think, write, and talk about the environment has important implications for how it is governed" (Coffey 2016, 203), hegemony analysis will be applied because it provides interesting insights in trying to reconstruct how specific ideas and answers grow in the attempt to combat climate change. We will build on poststructuralist theory, which assumes that the answers offered in trying to manage climate change are formulated from the discursively produced identity of an entity. Power structures, or more precisely hegemonies, play an important role here, as it is they that make discursively negotiated responses to climate change appear more—or less—important and valid.

The last few decades of progress and research have produced a variety of different approaches and answers to counter and contain the climate catastrophe, that follow the idea of sustainable development (SD). Conceptually, a distinction can be made between different approaches of SD: A general distinction can be made between weak and strong sustainability, as the concept of SD "has resisted a single, accepted interpretation" (Dietz and Neumayer 2007, 617). The concept of weak sustainability favours the central aspects of ecological modernisation such as technological innovation and increased efficiency in securing the earth's climate, assuming that the natural resources are to some extent substitutable (Neumayer 2013, 22–23). On the other hand, strong sustainability recognises the planetary boundaries, not expecting that the resources extracted for growth can be replaced. Accordingly, strong sustainability aims its focus on alternative economic approaches, rejecting the growth paradigm and relying on sufficiency (Zeigermann Forthcoming) and the precautionary principle (Neumayer 2013, 109–110, 176). Depending on which sustainability approach is followed, different responses to the climate crisis are favoured. One assumption as to why some responses to climate change are constructed as hegemonic, while others do not make it to the top of the global to-do list, is that the hegemonic order of our societies—structured inter alia by gender stereotypes discursively designed by drawing on ancient dichotomies—as elaborated by ecofeminism (MacGregor 2010), often criticises the current economic system both for dominating nature and for its androcentrism.

Although we know today that the consequences of climate change are gendered (Dankelman 2002; Denton 2002; Ahmed and Fajber 2009), as studies for instance show that women caring for the families' daily needs in the Global South are hit the hardest by climate change (Enarson 2012; Resurrección 2017), we are still in our infancy when it comes to understanding how structures that produce climate governance and policy follow "a gendered logic of appropriateness" (Kronsell et al. 2023, 226). As we live in a patriarchal world, which is characterised in particular by the so-called 'male gaze' through which also climate policy is formulated (Flavell 2023, 4–6), the patriarchy has an impact on climate policy (Tanyag 2023); some even speak of an androcentric climate policy (Spitzner et al. 2020) seeing masculinised perspectives as a benchmark in problem definition, the development of theoretical assumptions and empirical investigations (Spitzner et al. 2020, 50, 51). From these interconnections we can deduce that "governmental rationalities underpinning environmental decision-making and subjectification are gendered" (Foster 2017, 216). Pointing to "the hegemony of masculinity in (global) politics, including climate politics" (Wilson and Chu 2020, 1099), we can search within entities' politics and policy for these discursively produced hegemonies and their impact regarding global climate policy and answers formulated to combat climate change. One aim of this project is to reconstruct the identities discursively created by following an offensive-hegemonic strategy as employed by the two players.

But why are we focusing on India and the EU? The rationale behind this choice is rooted in the divergent yet to some extent interconnected nature of these entities. The two largest democracies in the world—although India has experienced a decline in democratic norms in the latter part of the period under review—are the third (India) and fourth (EU) largest producers of greenhouse gas (GHG) emissions after China and the USA (World Resource Institute 2023). Hence, their climate policy can have a strong impact on the overall global climate. Indeed, we are living in a time of some progress in climate policy and yet it is currently not enough to meet the 1.5-degree target. India, as a rapidly developing nation of the Global South, grapples with the challenge of reconciling economic growth imperatives with the pressing need for sustainable development practices while at the same time, ranking within the 10 countries most affected by climate change worldwide in 2020 (Eckstein, Künzel, and Schäfer 2021, 8). On the other side, the EU is increasingly affected by climate change within its own borders, while it has been pursuing an active climate policy for a long time and is trying to harmonise this with an enormous economic programme. As a regional organisation located in the Global North, the EU finds itself in complex terrain with its climate policy where unity and diversity converge in shaping a collective climate identity comprising diverse institutions and many other actors. In both cases, climate policy is to a large extent also part of foreign policy, which in the case of India is strongly characterised by the executive actors, and in the case of the EU by both legislative and executive actors.

To understand the complex and contested field of climate policy, the motives, justifications and ideas for the respective climate policy must be analysed and thus also the climate discourse, which itself is constituted by power relations and creates the respective identity of India and the EU. This research aims to dissect and analyse a specific aspect of the multifaceted dimensions of identity construction within the context of their respective climate policies: Their gendered structures. Both entities (in the case of the EU some of the future member states) have been participants in global climate talks since their inception in 1972, demonstrating a longstanding commitment to addressing environmental challenges. The EU, renowned as a frontrunner in climate action (see instead of many Oberthür and Dupont 2021), merits examination with a view to unravelling the strategies and policies that have positioned it at the forefront of global climate protection efforts. After all, the bloc's normative power (Manners 2002) may also materialise in climate policy. Simultaneously, India—historically an advocate for the interests of developing nations (Michaelowa and Michaelowa 2012, 576; Gieg 2023, 39–43)—has also evolved into a key player. New Delhi has assumed a crucial role owing to its transition into a major emerging economy and subsequently into an emerging power (Gieg and Lowinger 2015, 5) with the ambition to play "the role of leader of the developing world" (Gieg 2023, 280). Accordingly, both actors potentially have a significant impact on other players. In addition, India and the EU strive to consolidate their efforts, especially within the framework of their Strategic Partnership, established in 2004. Although this partnership of now two decades has not met expectations (Müller-Brandeck-Bocquet et al. 2021, 3), there nevertheless lies substantial potential for collaboration in the field of climate policy (Jayaram 2021; Jörgensen 2021), not least due to the actors' outlined common ground. The object of the study's relevance can therefore not only be derived from the significance of each individual case, but also from the bilateral dimension.

India and the EU are also interesting cases when it comes to the intertwining of gender and the environment or gender and climate. India can be seen as the cradle of ecofeminism, and various studies on the gender/environment relationship have already been carried out, analysing different topics in India (Agarwal 1992; Gururani 2002; Truelove 2011; Shrestha, Joshi, and Clément 2019; Michael et al. 2020). While research has long been absent on the intersection of gender and climate with regard to the EU (MacGregor 2010, 227),[1] the EU can be seen as a pioneer in terms of gender mainstreaming (Debusscher and Manners 2020), although there is criticism of the implementation (Vida 2021; Allwood 2014). As the analysis will show, these aspects are also important with regard to the gendered climate identity. However, the analysis of these actors not only offers insights into their climate policy and how it became what it is via powerful discourses but also contributes to a broader understanding of global efforts towards mitigating climate change, as the final chapter will show.

Thus, the relevance of this work is not solely derived from the very extensive discourse analyses—never before available in such detail—that untangle India's and the EU's identities in the context of their respective climate policies, but rather, also stems from the importance of the findings for the global stage. A further focus is the nexus of gender and climate, and the question of how seemingly omnipresent—yet at the same time only implicitly articulated—gender structures shape climate policy is another focus. Moreover, the application of Martin Nonhoff's hegemony analysis and its expansion to include Judith Butler's theory of subjectivation on the one hand and the ecofeminist lens on the other, represent an innovative approach to the topic. Following on from some studies on the discourse analysis of environmental and climate policy (cf. Sect. 1.3), the approaches are combined in this way for the first time, expanded and utilised in practice.

1.2 Research Questions

The aim of this study is to reconstruct the climate policy identities—conceptualised here as being constituted and represented by the respective societies (cf. Sect. 2.2.3)—of India and the EU and discuss them through an ecofeminist lens, trying to find out what impact gendered structures have on the entities' climate policies. Based on the poststructuralist assumption that the foreign policy identity of an entity such as a state or regional organisation determines its actions (Hansen 2012, 99), for instance its policy formulation, the reconstructed identity is the key to uncovering and critically reflecting India's and the EU's climate policy courses. The idea that gender structures play a significant role in the discursive process formulating potential answers in the fight against climate change is one of the theses pursued in this project. It necessitates a comprehensive reconstruction of

[1] Individual studies have emerged in the last ten years or so (cf. Allwood 2014; Magnusdottir and Kronsell 2015), but for a long time the 'classic' object of investigation was the Global South.

the two climate discourses, as we need to know the content of the discourses in order to be able to critically reflect on their structures (Nonhoff 2019, 100).

Three research questions are derived from the general interest in the gendered climate policy identities of India and the EU:

(1) What demands and strategemes are present in the climate discourses of India and the EU within the research period from 2010 to 2021 and how do they construct the respective climate policy identity?
(2) Which gender aspects are part of the respective climate discourse regarding both gendered subject positions and gendered patterns justifying the responses to climate change? Can typical or frequent patterns be identified?
(3) What is the significance of the gendered identities for the EU's and India's climate policies? What implications do they have at bilateral and global level?

In order to consider these questions within an appropriate time frame, the period from 2010 to 2021 was chosen. The research spans a duration of twelve years, driven by several considerations. As discourses normally change only slowly or due to disruptive stimuli (cf. Sect. 2.2.1), a twelve-year study period enables the deconstruction of medium-term developments, for example due to changes in government, newly emerging actors such as civil society movements or disruptive political developments on the global stage. It is therefore no coincidence that 2021 was chosen as the last year of the study period, ensuring that any potentially drastic changes in the discourse due to the global consequences of the Coronavirus pandemic can also be captured. As the study will show, there are further aspects of the domestic political debate about the right climate policy that make this period of investigation particularly interesting. On the EU side, the bloc's renewed ambitions caused by the European Parliament election's outcome and the huge impetus of the European Green Deal must be mentioned in this context, and on the Indian side New Delhi's increasing aspirations in the field of renewable energies—not least due to economic needs—and Union-level elections in 2014 which brought Narendra Modi to power. Of course, the internal and external issues of the respective climate discourse are almost endless; many of them are addressed in detail in the analysis, depending on their contribution to the gendered structures of the respective climate policy identity.

While the Indian climate discourse has already been analysed by different scholars with regard to its changing role in the time up to and in Copenhagen 2009 (though the following examples are not poststructuralist: Michaelowa and Michaelowa 2011; Betz 2012; Isaksen and Stokke 2014; Thaker and Leiserowitz

2014; Jäger 2018), the discourse up to Paris and beyond is only slowly becoming the subject of research (Sengupta 2019; Swarnakar, Shukla, and Broadbent 2022). It is therefore of particular interest to analyse the Indian discourse after its first impressive changes prompted by the COP15 events in 2009. Starting the research period in 2010 is also promising regarding the EU and its Copenhagen disappointment, as it was a turning point of the EU's climate discourse concerning both the international and the domestic stage. Therefore, subsequent changes in the EU's reasoning can be expected, including with regard to the implicitly anchored gender structures.

To analyse the discursively produced identities of India and the EU, Martin Nonhoff's poststructuralist hegemony analysis will be applied. Compared to other meta-theories such as constructivism, poststructuralism offers the advantage that it allows the "deep-seated" identities to be determined, detecting the general lines—in our case with the help of coding—instead of focusing on the concrete words used, as most approaches in constructivism do (Buzan and Hansen 2010, 199). In addition, poststructuralism recognises that the respective actor's identity is "not an entity that can be fully controlled" (Buzan and Hansen 2010, 199). On the one hand this makes the volatile and fragile character of an identity clear and on the other hand it sees a dynamic circle of actors shaping the identity, whereas constructivism usually sees a fairly stable group of actors (Buzan and Hansen 2010, 199).

Nonhoff is one of the well-known members and former students of the Essex School, which is famous for its poststructuralist discourse analysis revealing power relations and hegemonies. Trying to close the Essex School "methodology gap" (Nonhoff 2019, 64), Nonhoff (2006) has formulated a conceptual framework for how hegemony analysis can be implemented in practice through the deconstruction of the so-called strategemes, the smaller units of the offensive-hegemonic strategy. Accordingly, we assume a discursively ontologised world, producing both gendered answers regarding how to combat climate change and gendered subject positions, which are also able to have an impact on the formulated responses. Pursuing the second research question, which asks for the gendered subject positions that shape the discourse, the set of analyses is supplemented by Judith Butler's theory of subjectivation, enabling us to deconstruct this dimension of gendered discourses. But how exactly can we uncover the gendered discourse patterns gendering individuals or groups and responses to climate change and identify the underlying power structures as gender structures?

To achieve this, ecofeminist insights will resonate throughout the entire analysis, because this is the most promising way to reveal the underlying gender structures. As these structures are highly implicit, it is necessary to fall back on the realisation that the world has been constructed dichotomously, which also includes the dichotomisation of gender (cf. Sect. 2.3.3). Hence, I will use the ecofeminist "concept of dualisms [which] is an illuminating means of understanding and deconstructing rhetorical strategies in climate politics" (Flavell 2023, 58). To this end, we must first examine the understanding of gender on which the study is based.

"[G]ender is not just an attribute of individuals, but a system of signification and meaning" (Aggestam and True 2021, 392) and is therefore understood as being negotiated in social, or more precisely, discursive processes, which are indeed also negotiated in a space where power structures operate. Consequently, gender is conceptualised here as a system of signification and meaning determined by and building power structures which materialise in various discourses for sorting the world along different gendered (and mostly dichotomised) axes of power. Thus, gender "affects actors' self-understanding and perceived interests (…) defining identity, the home, public and private, and the nation-state" (Aggestam and True 2021, 392). Thus, a co-construction of gender must be assumed. Gender is discursively produced, while it also discursively influences various phenomena such as the identity of an entity. As will be shown later, when it comes to entities such as states or regional organisations, this happens most often in a highly implicit, almost automatic way, reproducing "the current gender order" (Kronsell 2019, 113) without naming it as such. Therefore, gender does not often become visible as a normative ideal, but as "a system structuring power relations" (Gaard 2015, 22) (re-)produced by society over time, often in the form of decades- or centuries-old stereotypes.

These mostly implicit patterns can only be traced by means of a theoretical approach that is aware of these social structures and makes it possible to both reveal and constructively criticise them. This is why ecofeminism is used as an additional foil to discuss the findings of the hegemony analysis. When using "the feminist tool of critical deconstruction of binaries" (Flavell 2023, 57) there is a particular danger that deconstructing the dichotomously structured world reproduces it. It is therefore all the more important to later focus on the responses formulated by ecofeminists which aim to overcome these dichotomies.

Applying ecofeminism as a normative foil and as part of the analytical set used in this project means, in concrete terms, to "frame the analysis of power within structures of patriarchy, colonialism and capitalism, and the role of institutions and policies in reproducing that power with an explicitly social, gender (…) and climate justice perspective" (Khoury 2023, 2).

1.3 State of Research and Own Contribution

Addressing the multidimensional research questions of this project, it is necessary to first analyse where research stands in the field of poststructuralist discourse analysis concerning both the climate policy identity of the cases examined and the major steps that have brought International Relations (IR) to the table. The research gaps and needs are identified by examining the current state of research and showing how this project contributes to the complex web of gendered climate policy identities of India and the EU. The state of research in ecofeminism is only mentioned in passing in this chapter as the theoretical framework of this work is later derived from its genesis and state of research (cf. Sect. 2.3).

With the emergence of poststructuralist studies in the course of the linguistic turn, but also in the course of a general critical turn in IR (Zehfuss 2013), some pioneering scholars in IR used this approach to generally criticise the then-prevailing neorealist approach as insufficient and neglecting the subjective and ideational aspects that shape international behaviour (Ashley 1984; DerDerian and Shapiro 1989; Walker 1990). Later poststructuralist analyses focused on the language and practices of diplomacy (Der Derian 1991), the connection between gender and international politics (Enloe 1989) or gender and international security practices (Shepherd 2008), poststructuralist methodologies in IR (Milliken 1999), and the constitution of identity (Campbell 1992; Herschinger 2011) and security in the context of foreign policy (Hansen 2006)—thereby establishing a robust strand of poststructuralist foreign policy analysis.

Today, poststructuralist research in IR encompasses a relatively broad field of topics, (Herschinger and Renner 2014; Çalkıvik 2017; Herschinger and Renner 2017) theories and methods (Angermuller et al. 2014; Herschinger 2017; Nabers 2023). Over time, poststructuralism has gained a certain amount of hard-won recognition in the discipline (Diez 2014, 382–85). Beyond foreign and security policy, researchers have increasingly turned their attention to environmental and climate policy, employing poststructuralist approaches. The first works were a critical reflection of ecological modernisation (Hajer 1995), the parallelism between a transforming governmentality and a changing environmental discourse (Oels 2005) or the international anti-whaling discourse (Epstein 2008). After about a decade of research, several reflexive works have emerged that provide an insight into this lively field (Feindt and Oels 2005; Hajer and Versteeg 2005). Later, studies regarding 'climate protection' as world politics' empty signifier (Methmann 2010), different case studies using interpretative approaches to global climate governance (Methmann, Rothe, and Stephan 2013) and discourse analytical research regarding the Anthropocene (Rothe 2020) all followed.

Today, poststructuralist discourse research on international and global environmental and climate issues is broadly anchored (Leipold et al. 2019; Methmann, Rothe, and Stephan 2013). Nevertheless, there remain research gaps relevant to this study: The Indian Union's climate policy identity has to date only been the subject of sporadic poststructural research (Pathak and Parris 2021; Prys-Hansen 2022). Hence, this study will contribute to this rather young field of research. On the other hand, the EU's foreign policy identity has been widely analysed (Wæver 1996; Neumann 1998; Diez 1999, 2004; Youngs 2004; Green 2007; Krzyżanowski 2010; Lönnendonker 2016; Sarısakaloğlu 2019). Its climate policy identity has also been analysed regarding both the time before (Krzyżanowski 2015) and after the Paris Agreement (Wendler 2022). But most of the studies examining the EU's role or identity in climate policy almost exclusively focus on the question of the EU's leadership in global climate governance (see instead of many Oberthür and Dupont 2021). However, none of the studies mentioned used an additional ecofeminist lens to research the EU's climate policy identity with a focus on gendered discourse patterns and reconstructing the potentially underlying gendered structures which produce and stabilise the identity. This study addresses both desiderata: The EU's climate policy identity beyond the question of leadership, and the missing ecofeminist lens. In the case of India, the situation is similar: Feminist perspectives are increasingly being used to analyse India's foreign policy (Gopi 2023; Khullar 2023), particularly when it comes to the implementation of the global Women, Peace and Security agenda (Kumar Singh 2010; Lahiry 2019; Manchanda 2020), but not regarding the climate policy (identity) of the Indian Union. In the Indian context, (eco-)feminist research refers rather to more narrow case studies such as specific regions or environmental problems (cf. Chap. 1.1). Addressing this shortcoming is also a goal of the project.

Furthermore, Sina Leipold et al. (2019, 456) note that there is a large research gap regarding the investigation of multiple cases within one study analysing entities' environmental or climate-change policies; this project aims to tackle that void. Although the project at hand is decidedly not aiming for a comparison of India's and the EU's gendered climate identities, I will nevertheless address this lacuna by examining whether the chosen theoretical approach is suitable for analysing the respective gendered identity and its meaning of both a nation state and a regional organisation. This is interesting because Nonhoff (2006, 231) states that relatively stable limits in the discourse are essential in order to install the successful offensive-hegemonic strategy, which we are searching for in the later analysis. Since the existence of these stable limits can be assumed within a nation state but not necessarily in the case of a regional organisation, it may become necessary in order to also make the ecofeminist hegemony analysis fruitful as an analytical framework in the case of the EU, to adapt or develop it further. At the same time there is a tension in how the approach is applied: Whereas hegemony analysis was conceived for liberal democracies, India has undergone contrary developments in this respect in recent years (Gieg and Müller-Brandeck-Bocquet 2024).

Another research gap identified by Leipold and her colleagues (Leipold et al. 2019, 457) concerns the combination of different theoretical approaches to environmental and climate discourses. She calls for the now-widespread approaches, which mostly follow either the Critical Discourse Analysis, the poststructuralist discourse analysis of Michel Foucault, or that of Ernesto Laclau and Chantal Mouffe, to be combined, while pointing out the danger of eclecticism (Leipold et al. 2019, 457). I will follow this recommendation insofar as I extend the hegemony analysis to include the poststructuralist ecofeminist approaches. Trying to avoid the danger of eclecticism, Nonhoff's entire set of analysis will nevertheless be applied (cf. Sect. 2.4.1), albeit supplemented by the ecofeminist lens and Butler's theory of subjectivation. This is precisely where I start with the following study, in order to be able to work out not only the hegemonies and the identity that arises from them, but also their gendered basic structure.

The combination of hegemony analysis and ecofeminism also enables a further innovation in research: The necessary concretisation of the concept of power in poststructuralist research, which has often been called for (Leipold et al. 2019, 445; Ziai 2012), is to be addressed by expanding it to include the gender perspective. Although this does not explicate the concept of power, I am taking important steps towards giving it a more concrete shape.

The state of research in ecofeminist studies will be reviewed later (cf. Sect. 2.3), however, at this point, it is important to note that ecofeminist research has so far not focused on state entities' identities (Aggestam and True 2021, 388) and that there is an urgent need for ecofeminist research on climate policy discourses and its power structures (MacGregor 2010)—not least as "previous research in the field on climate governance has made scant use of critical feminist work and paid insufficient attention to gender as a power relation an a social construct" (Kronsell 2019, 107). Together with other recent studies (Tanyag 2023; Flavell 2023), this project will serve this need for research, since "in climate change studies, gender is mostly handled in a men-versus-women dichotomy and little or no attention has been paid to power and social and political relations" (Djoudi et al. 2016, 248). Overall, Sherilyn MacGregor (2010, 229) identified a great need for such a research direction, stressing that "a feminist interrogation of how (…)

the politics of climate change are gendered seems to be an important addition to—and perhaps as a corrective to—the existing 'gender impacts' research", calling for the use of "the tools necessary for digging down and pulling up the deep roots of the discourses that frame gender and climate politics everywhere" (MacGregor 2010, 236). This study can be seen as an attempt to identify and apply the tools necessary to pull up the gendered nature of Indian and EU climate policy.

To sum up, this work stands on the shoulders of decades of ecofeminist research as well as poststructuralist endeavours in IR in general and the hegemony analysis approach in particular. In doing so, this project contributes to the further dissemination of the now widely recognised approach of poststructuralist discourse analysis by linking it to the recently rediscovered ecofeminist thinking and applying this combination of ecofeminist hegemony analysis for the first time—and in parallel—to two cases. This not only provides information on the significance of gendered identities for the respective climate policy but also allows the applicability of the approach to be delimited in relation to nation states and regional organisations. In addition to the fundamental findings on the question of how the respective identity is gendered, the two discourse analyses represent independent deconstructions of the respective climate discourse, which do not yet exist in this detail. In this respect, they also form an important contribution to discourse-analytical climate policy research.

1.4 Scientific Understanding and Positionality

In the pursuit of scientific understanding within this dissertation, the poststructuralist lens of hegemony analysis combined with ecofeminism stands at the forefront. Adopting a postpositivist stance, the applied hegemony analysis is not intended to reconstruct causalities, but rather to problematise and question the processes that have led to the current state of affairs. As aptly articulated by Thomas Diez (2013, 205):

> *"The purpose of the study of hegemony is to problematize how we got to where we are, and how certain concepts become seen as natural and logical. This then opens the debate for discussions about alternative futures (...) but it also serves to help understand better the present results of our past."*

The acknowledgment of discursive constructions and the absence of a deterministic approach characterise this work as an integral part of postpositivist theory-building, challenging established narratives and paving the way for critical discussions on alternative trajectories (Campbell 2010, 228) in the realm of climate policy.

In the pursuit of hegemony analysis within the context of my research, a primary focus lies on unravelling the intricate web of gender structures. Drawing on ecofeminist convictions that echo within the realms of poststructuralism, the investigation delves into the belief that the world operates on a dichotomous order (cf. Sect. 2.3.3). This order encompasses gendered dichotomies, thereby potentially creating an impression during the reconstruction of this order that the world is neatly divided into masculinised and feminised spheres. However, the overarching objective is to elucidate that the world is constructed in this manner rather than existing ontologically in such binary terms. It is crucial to navigate the peril inherent in a poststructuralist discourse analysis, where there is a risk that the interpretation may lean towards an assumption that I endorsed the inherent truth of this constructed world, rather than emphasising that the world is shaped through discourse and that certain groups benefit from intrinsic hegemonies. Hence, a foundational step involves the explicit identification and naming of phenomena, as it is only through this initial act of recognition that a subsequent phase of critical examination and ecofeminist problematisation becomes feasible. The mostly implicitly-articulated gender structures (Spitzner 2020, 17) must therefore be broken up in their dichotomy and assigned to the sphere connoted as male or female. However, this harbours two challenges: One, this exercise reproduces the dichotomies that I reject and yet that still persist. Two, these dichotomies which are derived from the time of the Western enlightenment (cf. Sect. 2.3.3) are presented as universally valid—although they of course are not, but were rather often imposed on other cultures and states in the course of colonisation, as in the case of India.

In undertaking this poststructuralist exploration, it is imperative to situate my own positionality within the broader context of the research. As a white, non-disabled Western woman engaging in this inquiry, I am acutely aware that I do not directly experience some of the power structures under scrutiny and, as Sherilyn MacGregor (2010, 227) put it, that "academic work can contribute to the silencing and misrepresentation of people in the developing world, despite our best intentions". Recognising this privilege and this pitfall, I approach the subject matter through an extensive analysis of diverse documents, a thorough review of secondary literature and with background talks. Employing poststructuralist perspectives, it is furthermore essential to acknowledge that this critical theory itself emerges predominantly from a Western theoretical tradition.

Given the potential limitations of such a perspective, I seek to mitigate potential biases and broaden the analytical lens by incorporating an ecofeminist approach. This lens is particularly vital when examining the interplay between gender, climate policies, and power dynamics. In doing so, I intentionally draw upon the work of non-Western scholars within the ecofeminist discourse to enrich and diversify the analytical approach of this study. This deliberate inclusion aims to counterbalance the inherent subjectivity of my viewpoint, recognising the necessity of incorporating multiple voices and perspectives.

It is crucial to emphasise that this work is inherently subjective, shaped by my positionality as a researcher. While striving for objectivity through rigorous analysis, the acknowledgment of my identity as a white Western woman underscores the necessity of interpreting this research within the context of its authorship. In embracing this reflexivity, I aspire to contribute to a more nuanced understanding of the complex interrelations between identity, power, and climate policy discourse in both the Indian and European contexts, and furthermore, to apply and make fruitful the perspective of ecofeminism in the pursuit of "global justice and ecological flourishing" (Cuomo 2002, 9), and hence also a good life for all (Hoppe 2019, 263; Gratz and Zürn 2020).

1.5 Outline of the Study

The structural arrangement of this dissertation may not overtly mirror the inductive-deductive methodology employed throughout the research project. However, as will be later exemplified when deducing the strategeme set, the work has been crafted through a continuous process of forward and backward movements applying poststructuralist discourse analysis with its ecofeminist filter in the sense of an explorative project.

The study is divided into five chapters in order both to approach the research questions (cf. Chap. 1.2) and to illuminate all the building blocks that are necessary to answer them. Chap. 2 provides a comprehensive overview of the theoretical foundation and conceptual framework, outlining the hegemony analysis. The multifaceted components of the theoretical framework, including the definition of demands and the understanding of the discourse shaping an entity's identity, subjectivation and ecofeminism, are delineated. Sect. 2.4 shows the concrete steps implementing the theoretical framework and provides a glimpse into the discourse corpus.

The first case, the Indian climate change discourse, is then analysed in Chap. 3. There, the results of the hegemony analysis are critically discussed with the help of the ecofeminist foil. Delving into the specifics of India's climate change governance, Sect. 3.1 first traces the genesis of India's role on the global stage up to 2010 (Sect. 3.1.1), coupled with an exploration of its national climate change governance (Sect. 3.1.2). The ecofeminist hegemony analysis of the Indian case will be applied in order to examine the manifold facets of India's climate discourse (Sects. 3.2.1 to 3.2.7), dissecting the nation's stance on sustainable development, climate justice, and its evolving role in international climate negotiations throughout the research period from 2010 to 2021 alongside the timeline structured by the COPs and the changing political constellations in India. At the end of this part, the deductive nature of the study becomes clear when the theoretical assumptions are expanded on the basis of the findings (Sect. 3.2.8). The gendered Indian-climate-identity will be reflected in the final part of Chap. 3, when I discuss my findings of the hegemony analysis against the backdrop of decades of ecofeminist theory building (Chap. 0).

The second case, the EU's climate discourse and resulting gendered identity is explored in Chap. 4. Before analysing the discourse within the research period, the European Union's climate governance will be reconstructed, tracing its policies and milestones up to 2010 to provide the necessary context for the consecutive analysis (Sects. 4.1.1 to 4.1.3). Subsequently, the EU's climate change discourse is meticulously analysed, spanning its (re-)orientation post-Copenhagen, engagement with the green economy and its positioning before and after Paris (Sects. 4.2.1 to 4.2.5). The final periods of the ecofeminist hegemony analysis deal with the climate discourse unfolding around the European Green Deal, the implementation of which is influenced by the Covid pandemic (Sects. 4.2.6 to 4.2.9). In Sect. 4.3, I will critically discuss the gendered facets of the EU's climate identity, deciphering its evolution through different (leadership) periods and amidst global crises such as the Covid pandemic, reflecting what a gendered identity actually means for the formulation of answers to the climate crisis.

Chap. 5 sums up the findings, synoptically answers the research questions and presents a final outlook. In Sect. 5.1, I will synthesise and contrast against each other the findings from the gendered climate change identities of India and the EU. Sect. 5.2 reflects on the significance of understanding these identities through an ecofeminist hegemony analysis. This entails asking whether the case studies can serve as blueprints for analysing further climate discourses as well as assessing the contribution of this study in light of the research gaps identified earlier. Moreover, the chapter will identify possible strands of further research. The final Sect., 5.3, concludes the study by addressing the significance of gendered climate identities at the bilateral and global levels.

Ultimately, the aim of this work is to show why it is rewarding to analyse the gendered components of an identity that influences the respective climate policy and to see how an understanding of these can contribute to a different global climate policy in the sense of strong sustainability formulating answers to tackle the climate catastrophe, while respecting the irreversibility of the ecological system.

References

Agarwal, Bina. 1992. "The Gender and Environment Debate: Lessons from India." *Feminist Studies* 18 (1): 119–58. https://doi.org/10.2307/3178217.

Aggestam, Karin, and Jacqui True. 2021. "Political leadership and gendered multilevel games in foreign policy." *International Affairs* 97 (2): 385–404. https://doi.org/10.1093/ia/iiaa222.

Ahmed, Sara, and Elizabeth Fajber. 2009. "Engendering adaptation to climate variability in Gujarat, India." *Gender & Development* 17 (1): 33–50. https://doi.org/10.1080/13552070802696896.

Allwood, Gill. 2014. "Gender Mainstreaming and EU Climate Change Policy." *European Integration online Papers*, 1–26. https://doi.org/10.1695/2014006.

Angermuller, Johannes, Martin Nonhoff, Eva Herschinger, Felicitas Macgilchrist, Martin Reisigl, Juliette Wedl, Daniel Wrana, and Alexander Ziem, eds. 2014. *Diskursforschung: Ein interdisziplinäres Handbuch.* Sozialtheorie. Bielefeld: Transcript-Verl.

Ashley, Richard K. 1984. "The poverty of neorealism." *International Organization* 38 (2): 225–86. https://doi.org/10.1017/S0020818300026709.

Betz, Joachim. 2012. "India's Turn in Climate Policy: Assessing the Interplay of Domestic and International Policy Change." GIGA Working Papers 190.

Buzan, Barry, and Lene Hansen. 2010. *The Evolution of International Security Studies.* Cambridge: Cambridge University Press.

Çalkıvik, Aslı. 2017. "Poststructuralism and Postmodernism in International Relations." In *Oxford Research Encyclopedia of International Studies*, edited by Aslı Çalkıvik: Oxford University Press.

Campbell, David. 1992. *Writing security: United States foreign policy and the politics of identity.* Minneapolis: University of Minnesota Press.

Campbell, David. 2010. "Poststructuralism." In *International relations theories: Discipline and diversity*, edited by Timothy Dunne, Milja Kurki, and Steve Smith. 2nd ed., 223–46. Oxford: Oxford University Press.

Coffey, Brian. 2016. "Unpacking the politics of natural capital and economic metaphors in environmental policy discourse." *Environmental Politics* 25 (2): 203–22. https://doi.org/10.1080/09644016.2015.1090370.

Cuomo, Chris. 2002. "On Ecofeminist Philosophy." *Ethics and the Environment* 7 (2): 1–11. https://www.jstor.org/stable/40339033.

Dankelman, Irene. 2002. "Climate Change: Learning from Gender Analysis and Women's Experiences of Organising for Sustainable Development." *Gender and Development* 10 (2): 21–29.

Debusscher, Petra, and Ian Manners. 2020. "Understanding the European Union as a Global Gender Actor: The Holistic Intersectional and Inclusive Study of Gender+ in External Actions." *Political Studies Review* 18 (3): 410–25. https://doi.org/10.1177/1478929919899345.

Denton, Fatma. 2002. "Climate Change Vulnerability, Impacts, and Adaptation: Why Does Gender Matter?" *Gender and Development* 10 (2): 10–20.

Der Derian, James. 1991. *On diplomacy: A genealogy of Western estrangement.* Oxford: B. Blackwell.

DerDerian, James, and Michael J. Shapiro, eds. 1989. *International, intertextual relations: Postmodern readings of world politics.* Lexington: Lexington Books.

Dietz, Simon, and Eric Neumayer. 2007. "Weak and strong sustainability in the SEEA: Concepts and measurement." *Ecological Economics* 61 (4): 617–26. https://doi.org/10.1016/j.ecolecon.2006.09.007.

Diez, Thomas. 1999. "Speaking 'Europe': the politics of integration discourse." *Journal of European Public Policy* 6 (4): 598–613. https://doi.org/10.1080/135017699343496.

Diez, Thomas. 2004. "Europe's others and the return of geopolitics." *Cambridge Review of International Affairs* 17 (2): 319–35. https://doi.org/10.1080/0955757042000245924.

Diez, Thomas. 2013. "Normative power as hegemony." *Cooperation and Conflict* 48 (2): 194–210. https://doi.org/10.1177/0010836713485387.

Diez, Thomas. 2014. "Bedeutungen und Grenzen: Anmerkungen zur Diskursforschung in den deutschsprachigen Internationalen Beziehungen." In Herschinger and Renner 2014, 381–98.

Djoudi, Houria, Bruno Locatelli, Chloe Vaast, Kiran Asher, Maria Brockhaus, and Bimbika Basnett Sijapati. 2016. "Beyond Dichotomies: Gender and Intersecting Inequalities in Climate Change Studies." *Ambio* 45 (Suppl 3): 248–62. https://doi.org/10.1007/s13280-016-0825-2.

Eckstein, David, Vera Künzel, and Laura Schäfer. 2021. "Global Climate Risk Index 2021." https://www.germanwatch.org/sites/default/files/Global%20Climate%20Risk%20Index%202021_2.pdf.

Enarson, Elaine P. 2012. *Women Confronting Natural Disaster: From Vulnerability to Resilience.* Boulder, Colo: Lynne Rienner Publishers. http://search.ebscohost.com/login.aspx?direct=true&scope=site&db=nlebk&db=nlabk&AN=552040.

Enloe, Cynthia. 1989. *Bananas: Making Feminist Sense of International Politics.* Berkeley: University of California Press.

Epstein, Charlotte. 2008. *The power of words in international relations: Birth of an anti-whaling discourse.* Politics, science, and the environment. Cambridge, Mass. MIT Press.

Feindt, Peter H., and Angela Oels. 2005. "Does discourse matter? Discourse analysis in environmental policy making." *Journal of Environmental Policy & Planning* 7 (3): 161–73. https://doi.org/10.1080/15239080500339638.

Flavell, Joanna. 2023. *Mainstreaming Gender in Global Climate Governance.* London: Routledge.

Foster, Emma A. 2017. "Gender, Environmental Governmentality, and the Discourses of Sustainable Development." In MacGregor 2017, 216–28.

Gaard, Greta. 2015. "Ecofeminism and climate change." *Women's Studies International Forum* 49 (1): 20–33. https://doi.org/10.1016/j.wsif.2015.02.004.

Gieg, Philipp. 2023. *India's Africa Policy: Challenges of a Millennia-Old Relationship.* Africa's global engagement: perspectives from emerging countries. Singapore: Palgrave Macmillan.

Gieg, Philipp, and Timo Lowinger. 2015. "Introduction: Exploring Emerging India." In *Exploring Emerging India: Eight Essays.*, 4–9. Würzburger Arbeitspapiere zur Politikwissenschaft und Sozialforschung 7.

Gieg, Philipp, Timo Lowinger, Manuel Pietzko, Anja Zürn, Ummu S. Bava, and Gisela Müller-Brandeck-Bocquet, eds. 2021. *EU-India Relations.* Cham: Springer International Publishing.

Gieg, Philipp, and Gisela Müller-Brandeck-Bocquet. 2024. "Navigating Norms and Pragmatism in EU-India Relations. Multilateralism, Connectivity and the Future of the Strategic Partnership." In *The European Union's Geopolitics. The Lackluster World Power*, edited by Mathias Jopp and Johannes Pollak. Cham: Springer.

Gopi, Rashmi. 2023. "National Security in India's Foreign Policy: A Feminist Reading on Developments from 2004 to 2020." In Ojha and Jaiswal 2023, 83–103.

Gratz, Mandy, and Anja Zürn. 2020. "Naturverhältnisse feministisch denken: Anforderungen an einen aktuellen Ökofeminismus." *Forum Wissenschaft* 37 (3): 22–25.

Green, David M. 2007. *The Europeans: Political identity in an emerging polity.* Studies on the European Polity. Boulder: Lynne Rienner Publishers.

Gururani, Shubhra. 2002. "Forests of Pleasure and Pain: Gendered practices of labor and livelihood in the forests of the Kumaon Himalayas, India." *Gender, Place & Culture* 9 (3): 229–43. https://doi.org/10.1080/0966369022000003842.

Hajer, Maarten, and Wytske Versteeg. 2005. "A decade of discourse analysis of environmental politics: Achievements, challenges, perspectives." *Journal of Environmental Policy & Planning* 7 (3): 175–84. https://doi.org/10.1080/15239080500339646.

Hajer, Maarten A. 1995. *The Politics of Environmental Discourse: Ecological Modernization and the Policy Process.* Oxford: Clarendon Press.

Hansen, Lene. 2006. *Security as Practice: Discourse Analysis and the Bosnian War.* New York, NY: Routledge.

Hansen, Lene. 2012. "Discourse Analysis, Post-Structuralism, and Foreign Policy." In *Foreign Policy: Theories, Actors, Cases*, edited by Steve Smith, Amelia Hadfield, and Timothy Dunne. 2. ed., 94–109. Oxford: Oxford Univ. Press.

Herschinger, Eva. 2011. *Constructing Global Enemies: Hegemony and Identity in International Discourses on Terrorism and Drug Prohibition.* London: Routledge.

Herschinger, Eva. 2017. "Hegemonie und Identität: Ernesto Laclaus Werk in den Internationalen Beziehungen." In *Ordnungen des Politischen: Einsätze und Wirkungen der Hegemonietheorie Ernesto Laclaus*, edited by Oliver Marchart, 127–43. Wiesbaden: Springer VS.

Herschinger, Eva, and Judith Renner, eds. 2014. *Diskursforschung in den Internationalen Beziehungen.* Baden-Baden: Nomos.

Herschinger, Eva, and Judith Renner. 2014. "Einleitung: Diskursforschung in den Internationalen Beziehungen." In Herschinger and Renner 2014, 9–36.

Herschinger, Eva, and Judith Renner. 2017. "Diskursforschung in den Internationalen Beziehungen." In *Handbuch Internationale Beziehungen*, edited by Frank Sauer and Carlo Masala, 313–37. Wiesbaden: Springer.

Hoppe, Katharina. 2019. "Donna Haraways Gefährt*innen: Zur Ethik und Politik der Verwobenheit von Technologien, Geschlecht und Ökologie." *Feministische Studien* 37 (2): 250–68. https://doi.org/10.1515/fs-2019-0026.

Isaksen, Kari-Anne, and Kristian Stokke. 2014. "Changing climate discourse and politics in India. Climate change as challenge and opportunity for diplomacy and development." *Geoforum* 57: 110–19. https://doi.org/10.1016/j.geoforum.2014.08.019.

Jäger, Manuela. 2018. *Indien in den internationalen Klimaschutzverhandlungen*. Marburg: Metropolis.

Jayaram, Dhanasree. 2021. "India's Climate Diplomacy Towards the EU: From Copenhagen to Paris and Beyond." In Gieg, Lowinger, Pietzko, Zürn, Bava, and Müller-Brandeck-Bocquet 2021, 201–26.

Jörgensen, Kirsten. 2021. "Low-Carbon Development: An Idea Whose Time Has Come—Unlocking Climate Cooperation Between India and the EU." In Gieg, Lowinger, Pietzko, Zürn, Bava, and Müller-Brandeck-Bocquet 2021, 185–99.

Khoury, Stefanie. 2023. "A 'lifeline out of the COVID -19 crisis'? An ecofeminist critique of the European Green Deal." *Law & Policy*. https://doi.org/10.1111/lapo.12211.

Khullar, Akanksha. 2023. "Re-envisioning South Asia's Foreign Policy from a Feminist Perspective." In Ojha and Jaiswal 2023, 105–31.

Kronsell, Annica. 2019. "The contribution of feminist perspectives to climate governance." In *Understanding climate change through gender relations*, edited by Susan Buckingham and Virginie Le Masson, 104–20. London: Routledge.

Kronsell, Annica, Gunnhildur L. Magnusdottir, Nanna Rask, and Benedict E. Singleton. 2023. "An Intersectional Exploration of Climate Institutions." In *The Oxford Handbook of Comparative Environmental Politics*, edited by Jeannie Sowers, Stacy D. VanDeveer, and Erika Weinthal, 223–39: Oxford University Press.

Krzyżanowski, Michał. 2010. *The discursive construction of European identities: A multi-level approach to discourse and identity in the transforming European Union*. Frankfurt am Main: Peter Lang.

Krzyżanowski, Michał. 2015. "Discourse analysis, policy analysis, and the borders of EU identity." *JLP* 14 (1): 110–33. https://doi.org/10.1075/jlp.14.1.06krz.

Kumar Singh, Dinesh. 2010. "Women, Security, Peace and Conflict in South Asia." *The Indian Journal of Political Science* 71 (2): 651–61.

Lahiry, Sujit. 2019. "Conflict, Peace and Security: An International Relations Perspective with Special Reference to India." *Millennial Asia* 10 (1): 76–90. https://doi.org/10.1177/0976399619825691.

Leipold, Sina, Peter H. Feindt, Georg Winkel, and Reiner Keller. 2019. "Discourse analysis of environmental policy revisited: traditions, trends, perspectives." *Journal of Environmental Policy & Planning* 21 (5): 445–63. https://doi.org/10.1080/1523908X.2019.1660462.

Lönnendonker, Julia. 2016. *Konstruktionen europäischer Identität*. Köln: Herbert von Halem.

MacGregor, Sherilyn. 2010. "'Gender and climate change': from impacts to discourses." *Journal of the Indian Ocean Region* 6 (2): 223–38.

MacGregor, Sherilyn, ed. 2017. *Routledge international handbook of gender and environment.* Routledge international handbooks. London: Routledge.

Magnusdottir, Gunnhildur L., and Annica Kronsell. 2015. "The (In)Visibility of Gender in Scandinavian Climate Policy-Making." *International Feminist Journal of Politics* 17 (2): 308–26. https://doi.org/10.1080/14616742.2014.896661.

Manchanda, Rita. 2020. "Difficult Encounters with the WPS Agenda in South Asia: Re-scripting Globalized Norms and Policy Frameworks for a Feminist Peace." In *New Directions in Women, Peace and Security*, edited by Soumita Basu, Paul Kirby, and Laura J. Shepherd, 61–82: Bristol University Press.

Manners, Ian. 2002. "Normative Power Europe: A Contradiction in Terms?" *JCMS: Journal of Common Market Studies* 40 (2): 235–58. https://doi.org/10.1111/1468-5965.00353.

Methmann, Chris, Delf Rothe, and Benjamin Stephan, eds. 2013. *Interpretive approaches to global climate governance: (de)constructing the greenhouse.* Interventions. London: Routledge.

Methmann, Chris P. 2010. "'Climate Protection' as Empty Signifier: A Discourse Theoretical Perspective on Climate Mainstreaming in World Politics." *Millennium* 39 (2): 345–72. https://doi.org/10.1177/0305829810383606.

Michael, Kavya, Manish K. Shrivastava, Arunima Hakhu, and Kavya Bajaj. 2020. "A two-step approach to integrating gender justice into mitigation policy: examples from India." *Climate Policy* 20 (7): 800–814. https://doi.org/10.1080/14693062.2019.1676688.

Michaelowa, Katharina, and Axel Michaelowa. 2011. "India in the international climate negotiations: from traditional nay-sayer to dynamic broker." 70. Accessed January 12, 2022. https://www.files.ethz.ch/isn/133820/WP70.pdf.

Michaelowa, Katharina, and Axel Michaelowa. 2012. "India as an emerging power in international climate negotiations." *Climate Policy* 12 (5): 575–90. https://doi.org/10.1080/14693062.2012.691226.

Milliken, Jennifer. 1999. "The Study of Discourse in International Relations." *European Journal of International Relations* 5 (2): 225–54. https://doi.org/10.1177/1354066199005002003.

Müller-Brandeck-Bocquet, Gisela, Philipp Gieg, Timo Lowinger, Manuel Pietzko, and Anja Zürn. 2021. "India and the European Union: A Growing Responsibility to Cooperate in a Changing World." In Gieg, Lowinger, Pietzko, Zürn, Bava, and Müller-Brandeck-Bocquet 2021, 3–16.

Nabers, Dirk. 2023. "Poststructuralist theories: making space for a linguistic analysis of political discourse." In *Handbook of political discourse*, edited by Piotr Cap, 66–78. Cheltenham, UK: Edward Elgar Publishing.

Neumann, Iver B. 1998. *Uses of the other: "The East" in European Identity Formation.* Minneapolis: University of Minnesota Press.

Neumayer, Eric. 2013. *Weak Versus Strong Sustainability: Exploring the Limits of Two Opposing Paradigms.* Fourth edition. Cheltenham: Edward Elgar.

Nonhoff, Martin. 2006. *Politischer Diskurs und Hegemonie: Das Projekt »Soziale Marktwirtschaft«.* Bielefeld: transcript Verlag.

Nonhoff, Martin. 2019. "Hegemony Analysis: Theory, Methodology and Research Practice." In *Discourse, Culture and Organization: Inquiries into Relational Structures of Power*, edited by Tomas Marttila, 63–104. Cham: Springer International Publishing.

Oberthür, Sebastian, and Claire Dupont. 2021. "The European Union's international climate leadership: towards a grand climate strategy?" *Journal of European Public Policy* 28 (7): 1095–1114. https://doi.org/10.1080/13501763.2021.1918218.

Oels, Angela. 2005. "Rendering climate change governable: From biopower to advanced liberal government?" *Journal of Environmental Policy & Planning* 7 (3): 185–207. https://doi.org/10.1080/15239080500339661.

Ojha, Abhiruchi, and Pramod Jaiswal, eds. 2023. *South Asian Women and International Relations*. 1st ed. Springer Singapore.

Pathak, Swapna, and Christie Parris. 2021. "India's diplomatic discourse and development dilemma in the international climate change regime." *India Review* 20 (1): 1–28. https://doi.org/10.1080/14736489.2021.1875699.

Prys-Hansen, Miriam. 2022. "Politics of Responsibility: India in Global Climate Governance." In *Great powers, climate change, and global environmental responsibilities*, edited by Robert Falkner and Barry Buzan, 139–63. Oxford: Oxford University Press.

Resurrección, Bernadette P. 2017. "Gender and Environment in the Global South." In MacGregor 2017, 71–85.

Rothe, Delf. 2020. "Governing the End Times? Planet Politics and the Secular Eschatology of the Anthropocene." *Millennium* 48 (2): 143–64. https://doi.org/10.1177/0305829819889138.

Sarısakaloğlu, Aynur. 2019. *Europas Identität und die Türkei: Eine länderübergreifende Framing-Analyse der Mediendebatte über den EU-Beitritt der Türkei.* Critical studies in media and communication Band 22. Bielefeld: transcript. http://www.transcript-verlag.de/978-3-8376-4626-9.

Sengupta, Sandeep. 2019. "India's Engagement in Global Climate Negotiations from Rio to Paris." In *India in a Warming World: Integrating Climate Change and Development*, edited by Navroz K. Dubash, 114–41. Oxford: Oxford University Press.

Shepherd, Laura J. 2008. *Gender, violence and security: Discourse as practice*. London: Zed Books.

Shrestha, Gitta, Deepa Joshi, and Floriane Clément. 2019. "Masculinities and hydropower in India: A feminist political ecology perspective." *Int J Commons* 13 (1): 130. https://doi.org/10.18352/ijc.920.

Spitzner, Meike. 2020. "Transformative Klimapolitik. Genderwirkungsprüfung und emanzipative Suffizienz-Rechte." *Forum Wissenschaft* 37 (3): 17–21.

Spitzner, Meike, Diana Hummel, Immanuel Stieß, Gotelind Alber, and Ulrike Röhr. 2020. "Interdependente Genderaspekte der Klimapolitik: Gendergerechtigkeit als Beitrag zu einer erfolgreichen Klimapolitik: Wirkungsanalyse, Interdependenzen mit anderen sozialen Kategorien, methodische Aspekte und Gestaltungsoptionen." 30. Accessed December 17, 2023. https://www.umweltbundesamt.de/sites/default/files/medien/1410/publikationen/2020-02-06_texte_30-2020_genderaspekte-klimapolitik.pdf.

Sturgeon, Noël. 2017. "Facing the future, honouring the past: whose gender? Whose nature?" In MacGregor 2017, 21–22.

Swarnakar, Pradip, Rajshri Shukla, and Jeffrey Broadbent. 2022. "Beliefs and Networks: Mapping the Indian Climate Policy Discourse Surrounding the Paris Climate Change Conference in 2015." *Environmental Communication* 16 (2): 145–62. https://doi.org/10.1080/17524032.2021.1973528.

Tanyag, Maria. 2023. "Feminist Governance and Climate Change." In *Handbook of Feminist Governance*, edited by Marian Sawer, Lee A. Banaszak, Jacqui True, and Johanna Kantola, 262–73. International handbooks on gender. Cheltenham, UK: Edward Elgar Publishing.

Thaker, Jagadish, and Anthony Leiserowitz. 2014. "Shifting discourses of climate change in India." *Climatic Change* 123 (2): 107–19. https://doi.org/10.1007/s10584-014-1059-6.

Truelove, Yaffa. 2011. "(Re-)Conceptualizing water inequality in Delhi, India through a feminist political ecology framework." *Geoforum* 42 (2): 143–52. https://doi.org/10.1016/j.geoforum.2011.01.004.

Vida, Bianka. 2021. "Policy framing and resistance: Gender mainstreaming in Horizon 2020." *European Journal of Women's Studies* 28 (1): 26–41. https://doi.org/10.1177/1350506820935495.

Wæver, Ole. 1996. "European Security Identities." *JCMS: Journal of Common Market Studies* 34 (1): 103–32. https://doi.org/10.1111/j.1468-5965.1996.tb00562.x.

Walker, R. B. J. 1990. *Inside/outside: International relations as political theory.* Reprinted. Cambridge studies in international relations 24. Cambridge: Cambridge University Press.

Wendler, Frank. 2022. *Framing Climate Change in the EU and US After the Paris Agreement.* Cham: Springer International Publishing AG.

Wilson, Joanna, and Eric Chu. 2020. "The embodied politics of climate change: analysing the gendered division of environmental labour in the UK." *Environmental Politics* 29 (6): 1085–1104. https://doi.org/10.1080/09644016.2019.1629170.

World Resource Institute. 2023. "The Top 10 GHG Emitters Contribute Over Two-Thirds of Global Emissions." Accessed December 25, 2023. https://www.wri.org/insights/interactive-chart-shows-changes-worlds-top-10-emitters.

Youngs, Richard. 2004. "Normative Dynamics and Strategic Interests in the EU's External Identity." *JCMS: Journal of Common Market Studies* 42 (2): 415–35. https://doi.org/10.1111/j.1468-5965.2004.00494.x.

Zehfuss, Maja. 2013. "Critical theory, poststructuralism, and postcolonialism." In *Handbook of International Relations*, edited by Walter Carlsnaes, Thomas Risse, and Beth Simmons. London: SAGE.

Zeigermann, Ulrike. "Im Spannungsfeld zwischen Wissen und Werten: Über Wesen und Ziel sozialwissenschftlicher Nachhaltigkeitsforschung.", 141–168 (forthcoming).

Ziai, Aram. 2012. "Postkoloniale Studien und Politikwissenschaft: Komplementäre Defizite, Stand der Forschung und Perspektiven." *Politische Vierteljahresschrift* 53 (2): 291–322. https://doi.org/10.5771/0032-3470-2012-2-291.

Research Design 2

2.1 Overview of Theoretical Foundation and Conceptual Framework

Hegemony analysis and ecofeminism serve as the theoretical foundation of this study.[1] Enhancing the hegemony analysis of Martin Nonhoff (2019) (cf. Sect. 2.2) through ecofeminism (cf. Sect. 2.3) in its poststructuralist form, which serves as a conceptual filter focusing on the nexus of environment and gender, enables the parallel exploration of the gendered climate policy identities of both India and the European Union (EU). Following this path, gendered discourse elements concerning specifically-gendered subject positions as well as gendered justification strategies can be systematically deconstructed. The findings of such an ecofeminist hegemony analysis allow us to understand the use of gendered discourse patterns to justify the specific climate policies and targets of both India and the EU.

As Nonhoff grounds his hegemony analysis on the founders of the poststructuralist Essex School of discourse analysis, Ernesto Laclau and Chantal Mouffe (1985), their work plays an important role for this project (cf. Sect. 2.2.1 below). As a consequence, *ontologically*, this study will shed light on the discourses which in poststructuralist thinking create reality (Schieder and Spindler 2010, 25). Thus, poststructuralist *epistemology* emphasises that knowledge is constructed

[1] Since discourse analysis must be seen as an integral part of the critical poststructuralist theory, the boundaries between theory and method are fluid when it comes to poststructuralism (Herschinger and Renner 2014, 12).

A. Zürn, *Of Climate and Gender*, https://doi.org/10.1007/978-3-658-46774-6_2

and not objective or universal, given that discourses, power relations and hegemonies shape what is accepted as knowledge. Therefore, *methodologically*, the deconstruction of the discourse which seeks to reveal meanings, assumptions, power dynamics and prevailing hegemonies is at the heart of this study. *Normatively*, ecofeminist thinking—in addition to its function as a conceptual lens that sets the focus on the environment-gender nexus—will be used in the sense of a foil (Howarth, Glynos, and Griggs 2016, 102) to collate the findings from the hegemony analysis with poststructuralist ecofeminism.[2]

Of course, hegemony analysis particularly addresses the dimension of the respective *society* producing and reproducing meaning; adding the lens of ecofeminism, the focus is on meaning regarding climate change and ways to deal with it or fight it. However, society—while at the same time appearing to be united—is fragmented and consists of manifold *subject positions* with innumerably political interests. To account for the subject positions of groups and single persons, and hence to meet the second research question, the research design requires the addition of Judith Butler's theory of subjectivation (cf. Fig. 2.1). The unique combination of ecofeminist thinking and hegemony analysis will thus be complemented by Butler's focus on the process of becoming a subject through being accepted for specific subject positions. This theoretical background allows us to consider group- and single-subject positions such as *women's groups* or *women as care takers* to concretise whether and how they are gendered and therefore accepted in the discourse solely for specific themes.

Building on this theoretical foundation—and having in mind that the research questions invoke India and the EU's foreign and climate policy identities and aim at illuminating their construction by discourses, demands and subject positions of groups and single persons—the conceptual framework of this study depicts the interconnectedness of these elements (cf. Fig. 2.2).

---

[2] Poststructuralist approaches often face the critique that they would contradict themselves if they became normative, because normative critique is a vital part of the discourse that must itself be deconstructed. A critical examination of the findings from the discourse analysis and the recommendations to be derived from them would thus be difficult. One answer to this problem is to insert the foil as described above making the results of poststructuralist analysis fruitful for further considerations beyond the deconstruction of the discourses. Tomas Marttila (2019) discusses in detail the problem of the discursive nature of normative critique or demands and what follows for poststructuralist projects.

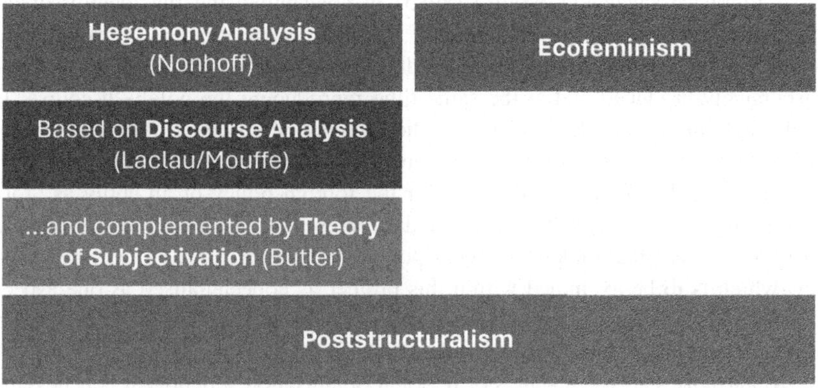

Fig. 2.1 Building blocks of the study's theoretical foundation

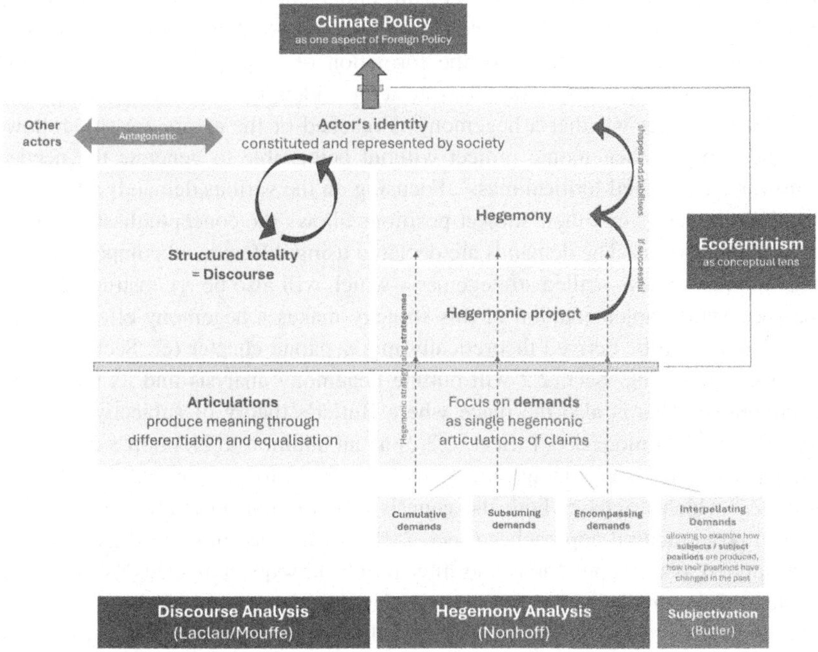

Fig. 2.2 Conceptual framework

It follows Laclau and Mouffe (1985) in conceptualising the identity of an international actor (such as a state or a regional organisation) as constituted by society. Society encompasses manifold different subject positions which are both discursively-produced and at the same time reproducing the policy discourse. It is through and within these societies where policy-determining identity is established, not least by the self-differentiation from the outside (Parker et al. 1992, 5). Accordingly, this identity is decisive for the foreign policy of an entity as "foreign policies are dependent upon particular representations of [...] the national or institutional Self that undertakes these policies" (Hansen 2012, 95). Climate policy, which is to be examined within this project, is conceptualised as one aspect of the foreign policy of an entity such as the EU (Müller-Brandeck-Bocquet and Rüger 2015, 11) or a state (McDonald 2012, 385).

To partially stabilise a state's or regional organisation's identity it is necessary to implement the discursive formation of a hegemony. For a hegemonic project to be successful, it must ultimately be supported by many different actors; especially by those who are recognised as powerful, either due to their political standing, their resources in terms of knowledge or finance (Nonhoff 2006, 21), such as socio-political actors. Since only the formation of a hegemony can shape and also—at least temporarily—stabilise an actor's identity, the study at hand will particularly detect whether a hegemony is reached or the actors are still following their specific hegemonic project without being able to generate the needed followers, a so called 'critical mass'. Focusing on the various demands articulated by different actors and their subject positions allows the conceptualisation of the hegemony analysis. The demands are declared using different subcomponents of a hegemonic strategy—called strategemes—which will also be reconstructed. Only the successful implementation of this strategy makes a hegemony effective. The strategemes will be derived theoretically in a separate chapter (cf. Sect. 2.2.4).

In the following, Sect. 2.2 will outline hegemony analysis and its theoretical assumptions. This is also the place where Butler's theory of subjectivation will be introduced in more detail (Sect. 2.2.2) as an addition to Nonhoff's hegemony analysis. The second building bloc of the theoretical foundation will be added in Sect. 2.3 which discusses both the genesis of ecofeminism (Sect. 2.3.1) and its turn to poststructural approaches (Sect. 2.3.2) before outlining how ecofeminism will applied as conceptual lens and filter in the subsequent analysis (Sect. 2.3.3).

Sect. 2.4 completes the research design by laying out in detail how the ecofeminist hegemony analysis is conducted (Sect. 2.4.1) and explaining the data selection process and thus the construction of discourse corpus (Sect. 2.4.2).

2.2 Hegemony Analysis

Nonhoff's (2019) hegemony analysis serves as the first building block of the study's theoretical foundation. Given its deep grounding in Laclau and Mouffe's poststructuralist discourse analysis, their concepts of discourse and articulation will be outlined in Sect. 2.2.1, as will Nonhoff's adaptation in the form of his focus on demands. Sect. 2.2.2 will then summarise Butler's theory of subjectivation as a subject-related addition to the research design. After turning to theoretical assumptions on the identity construction of states and regional organisations (Sect. 2.2.3)—whose reconstruction in this study is key to answering the research questions later—the offensive-hegemonic strategy and its strategemes will be outlined (Sect. 2.2.4).

2.2.1 Discourse, Articulations and Demands

Nonhoff's hegemony analysis heavily draws on Laclau and Mouffe's discourse analysis who deduce their conceptualisation of *discourse* from *articulations*:

> „*We will call articulation any practice establishing a relation among elements such that their identity is modified as a result of the articulatory practice. The structured totality resulting from this articulatory practice, we will call discourse*" (Laclau and Mouffe 1985, *105)*.

Discourse thus comprises various facets (Nonhoff 2019, 67–68). It embraces the smallest units, *articulations*. These can produce meaning through a process of differentiation or equalisation. Out of this *articulatory practice* a structure can emerge, which Mouffe and Laclau describe as *structured totality* (cf. Fig. 2.2). This understanding of discourse is particularly relevant for the hegemony analysis used in this project. It shows that discourses comprise of and address different dimensions.

What is meant by discourse in an ontological sense? The Essex School reaches beyond just the spoken or written word as discourse, by also including discursively-constructed social practices (Laclau and Mouffe 1985, 107; Nabers 2015, 131). In the words of Herschinger (2012, 71), they conceptualise discourse as "a system of meaningful practices, which relates differences to establish their meaning". Emphasising that discourses define the social per se, Laclau (1980, 87) makes clear that discourses are "the ensemble of the phenomena in and through which social production of meaning takes place, an ensemble which constitutes a society as such". Consequently, no society can exist without discursive structures, as discourses are made by the society—and vice versa. Hence, society can never be fully constituted because of its relational construction (Demirović 2007, 62). This aspect becomes particularly important regarding the discursively-produced identity of an actor which is represented by the respective unfulfilled society (cf. Sect. 2.2.3).

The outlined understanding of discourse is shared by various scholars, working in different disciplines (Angermuller et al. 2014). Nonhoff (2019, 69–70) also follows the Essex School, drawing special attention to *demands* which he defines as the smallest discursive element of a political discourse (Nonhoff 2010, 309). Laclau (2005, ix) describes demands as the process of articulating claims by several subject positions to build a group, even though those demands may vary or follow different interests—an important aspect for the upcoming analysis. Nonhoff (2019, 70) argues that especially for methodological reasons, it is fruitful to think of articulations as demands, as these are a vital part of political daily life and are therefore graspable. In particular when it comes to the debate about how a society wants to live in the future, but also regarding reference points lying in the past, the added value of demands becomes tangible (Nonhoff 2019, 70). In today's political discourses climate- and environment-related demands are highly present also shaping debates of the common future. Conceptualising demands as a "quintessentially political form" (Glynos and Howarth 2007, 116), they can call into question—as well as support—widespread practices or norms represented by the government. Furthermore, demands can lead to the adoption of a specific paradigm or the leadership of particular groups or persons (Nonhoff 2010, 309). In fact, demands are the material used to pursue one's own agenda with the help of the possible hegemonic formations.

To understand the different discursive formations necessitates distinguishing between various types of demands (Nonhoff 2019, 75–76): (1) *Cumulative demands* concretise one specific aspect of the universal, a necessary precondition to its completion. Cumulative demands are open for amplification with further demands which are also oriented towards the desirable common good, as cumulative demands do not automatically go hand-in-hand with all other demands of the chain of equivalence (Nonhoff 2006, 119). (2) A *subsuming demand* fulfils other thematically-related demands which also serve the supposed achievement of the common good (Nonhoff 2010, 309). (3) *Encompassing demands* "articulate[s] a sufficient condition for overcoming the lack of the universal" (Nonhoff 2019, 75) and can therefore be understood as a maximization of cumulative demands.

All these types of demands can be expected to be present in debates on climate- or environmental protection. To identify whether one of the demands of the chain of equivalence became the encompassing demand will be one of the goals of the later analysis, as this is the necessary type to reach a hegemony.

Aiming to shed greater light on the discursive making of (gendered) subject positions such as *women as victims*, the set of different types of demands conceptualised by Nonhoff will be extended with (4) *interpellating demands,* following Judith Butler's theoretical assumptions (cf. Sect. 2.2.2). While interpellating demands follow the mission to establish one's subject position, they operate on the first dimension of a discourse (cf. Sect. 2.3.3) cumulative, subsuming and encompassing demands are oriented towards the common good, and therefore strive towards the *empty signifier* (cf. Sect. 2.2.3) operating on the second dimension of a discourse (cf. Sect. 2.3.3). Depending on their impact these demands are able to create a specific hegemonic discursive formation such as a hegemonic project or a hegemony. These will be deconstructed later in the study to reveal whether a hegemonic project of one or more actors is successfully enforced and thus able to become the society's hegemony (cf. Table 2.1) and therefore the EU's or India's climate identity.

Table 2.1 Three levels of success of a hegemonic practice (based on Nonhoff 2019, 76)

(3) Hegemony = Successfully enforced hegemonic project
(2) Hegemonic project
(1) Single hegemonic articulation

The success or failure of a demand, and the question of whether it will contribute to the identity of an entity or whether it is a vital part of the hegemonic project, can depend on who articulates it and how the person is positioned. The subject, or rather, the subject *position* is also produced discursively. How is the process of subjectivation linked to demands and how are subject positions produced?

2.2.2 Subjectivation and Gendered Subject Positions

In 2.3.3, different subject positions reconstructed by ecofeminist scholars will be presented. These subject positions, such as *women as victims of climate change* or *carers for nature*, are the product of the first dimension of gendered climate discourses. Nonetheless, it remains open how these subject positions, which can become central actors within a discourse, emerge, or become gendered or recognised as legitimate subjects shaping a discourse.

When Laclau and Mouffe use the term subject, it is usually "in the sense of 'subject positions' within a discursive structure" (Laclau and Mouffe 1985, 115). Although a subject is characterised by many different aspects, its *position* is most often recognised exclusively for one aspect. In fact, subject positions are dual-faceted; they are both an inherent part of the discourse yet also able to influence it. For this study, the individual subject position as well as group subject positions are of great interest, since they provide information on whether and how certain groups are shaped by the discourse—which might be explicitly gendered. Regarding the analysis of offensive-hegemonic strategies in climate politics, we are concerned with groups such as *the government* or individual subject positions, such as *the Commissioner for the Environment* or a *climate activist* trying to win other subject positions supporting his or her hegemonic project.

Asking how an individual or a group takes up a subject position and in particular how it becomes a subject allows us to search for gender stereotypes within the intersubjective and discursive process of *subjectivation.* Judith Butler specifically sheds light on the subject itself and—relying on Althusser—introduces the figure of *interpellation* (Butler 1997, 95),[3] but extends this concept with the aspect of *subjection.* Through this amplification it is possible to provide access to the discursively-produced gender roles, as Butler describes subjection as a condition of subjectivation (Villa 2012, 39), implicating the effect of power (Butler 1997, 2). The subject accepts the interpellation by subjugating him/herself to the power, such as the law or one's own conscience (Butler 1997, 115).[4]

Why should a subject submit to a power such as discursively imparted norms? Althusser based the submission of the subject on authority, but Butler sees the cause in the desire of the subject to be recognised or named as such (Butler

[3] Butler refers here to Althusser, as do Mouffe and Laclau (1985, 114–15) in their conceptualisation of subject positions, which is why the link with Butler is particularly promising. Butler, Laclau and Mouffe, however, all understand the interpellation, which is part of the process of subjectivation, not as an ideological process, as Althusser sees it. Laclau and Mouffe conceptualise interpellation as a discursive act (Glasze and Mattissek 2009, 162), and Butler emphasises the discursive character (Butler 1997, 111).

[4] In their later works, Butler revisited the character of power, showing that it is located in forms of action of discourses and norms which invoke the subject. For an in depth discussion of Butler's understanding and conceptualisation of norms and their functionality see Butler 2004, 43–56.

1997, 112).[5] If, in the context of climate, recognition as a woman is primarily associated with caring for the environment, the subject position of a woman is only recognised as such when she performs according to the norm of caring. The result of those processes is therefore a circular argument. The subject must submit themselves in order to be interpellated by other subjects and thus to be recognised. Even if a subject—which comprises more facets than the later discursively-produced subject position—were to contradict the interpellation, for example as a *climate change victim*, it becomes only subjectivated "through the performative effect of the interpellating demand" (Butler 1997, 99). Therefore, the subject has to accept the interpellation first before it can even change its position, because "[i]f one misrecognizes that effort to produce the subject, the production itself falters" (Butler 1997, 95). Consequently, the subject finds itself in a permanent loop of subjectivation and is therefore confronted with pressure to perform in line with the expectations faced by the subject position. Due to this infinite task the subject cannot be a stable one and a fixation cannot be achieved (Butler 1997, 99).

This opens up the space for the modification of the subject in the sense of subversion, which is in line with the critical epistemology of the study. Or with the words of Catriona Sandilands (1997, 21) "one is left with the possibility of a parody based on the inevitable disjuncture between subject and subject position". Through the process of subversion, a new understanding of a subject can develop. For example, if the subject *farmer* in the EU tends to be seen solely in the subject position of an *environmental exploiter*, subversive or parodistic reiteration can slowly change the understanding of the subject position towards *friend of the environment*. The attempt to change one's subject position can of course also fail. Overall, misrecognition of a subject is possible (Butler 1997, 96). In this study this is expressed, for example, if a subject is not recognised for its position, or if it does not receive the recognition for it by other subject positions such as influential or powerful actors.

With the help of Butler's approach of subjectivation, the way in which subjects are produced, as well as how their positions have changed in the past, can be examined. This is why Butler's interpellating demands are a necessary addition

[5] Butler (1990) illustrates this mechanism with the norm of compulsory heterosexuality. Following the approach of subjectivation, people submit to this norm, they behave like *typical* women or men, adopting specific subject positions regarding their sex, gender and desire to be accepted by other subjects against the background of the norm. People follow role models and examples, such as videos, their parents or advertising, when it comes to *typical* sports, haircuts or status icons (Villa 2012, 77) in the sense of heteronormativity and a binary gender system, as a prediscursive blueprint of the norm doesn't exist.

to Nonhoff's analytical strategy (see Sect. 2.2.4 below) in order to scrutinise the different aspects of subjectivation within the empirical sources.

2.2.3 The Discursive Construction of International Actors' Identities

The theoretical relationship between the subject position, the society and the common identity must be clarified. Following Nonhoff in drawing on Laclau and Mouffe (1985), this study conceptualises the identity of international actors such as states and regional organisations as constituted by societies. As has been shown, however, these societies encompass manifold different subject positions which are both discursively-produced and at the same time reproducing the policy discourse. Ultimately, as Epstein (2010, 343) puts it, "subject-positions constitute identities". Therefore, this common identity is not fixed, as "the theory of hegemony builds on the idea of a constitutive lack, and thus the continuous mobility and unfixity of any totality" (Nonhoff 2019, 68). But it still comprises a society which can stand for a state or regional organisation. This is not necessarily a contradiction to the fact that there can be no homogeneous society, because "the state is the domain where difference is translated into uniformity" (Inayatullah and Blaney 2004, 20) by hegemonies that are discursively produced through so-called chains of equivalence. Differences and dissent within a society are overformed by a common narrative depicting the identity of a society. It is through and within these societies where policy-determining identity is established, not least by the self-differentiation from the outside (Parker et al. 1992, 5). Hence, as already described, for the foreign policy of an actor, in our case its climate change policy, the identity is constitutive (Hansen 2012, 95). Since discursive formations operate in both directions, foreign policy also shapes the identity of an entity, trying to reach the so called universal, an expression of the common good of a society.

Conceptualising the discursive production of a state's or regional organisation's identity, this study will thus also follow Nonhoff in building on the work of Laclau and Mouffe:

(1) To gain a common identity, they emphasize the antagonistic character, which is necessary to differentiate a society from the outside (Laclau and Mouffe 1985, 122). This process of differentiation via antagonisms underlines the political character of the climate discourses with its inherent demands.

(2) Furthermore, inside a society, it is necessary to draw a picture of the various, mostly synchronous processes within discourses. Different demands are (re-) produced by subject positions and often articulated antithetically to other demands within the discourse.

For example, *environmental protection* and *energy security* could be positioned as opposed to each other. This mechanism is called the logic of difference (Laclau and Mouffe 1985, 129). The logic shows how divergent the various demands within a discourse can be. At the same time, however, the logic of equivalence can also have an effect within the discourse, when all discursively-produced differences are over-formed by one of these demands through collective reference "to something external" (Laclau and Mouffe 1985, 127). Thus, the logic of equivalence leads to the equivalisation of diametrically opposed demands, which is a condition conducive to being perceived as a community trying to reach the universal.

Mouffe and Laclau visualise these aspects of their theory by pointing to a colonialised country. There, one can find a lot of differentiations between the colonising power and the colonialised society, concerning diverse facets such as dress, languages or customs (Laclau and Mouffe 1985, 127). Even though those contents are highly distinct within the colonised society, the facets become equivalent through the differentiation from the external, in this case the colonising power. The differences repeal each other while commonly referring to the outside, or the so called "radical other", which doesn't have to be that radical in its otherness, but palpably different enough to enable the inside to define itself (Torfing 2005, 11).

To achieve such a separation from the outside, a common placeholder epitomising the universal or common good is discursively produced. This 'lowest common denominator' is represented by the *empty signifier*. Becoming the empty signifier means to be able to represent all equivalent demands and aspects of a discourse (Laclau 1994, 174, 1996). The signifier is almost emptied insofar as it can unite many different positions and accordingly appears completely blurred in its meaning. It must therefore be a 'catch-all term' representing a common point of reference. The empty signifier is one of the various *nodal points* (Laclau 2005, 69–72) or "organizing metaphors" (Rogers 2009, 838) of a discourse, such as *democracy* or *liberal world order*. Nodal points are those cumulative, subsuming or encompassing demands (cf. Sect. 2.2.1) which are approved by a critical number of different subject positions and are therefore a part of the so-called chain of equivalence. In the case of the empty signifier *liberal world order*, a variety of nodal points could be *hegemonic dominance by a specific state, club governance* or *multilateralism*. Nodal points serve to partially fix the meaning (Laclau and Mouffe 1985, 112) and therefore stem the tide of discursively-produced differences (Howarth 2010, 331).

Thomas Jacobs (2018, 303) visualises the fixation of meaning—which is in fact the formation of the chain of equivalence—in the metaphor of a spider's web (cf. Fig. 2.3). This network tracks different paths of nodal points and their various relationships to each other, to the centre stage of the web. There, we can find the encompassing demand equivalising all nodal points. These consist of cumulative or subsuming demands which go hand-in-hand with the encompassing demand, if this is the successfully established hegemonic project (Nonhoff 2019, 75–76). Hence, this encompassing demand is the empty signifier representing the actor's identity.

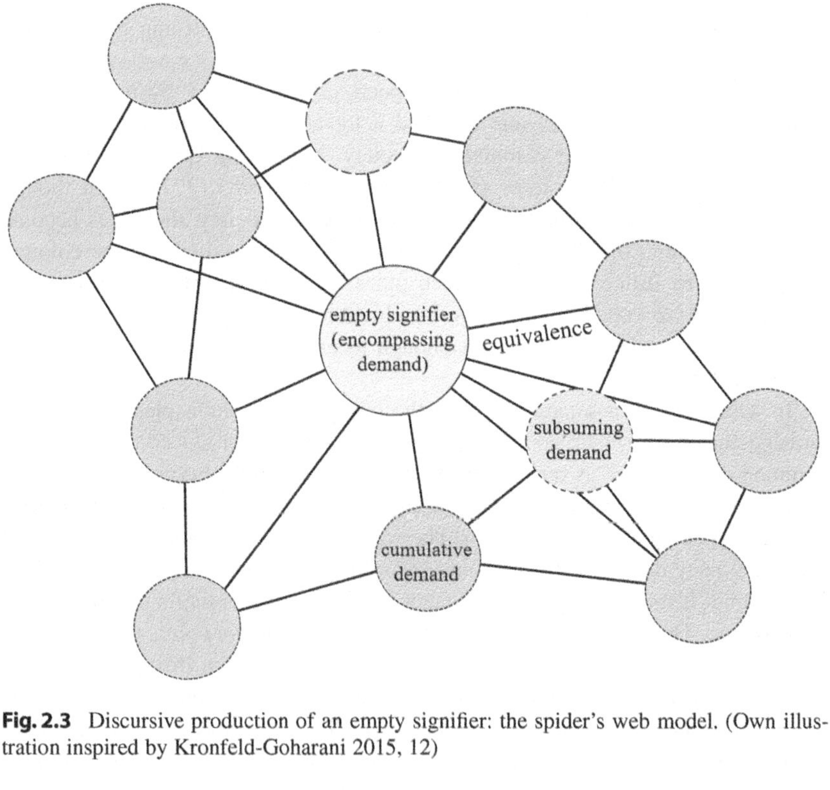

Fig. 2.3 Discursive production of an empty signifier: the spider's web model. (Own illustration inspired by Kronfeld-Goharani 2015, 12)

The aim of the later analysis is to reconstruct the identity and the strategy behind it. For this reason, the distinction between the different types of demands will not be pursued further at this point, but rather the encompassing demand will be specifically sought. It is important to note that the model of the spider's web depicts the demands articulated in the discourse and their equivalent relationship to each other. It illustrates how the chain of equivalence leads to an encompassing demand that resolves any differences between the various demands, because all demands see themselves represented to a greater or lesser extent by the encompassing demand, the empty signifier. Thus, the spider's web only concerns the *inner-societal* processes with lead to the production of an identity.

A more commonly used visualisation of the identity-producing discourse is the *chain of equivalence.* As already mentioned, it is the main task of the upcoming analysis to deconstruct this *chain of equivalence,* given that it plays an important role in the process of constructing a (state) identity (Howarth 2010, 318). This model integrates both the inner-societal processes—albeit in a less detailed manner than the spider's web—*and* the process of differentiation from the outside via antagonisms. The graphical illustration (cf. Fig. 2.4) can thus be understood as the less complex schematic representation of the chain leading to the 'heart of the spider's web', that is: the empty signifier. But the visualisation also shows an antagonistic space separating chain of equivalence X (which I will deconstruct in the later analysis) from its antagonistic counterpart Y. Since identity is formed in demarcation from the 'radical other', Fig. 2.4 depicts these parallel and constantly reproducing discursive processes. Precisely what concrete mechanisms and strategy lie behind these processes will be delineated in the following chapter.

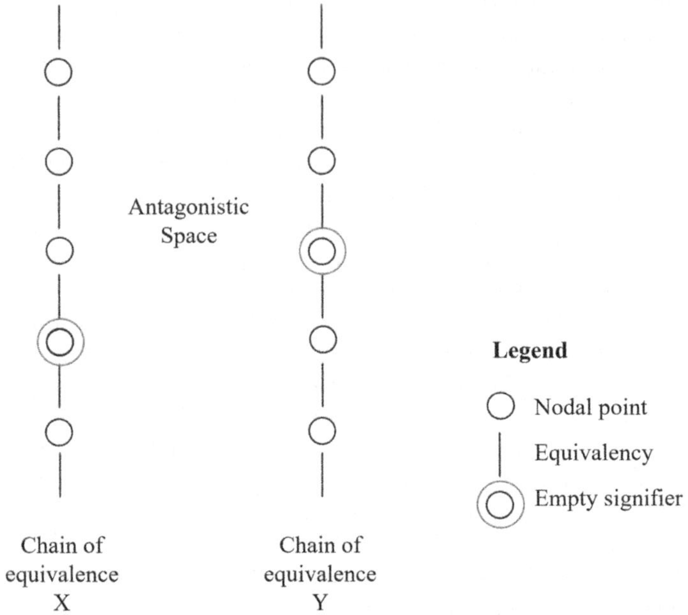

Fig. 2.4 Two opposing chains of equivalence X and Y (based on Nonhoff 2019, 90)

2.2.4 The Offensive-Hegemonic Strategy and its Strategemes

A successfully established hegemony, leading to a common identity, is based on a hegemonic strategy. Nonhoff (2019, 95–96) distinguishes the offensive-hegemonic and the anti-hegemonic strategy. While the former is intended to achieve a successfully established hegemonic project, hence a hegemony, the anti-hegemonic strategy is a counter-attempt, a strategy to contest and change the hegemony of a discourse. Even though I will occasionally deal during the project with anti-hegemonic articulations or spaces of the perceived opposition to the hegemony, the central aim of this study is to examine the mechanisms creating the identity of India and the EU. This allows a search for the strategic use of gendered discourse patterns, which might support the hegemonic project of powerful actors establishing their reading of India's or the EU's identity. Thus, I will focus on the offensive-hegemonic strategy.

An offensive-hegemonic strategy consists of various strategemes; these are smaller strategic components (Nonhoff 2010, 312). Trying to establish a hegemony, different actors use these strategemes. Hence, they must be identified and examined in their respective forms in order to reconstruct the drivers of an established hegemony—and thereby the discursive construction of the EU's and India's respective identities with a special interest concerning gendered components.

Nonhoff distinguishes between nine different strategemes (2019, 80). This number is not static, as he accentuates the expandability of the strategemes due to the fact that they are derived in two manners: inductively from the theory of Laclau and Mouffe and deductively from the findings of his study.[6] This already indicates that the set of strategemes is strongly dependent on the research subject (Nonhoff 2010, 312), and thus of course open for new strategemes emerging from the study at hand. Potential new strategemes will therefore be taken up in the later analysis. By paying attention to gendered discourse patterns, which most often emerge as specifically gendered demands or the recognition of particularly gendered subject positions, the deconstruction of the strategemes in this study are informed by ecofeminism (cf. Sect. 2.3 below) and by the Butler's approach of subjectivation (cf. Sect. 2.2.2 above).

The theoretically grounded strategemes identified by Nonhoff (2010, 313) are divided into four different groups as they differ in both their scope and frequency (see Table 2.2).

Table 2.2 Strategemes of the offensive-hegemonic strategy (Nonhoff 2019, 80)

A. **Core strategemes of the offensive-hegemonic strategy**
I. Articulation of equivalences between different demands made with regard to the universal
II. Antagonistic division of the discursive space
III. Representation
B. **Basic strategeme**
IV. Super-differential demarcation of a specific discourse's limits
C. **Supplementary hegemonic strategemes**
V. Gradual constitution of greater leeway for interpreting the symbolic equivalent of the universal
VI. (Re-)Articulation of subject positions for politico-societal forces
VII. Occasional but deliberate breach of the antagonistic line of division
D. **Secondary hegemonic strategemes**
VIII. Strategeme of true championship
IX. Strategeme of true meaning

[6] I refer back to Nonhoff's project on social market economy (Nonhoff 2006).

The first three strategemes, which are subsumed in the group of *core strategemes*, characterise almost every hegemonic practice introduced so far (cf. Sect. 2.2.1), while the other strategemes just occur from time to time and in a smaller amplitude. The analytical focus will therefore be on the core strategemes which are most important for successfully implementing a hegemonic project. As some of the other strategemes of groups B and C are used by different actors of both the Indian and the EU discourses on climate change—albeit with varying frequency—they will also be outlined. Due to their characteristics and the fact that the project at hand focuses on very volatile discourses, the strategemes of group D will not play a role in the later analysis.

(I) The first strategeme is characterised by its aim to establish a chain of equivalence, which makes the various demands appear to be equivalent and go "hand in hand" (Nonhoff 2010, 313). Following the logic of equivalence, strategeme I is characterised by equivalising possible differences within the discourse, such as *energy security* and *environmental protection*. For example, in relation to the encompassing demand *sustainable development*, both *energy security* and *environmental protection* are equivalent.

(II) It has already been emphasised above that equivalences arise in particular with regard to the so-called radical other, the outside. This relationship is represented by strategeme II, the *antagonistic division of the discursive space*. In order to help an offensive-hegemonic strategy to succeed, the hegemonic project follows the vital interest to discursively build this antagonistic space (Nonhoff 2019, 81), which symbolises the gap between the two chains of equivalence (cf. Fig. 2.4 above).

(III) Strategeme III illustrates the importance of *Representation*. The more universal an encompassing demand is, the higher is the chance of the hegemonic project to succeed in constructing a common identity. While the various nodal points of chain of equivalence X are antagonistic to one or more nodal points of chain Y, one specific nodal point turns out to be particularly representative in the sense that it embodies the antagonism in relation to all nodal points of the opposite chain of equivalence. Thus, the third strategeme is symbolised by the empty signifier of chain X, our research subject. If these three strategemes can be successfully implemented, the following strategemes are used primarily to further stabilise the hegemonic project.

(IV) The fourth strategeme following Nonhoff (2019, 80) is the *super-differential demarcation of a specific discourse's limits* which builds the category B *basic strategeme.* It is basic because a super-differential demarcation of a discursive space is needed to formulate a hegemonic articulation (Nonhoff 2006, 232). However, the strategeme is an important component in differentiating one specific discursive matter from others.[7] The distinction of the arenas should not be confused with the antagonism, which is in fact the division of one specific discursive arena into two antagonistic spaces.

(V) To extend the outreach of the hegemonic formation is the goal of group C*: supplementary hegemonic strategemes* (Nonhoff 2019, 80). Extending the outreach means that more and more subject positions follow the hegemonic articulation or project, helping to make it a success (Nonhoff 2006, 233). However, strategemes V-VII of Group C cannot be traced in every hegemonic formation. Strategeme V: *gradual constitution of greater leeway for interpreting the symbolic equivalent of the universal,* means that the empty signifier which is ultimately supposed to represent the striving for the universal and therewith the identity of an entity, will be emptied even further in order to synchronously open up the scope of interpretation.

(VI) Strategeme VI, the *(re-)articulation of subject positions for politico-societal forces*, is used when actors who try to establish a hegemony create subject positions for politico-societal forces in the chain of equivalence (Nonhoff 2006, 234) in order to extend the discursive visibility and the durability of the hegemonic project (Nonhoff 2006, 234). This is because politico-societal forces such as political parties, federations, civil society groups or NGOs play an important role in the process of establishing a hegemony (Nonhoff 2006, 233). To achieve a hegemony, it is more important to win socio-political forces for the hegemonic project rather than as many people as possible, because the socio-political forces, operating as multipliers, give the hegemonic project additional credibility. Here, the addition of Butler's subjectivation approach to Nonhoff's theory comes into play: it is at this point

[7] Nonhoff (2006, 231) describes this mechanism by using the example of a controversy about the economic consequences of the closing of a military base in a specific region. Politicians and others might argue that the commercial losses are secondary, as security policy is not in charge of protecting the regional economy, which is the duty of economic policy. In this example the sharp demarcation of security and economy as two different discursive arenas is made clear.

that gendered subject positions can be identified. Indeed, most socio-political subject positions are not explicitly gendered, as, for example, when the chamber of commerce supports the hegemonic articulation demanding green growth; however, when the support for the demand becomes credible especially due to the associated gender role of a subject position, for example when women's groups give additional reach to the demand for adaptation measures to protect the subsistence economy, one can speak of gendered subject positions, as the gender roles are used to push the hegemonic project.

(VII) Strategeme VII, *occasional but deliberate breach of the antagonistic line of division*, seems paradoxical at first sight because the hegemonic project depends on a basic degree of stability and because the antagonistic boundary is articulated as fixed and as a clear distinction between "good" and "bad" (Nonhoff 2006, 234). However, it may be helpful for the actors trying to establish a hegemonic project to place individual subject positions or demands outside its 'own' borders or to integrate an antagonistic demand or subject position into its 'own' chain of equivalence to strengthen its significance for instance by demonstrating your own power to set priorities for others.

(VIII–IX) Group D consists of *secondary hegemonic strategemes*. They come into play only on the second level of a hegemony—or more precisely in the case of an already relatively successful hegemonic project (Nonhoff 2006, 235).[8]

This overview of the different strategemes of an offensive-hegemonic strategy shows that many different functions of a discourse can be used to make a hegemonic project a successfully established hegemony. Adopting the ecofeminist hegemony analysis—conceived in one of the following chapters—therefore especially means deconstructing the discourse so that the strategemes used can be

[8] Here, the *strategeme of true championship* and the *strategeme of true meaning* act and consequently re-implement the mechanisms of strategemes I–VII. Therefore, it is hardly possible that the two strategemes are actually carried out—on the contrary, their success leads back to the first level of a hegemony. Especially at the beginning of a hegemonic project, it is very unlikely that these strategemes are detectable (Nonhoff 2006, 250) as the actors don't use them at this point of time. Since the later analyses will show that in both the Indian and the EU- climate discourses the hegemonic projects vary depending on temporal developments and are therefore often rather recent, these strategemes will not be examined in depth here, as they cannot be proven later due to the nature of the analysed discourses. For an introduction and discussion of the theoretical background of these strategemes see Nonhoff 2006, 234–38.

reconstructed. Only then can the various functions of the discourse—especially the use of gendered patterns in the context of the research interest—be uncovered, critically questioned and evaluated.

2.3 Ecofeminism: on the Nexus of Gender and Nature

To focus on gender in the framework of the reconstruction of hegemonic discourse patterns and to compare the findings of this study with ecofeminist insights, I will turn to ecofeminist approaches. As ecofeminism is only slowly being rediscovered and has a shadowy existence, if at all, in the discipline of international relations, this theoretical field will be introduced in detail. Interestingly, ecofeminism has been accompanied by a poststructuralist critique for several years, or rather decades. In order to break up this criticism and to work out the multifaceted connecting points with poststructuralist hegemony analysis, the genesis of ecofeminism will be elaborated (cf. Sect. 2.3.1) as will its poststructuralist approaches (cf. Sect. 2.3.2), which will lead to the deconstruction of gendered discourses (cf. Sect. 2.3.3).

Addressing the exploitation of women and the simultaneous or rather interlinked exploitation of nature was the starting point of ecofeminist thinking in the 1970s, while today's approaches widen the focus to the exploitation of marginalised groups and their experiences of discrimination. The theoretical and activist engagement with these realities has been closely intertwined and have influenced each other. Therefore, Heather Eaton and Lois Ann Lorentzen (2003, 1) speak of ecofeminism as "an intellectual and activist movement".

An important landmark of feminist thinking in the context of environmental as well as development studies was the analytical focus on the disproportionate impacts of climate and environmental change on rural women of the Global South (Resurrección 2017, 72). Scholars and scholar activists tackled inter alia the significance of female forestry- and farm work in India (Shiva 1988), the vulnerability of women due to the effects of droughts (Schroeder 1987) and floods (Cannon 2002), or the specifically gendered consequences of disasters (Reyes 2002; Neumayer and Plümper 2007). These works are flanked by contemporary theoretical and philosophical thoughts, which was laid the foundations for by various female theorists in the 1970s and 1980s (Merchant 2005, 194–95). Thus, ecofeminism is also referred to as a philosophical line of thought (Kennedy and Dzialo 2015, 924). In one of the sub-disciplines of environmental humanities—environmental sociology—considerations of the relationship between environment and gender have long been absent, wherefore especially the critical perspective "has much room to grow" (Kennedy and Dzialo 2015, 921). For some time now, scholars have also been looking at gender-nature relations from an economic viewpoint. Ecofeminism therefore combines different disciplinary backgrounds as well as multiple facets and feminist schools of thought. Reflecting these developments, Sherilyn MacGregor (2009, 126) frames ecofeminism as a "field", as do Eaton and Lorentzen (2003, 3).

To form an overview of this field, the theoretical progress and empirical findings of eco-feminist thinking will be outlined against the background of the hegemony analysis to incorporate the ecofeminstically-deconstructed and mechanisms of powerful gendered discourse patterns in the framework of environmental- and climate politics. Therefore, central works of the first ecofeminist debates will be briefly presented. These were often criticised for painting an essentialist picture of women—essentialism in the sense of attributing supposedly unchangeable characteristics to a clearly defined and therefore seemingly homogeneous group. This triggered a diversification of ecofeminist thinking into different strands, one of which is formed by poststructuralist informed approaches.

Poststructuralism has enabled the deconstruction of the historically contingent dichotomies articulated in climate discourses. Its introduction to ecofeminism has allowed to unfold the power structures shaping specific ideas of climate protection and measures to fight the climate crisis as well as socially constructed gender roles related to climate change. Years of poststructuralist theorising in ecofeminism and the resulting insights can thus be harnessed for the normative dimension of this study's theoretical foundation: as the normative foil interpreting the findings of the hegemony analysis.

2.3.1 The Genesis of Ecofeminism

Alongside development debates, feminist theory-building has been crucial to the evolution of ecofeminism and to this day is a catalyst informing the field (Gratz and Zürn 2020, 22). Simultaneously, it must be noted that the relationship has been precarious to some extent (Merrick 2017, 110), since it only seems that ecofeminism was initially dominated by conceptualising the category of gender as essentialistic. Although this idea of gender is later strongly criticised, the works stemming from this concept are indispensable to understanding the genesis of ecofeminist thought. Or, as Greta Gaard (2011, 44) puts it: "If there is to be a future for a 'New Eco-feminism', it will need to be more cognizant of its rich and prescient history". Barbara Holland-Cunz (2014, 113) points to

Françoise d'Eaubonne (1974), who first mentioned the term "ecofeminism" in her essay with the meaningful title "Le Féminisme ou la mort". D'Eaubonne considers feminism as the only answer to avert the patriarchal structures and consequences of—in her eyes—the key threads of these times, overpopulation[9] and environmental change.[10]

Another groundbreaking work is Rachel Carson's "Silent Spring", published in 1962, which was pivotal for both the US environmental movement and the debate on environmental change and gender.[11] The biologist dealt with the effects of various pesticides on flora and fauna by means of numerous studies and individual fates. The pattern she identified of "masculinised" control over the environment, for example by aggressive representatives of the chemical industry, was later used by various scholars such as Vandana Shiva, Maria Mies and Carolyn Merchant as a central reference for their ecofeminist thinking (Saeger 2017, 28). In particular the Indian *scholar activist* Shiva was a prominent early ecofeminist who remains engaged to this day both scientifically and as an activist, bringing together a well-founded analysis of different environmental threats such as desertification, pesticides and water scarcity with a scientific-feminist analysis.[12] In her work "Staying alive: Women, Ecology and Development", Shiva (1988) outlines an ecofeminist vision inspired by local women of the Global South to answer the oppression of women and environment. The Alternative Nobel Prize winner Shiva later continued on her work together with the German sociologist Maria Mies

[9] Overpopulation is, however, a very truncated way to answer the complex questions of climate change and sustainable development. The debate can be neocolonialised, for example, if—in particular—women of the Global South are asked to have fewer children (apart from the fact that their CO_2 footprint is currently very low). Ryberg (1998) also points out that it is not clear what is meant by ‚overpopulation' and at what point it would be reached. Furthermore, in the context of development, it is asked whether the image of overpopulation is merely used to justify specific measures (Aufhauser and Diketmüller 2001).

[10] In the 1970s further scholars dealt with the significance of nature in feminism, including Sherry Ortner (1972), Rosemary Ruether (1975), Mary Daly (1978) and Susan Griffin (1978). For a detailed contextualisation of some of the mentioned critical studies as well as partly radical demands, see the overview by Holland-Cunz (2014, 112–18).

[11] As a female biologist in the 1960s, who was one of the first to clearly name the dangers of pesticides, she was confronted with broad attacks from the chemical industry because of her work. The arguments against her and her study results were unsurprisingly very often sexist, her femininity was belittled and she was called hysterical or a witch (Saeger 2017, 38–40), reproducing the stereotypes Rachel Carson criticised in the context of her study.

[12] In this chapter, the focus is on Shiva's early work. Currently Shiva addresses environmental destruction due to the production of oil (2008), the influence of different environmental crises on world peace and how to create peace with the earth (2013) or how to gain back control over the commons in the case of seeds and farming knowledge (2020).

(1993),[13] who was one of the three female founders of the *Bielefelder Ansatz*. This school was an important ecofeminist and capitalism-critical voice of the 1980s, introducing the concept of *housewifisation* (Mies, Bennholdt-Thomsen, and Werlhof 1988). Showing that the global division of labour and especially unpaid care work—which includes environmental protection—is primarily to the detriment of women, the *Bielefelder Ansatz* demands the appreciation of the subsistence economy (Baier 2008, 78). This analysis is often understood as the construction of a special affinity and duty of care for the environment on the part of women—a picture which additionally strengthens the accusation of essentialism.

In 1997, Ariel Salleh (2017) conceptualises in her book "Ecofeminism as Politics" the nature-woman-labour relationship in a comparable way to that of Shiva and Mies. By contrast, Salleh adopts a Marxist-informed analysis and focuses on motherhood as a prestigious role in the Global South allowing women as well as men to have a political consciousness (Bauhardt 2019b, 472). She emphasises the added value of ecofeminism, as it "shifts the economic analysis towards an ecological problematic" (Salleh 2017, 139), since ecofeminist analysis contradicts capitalist patriarchal patterns, demanding a new and less capitalist economic order.

The approaches described so far have in common being confronted with the accusation of both heteronormativity and of essentialism. Bauhardt (2019b, 473) instead states that the ecofeminism of this time was more concerned with the practices of care work than with an essentialisation of the female body. The impression of essentialism may have been created because care work is primarily attributed to women, not least because of the insinuation of women as mothers due to social norms and discourses.

Taking the social construction further into account, Carolyn Merchant had already dealt with the structures lying behind the nature-gender relationship during the initial phase of ecofeminism. It is therefore paramount to give recognition to Merchant's work (1990) "The Death of Nature", which was first published in 1980. The ecofeminist philosopher and historian points to the fact that for many years women have been written in various cultures as being very close to nature. Merchant rejects the essentialist conception of women by addressing the construction of both the simultaneous dominance over nature and woman as well as the proximity of the two. Considering the structural dominance over nature and women, not least for capitalist and patriarchal reasons, she discloses these

[13] For an in depth discussion of Shiva's and Mies' work and related books, see Molyneux and Steinberg 1995.

structures through historical observations, using methods reminiscent of poststructuralist informed analysis. Her "ecological approach to history" (Merchant 1989, 7) focuses on historical contingency regarding intersubjectively produced narratives, as "Each environmental problem [...] needs to be examined in the context of its own specific history as well as its linkages to global political economies" (Merchant 1990, 10). Also regarding the "interaction between science and culture", Merchant's method can be seen as "poststructuralist in spirit" (Thompson and MacGregor 2017, 50).

Having in mind that ecofeminism has always developed against the backdrop of feminist theoretical progress, the 1990s were highly influenced by the *linguistic turn* (Rorty 1975), especially with the reception of Judith Butler (1990). Feminist theorizing turned to postmodern and therewith poststructuralist thoughts in the wake of the so-called third-wave feminism (Snyder 2008), which in turn led to a certain amount of disdain towards ecofeminism and the disappearance of its warm reception (Thompson 2006, 509). Holland-Cunz (2014, 134) states that postmodernism caused ecofeminism to almost disappear in the Global North for more than two decades, notwithstanding that the debates and political actions were still present in the Global South.[14]

Although some ecofeminists in the 1990s or later opposed most of the described approaches criticising their supposed essentialism, the scholars differed in their ensuing theorising. Indeed, many avoided calling themselves ecofeminists because it carried a certain stigma, as ecofeminists were in danger of being associated with essentialism and therewith esoteric concepts. A lot of thinkers therefore linked themselves with related approaches or searched for alternative labels, such as "ecological feminism" or "feminist environmentalism" (Gaard 2011, 27). Stemming from the poststructuralist critique of ecofeminism it could be assumed that some of the ecofeminists subscribed to this school of thought, but astonishingly this is not often the case; although some approaches are characterised as poststructuralist, only a few of the theorists also locate themselves in poststructuralism by name. However, factional disputes in feminist thinking of recent decades also left its mark on ecofeminist theorising. In the following, both outlined cases of poststructuralist thinking in ecofeminism will be examined; the decidedly poststructuralist contents of ecofeminism, as well as patterns that can be decoded as such.

[14] Eaton and Lorentzen (2003) outline multiple examples of ecofeminist thinking of that time and its institutionalisation in countries of the Global South such as in India, Taiwan, Chile or Kenya.

2.3.2 Towards Poststructural Approaches in Ecofeminism

From the branding of ecofeminism as essentialistic, and pointing to the fact that women are not a monolithic group, due to their different material backgrounds, Bina Agarwal exemplifies the structural construction of the environment-woman relationship (Resurrección 2017, 74). Decisively dealing with and criticising Shiva's thinking (Agarwal 1992, 124–26) she works out the "gender and class (…) organization of production, reproduction, and distribution" (1992, 127). Instead of biologising women and their subject positions, Agarwal takes a twofold approach: On the one hand, she emphasises the materialist basis (concerning aspects such as class or caste) forming people and their roles; on the other hand, Agarwal shows the construction of the woman-nature relationship, which is structured through the assumed typical roles of women—not least due to their reproductive capacities and traditional models of gendered labour division. This observation derives from a consideration of various dichotomies shaping feminist thought. Namely, the juxtaposition of the public vs. the private (Prügl 2021, 296; MacGregor 2006) or culture vs. nature (Tickner 1993, 61; Hekman 1990, 5), the former being constructed as masculine, with the latter as the feminine sphere (Flavell 2023, 58).

This is why Agarwal (1998) warns against too traditional a view on women, that ignores their different material backgrounds, as it might additionally weaken their already marginalised position. In this case, solely enhancing the position of the people protecting nature is therefore the solution. However, Agarwal's approach bears the danger that the women's subject position is drawn in a very one-dimensional way (Sandilands 1999, 61) given that women especially are thought of as nurturing the environment in her case study of Indian grassroot initiatives. Nevertheless, rejecting the essentialist automatism of the construction of gendered roles and pointing to their materialistic basis, Agarwal lays the groundwork for an awareness of the structural and dichotomous construction of gender-nature relations. To meet the materialistic conditions of development in the context of environment and gender, Agarwal (1992, 126) evolved the theoretical strand *feminist environmentalism*, combining materialistic and social constructivist thinking.[15] The latter can be perceived as an important step towards a poststructuralist ecofeminism.

[15] Her current work builds on these thoughts and turns to empirical cases of gender relations in forestry (Agarwal 2000; 2010, 2015) and in food sovereignty (Agarwal 2014) and farming (Agarwal 2020).

Another productive critique of the pioneers of ecofeminism was explicitly poststructuralist, and was formulated by Donna Haraway (1987). With the figure of the *cyborg* she tries to break the dichotomous thinking of "animal and human, organism and machine, public and private, nature and culture, men and women" (Haraway 1987, 18). Deconstructing the dichotomies means, for her, undermining old certainties, as well as rejecting these centuries-old structures. Therefore, a multiplicity of possible subject positions arises. Moreover, the cyborg's subject position is fragmented and thus aware of its entanglement in multifaceted power relations (Hoppe 2019, 254), which is a key poststructuralist aspect of this study (cf. Sect. 2.2.1). Later, Haraway introduces the term *naturecultures,* accentuating the simultaneity and relatedness of the concepts of nature and culture (Haraway 2012, 12), which she wants to deconstruct by introducing the term. So overall, the poststructuralist Haraway, locating herself within the realm of ecofeminism, is concerned with thinking about the good life for all, rather than formulating clear recommendations for action, as Hoppe (2019, 263) sums it up. This search for a good life for all can be described as an important common feature of ecofeminist scholars.

Haraway (2012) also introduced the *compagnions*, who are meant to symbolize a queer family in the context of interacting environment, technology, and humanity.[16] Queerness and the resulting breakup of dichotomous thinking has also been addressed by Greta Gaard (1997) combining ecofeminist and queer theories. Referring to Val Plumwood's (1993) critique of Western philosophy, Gaard (1997, 117) shows that up till that point queerness had been given insufficient consideration by ecofeminism. The inclusion of queers in ecofeminist thinking enables the deconstruction of the created and at the same time determining dichotomies (Gaard 1997, 119), which otherwise lead to the construction of a "colonialist identity" (Gaard 2017, xxiv), comprising the "master model" (Plumwood 1993, 23) which subordinates other actors, as well as nature and therefore also reproduction. Gaard (1997, 120) states that queer people are also part of the oppressed, in contrast to the master model, described as "a white, largely male elite", which is socially perceived as "a *human* model, while the

[16] Currently, Haraway is increasingly turning to new materialism, which is partly discussed within the framework of this theoretical approach derived from the compagnions. Although new materialism engages with some of poststructuralist aspects, while at the same time attempting to give more weight to materiality, this strand will not be taken up here. Criticising poststructuralist thinking for seemingly negating materiality at all, Karen Barad (2003, 2007) might be the best known scholar of new materialism in ecofeminism. Stacey Alaimo (2010) is also a representative of this school of thought; together with Susan Hekman (Alaimo and Hekman 2008) she published one of the first volumes offering diverse insights in the theoretical strand.

feminine is seen as a deviation from it" (Plumwood 1993, 23; emphasis in original). The master model is a blueprint for the rational, culturally anchored white man who can subjugate both the environment and women, constructing them as *the other*. Gaard (1997, 116) specifies the dichotomies of white/nonwhite and hetereosexual/queer,as those categories also significantly shape the master identity since being white and heterosexual is constructed as superior. Therefore, ecofeminism must think about the domination and subjugation of queers alongside the domination of women and the environment. The path to a just life lies in the "dismantling [of] the system itself" (Gaard 1997, 117) with its dichotomies and their inherent hierarchisation. Following Val Plumwood, these dichotomies are "linking postulates" (Plumwood 1993, 45):

> *"Linking postulates are assumptions normally made or implicit in the cultural background which create equivalences or mapping between the pairs. For example, the postulate that all and only humans possess culture maps the culture/nature pair on to the human/nature pair; the postulate that the sphere of reason is masculine maps the reason/body pair to the male/female pair."*

If these theoretical reflections by Plumwood and Gaard are taken further in a methodological sense, poststructuralist thinking and work can be identified here, regarding the deconstruction of powerful dichotomies, which otherwise primarily serve the hegemonic anchoring of particular interests of powerful actors. The hierarchisation of linking postulates with male connotations can be traced back to the era of Western enlightenment philosophy (Tickner 1993, 60–63) paving the way for both dichotomised hierarchical thinking and thereby justifying the way for colonialism in an intellectual way. This fact can explain why these dichotomies were spread by Western cultures and societies into colonised societies such as India.

To envision some of the dichotomies, the following Table 2.3 shows Val Plumwood's enumeration, supplemented by Greta Gaard.

Table 2.3 Overview of the dichotomies (based on Plumwood 1993, 43 and Gaard 1997, 116)

culture	nature
reason	nature
male	female
mind	body (nature)

(continued)

Table 2.3 (continued)

master	slave
reason	matter
rationality	animality (nature)
reason	emotion (nature)
mind, spirit	nature
freedom	necessity (nature)
universal	particular
human	nature (non-human)
civilised	primitive (nature)
production	reproduction (nature)
public	private
subject	object
self	other
white	non-white
empowered	impoverished
heterosexual	queer
reason	the erotic

These dichotomies and their effects as linking postulates within discourses will be taken up again in the next chapter, as they enable a preliminary identification of gendered discourse patterns in the context of climate change debates and policies, for which I am searching in the analysis of the climate identities of India and the EU.

Catriona Sandilands (1999) also undertook what she calls a postmodern project, examining ecofeminism as both a political theory and democratic practice. By using the identity-focused poststructuralist theory of Mouffe and Laclau (1985), Sandilands (1999, xvii) searches for the development and positions of ecofeminism's subjects, nature and women, to finally find out how to build a future democratic project comprising ecofeminist findings. In a precise theoretical examination, Sandilands (1999) incorporates Plumwood's and Gaard's thoughts on dichotomies as well as Haraway's and Butler's approaches engaging with subject positions (which are an important take of this study, (cf. Sect. 2.2.2) and queerness (Sandilands 1997), which is why Sandilands' oeuvre can also be labelled as poststructuralist. Interestingly, her work questions Plumwood's standpoint explaining the associated nurturing of nature by women with their socially constructed position, which thus seems to be somewhat one-dimensional (Sandilands 1999, 65). In contrast, Sandilands (1997, 19; emphasis in original)

> *"argue[s], that identities have never been (and are certainly not now) as monolithic, stable, or singly positioned as many ecofeminists seem to argue, and that it is a dreadful mistake to require that they become so in order to create a narrative of future liberation or sustainability".*

Pointing to manifold and simultaneously possible subject positions of people or groups, discovering their complexity as well as the necessity to anticipate it to serve environmental sustainability in its multiple facets is the reason that politics can change. Hence, if solely a one-dimensional picture of nature or women is painted instead, it serves only a particular interest, which might stay diametrically opposed to environmental protection or include just a limited part of it. That's why the struggle of many different ideas, based on a variety of subject positions (especially concerning environment), becomes necessary in order to make, for example, the current economic order and further politics contentious (Sandilands 1999, 86). Part of this can include the parody of the supposedly lucid subject positions (Sandilands 1997) being able to lead to a multiplicity of identities and thus a variety of demands trying to answer the climate catastrophe.

Andrea Nightingale (2006) also gave a central role to the process of subjectivation and the structuring power relations in one of her early works.[17] In her well-known study of a community forestry village in Nepal, she decisively analysed the processual character of the emergence and performance of women's subject positions which she found are co-constituted by space and ecological conditions, namely the dimension of the forest, which represents far more than simply a part of their livelihood (Nightingale 2006, 171–80). Due to the contradictions between environmental protection and self-sufficiency, the forest becomes a space of negotiation of ecological conditions where power relations as well as gender roles are debated and recalibrated (Nightingale 2006, 181). Thus, neither subject positions nor gender roles are fixed but still influence a group's decisions how to shape the human-nature relations. Depending on the context and power structures, multiple subject positions exist from which different demands concerning forest management are discursively produced.

For instance, the forest acquires a different meaning when considered in relation to men and hunting (Gururani 2002, 239), as opposed to women and the gathering of fuel while entering a space to "consult[ed] each other, and spend[t] time in good company" (Gururani 2002, 238). These clear borrowings

[17] She still contributes to the importance of these power structures and subjectification in her current research, albeit with a greater focus on the socioenvironmental state and its position in transformative change rather than gender in the context of the environment (Nightingale 2018, 2017).

from poststructuralism were not only reflected in Nightingale's "deconstructivist approach[es]" (Hummel and Stieß 2017, 187) guiding her work, rather, they inform the Feminist Political Ecology (FPE) as a whole (Resurrección 2017, 76).[18] Moreover, Nightingale also recognised and integrated the diverse critiques of feminism of that time. Bell hooks (1988), Cherríe Moraga and Toni Bambara (1983) all criticised the almost west-centred white feminism focusing on middle-class women while not addressing racism or classism (Nightingale 2006, 167). An answer to the critique had to be found, as it made clear that ecofeminism, by focusing on essentialist assumptions, obscured the view on racism and other forms of discrimination. By drawing on poststructuralist methods, especially deconstruction, which leads to the disclosure and therefore criticality of power relations, Nightingale makes connections to intersectional thinking[19], which attempts to address the intertwined and mutually reinforcing forms of discrimination.

Although I found that the poststructuralist approaches in ecofeminism are fragmented to some extent, Helen Merrick (2017, 110) notes that what unites the various theorists like Plumwood, Gaard and Sandilands is "Haraway's desire to disrupt the nature/culture dualism". To sum up, it can be stated that poststructural approaches in ecofeminism at some points only represent borrowings from the poststructuralist toolbox, but—and this is important—this makes it possible to expose and deconstruct the dichotomies listed again and again in the most diverse ecofeminist theoretical strands, regardless of whether they were culturally, discursively or even materially justified in their existence, and thus make them accessible to a hegemony-critical analysis and to the question of how and for what reasons the climate change discourses of India and the EU are gendered. In order to be able to carry out this analysis which has already been theoretically derived in detail in Sect. 2.2, it is first necessary to consider what I understand by

[18] Examples of poststructuralist insights in FPE include inter alia the edited volumes of Dianne Rochelau (1996) or Christine Bauhardt and Wendy Harcourt (2019) focusing on FPE and care. Various case studies were conducted by Leila Harris (2006), Roberta Hawkins and Diana Ojeda (2011), Megan Carney (2014) as well as Rebecca Elmhirst et al. (2017). Furthermore, Yaffa Truelove (2011) and Gitta Shrestha, Joshi Deepa and Floriane Clément (2019) dedicated their research to Indian cases. Current works are a FPE centred critique of the Anthropocene narrative (Barca 2020) or the analysis of gender expertise in environment and development (Resurrección and Elmhirst 2021).

[19] Nightingale refers to the work of different scholars outlining intersectional strands (Mohanty, Russo, and Torres 1991; Butler 1997; Valentine 2007). Kimberlé Crenshaw laid the foundation for intersectionality as a concept in 1989. Nightingale turned more specifically to intersectionality and FPE in her 2011 article.

gendered discourse patterns, or those that have been identified to date by various theorists and ecofeminist projects and studies.

2.3.3 Gendered Discourses and Climate Change Policy

While the previous chapter may have given the impression that poststructuralist ecofeminism is primarily used to examine discursively-produced subject positions in terms of their gendered manifestation, this chapter introduces an additional and at the same time closely related dimension of gendered discourses, which are an important aspect of the study. To deconstruct the gendered character of an actor's identity, which can lead to a specific climate change policy, the second dimension of gendered discourses is of great interest. This dimension influences perceptions of adequate answers to climate change and produces the identity of a foreign policy actor. How this identity is produced concretely has already been outlined (cf. Sect. 2.2.3). At this point, I will distinguish the two dimensions of gendered discourses or discourse patterns and show whether and in what form they have already been identified by other scholars.[20] I will especially focus on the outcomes of gendered discourses and therefore distinguish the following two dimensions:

(I) Discourses *shaping subject positions*, such as women solely perceived as victims of climate change.
(II) Discourses *influencing perceptions of the problem of climate change, and therewith responses to it, which determine the policies of an actor*—such as India or the EU—via their discursively-produced identity (for instance, masculinised discourses on climate change exclusively demanding technical innovations to address the climate crisis).

The second dimension of gendered discourses is linked to the first dimension, particularly when specific subject positions are associated with specific climate change solutions or problems, which in turn are able to be part of the production

[20] Maria Tanyag (2023, 265) also distinguishes between two categories when she examines the nexus of gender and global governance. On the one hand, she distinguishes "gender in governance", which, similar to my first dimension, refers to the concrete thematisation of gender in climate governance, and on the other hand, "gender of governance", which refers to the gendered structure of climate governance as a whole. Since global governance research does not address the identity of states and further entities, however, I choose my own, hegemony-analytical approach to the different dimensions of gendered discourses.

of the subject positions.[21] The connection between the two dimensions can also be attested in the other direction: Dimension I can also be part of the discursive production of gendered discourses influencing understandings of climate change and an actor's identity (Dimension II). Combined with the first assumption that Dimension I can determine the second dimension of discourses; a permanent co-construction must be assumed. However, this is not necessarily always the case. Furthermore, poststructural understanding of discourses discards causalities so that Dimension I and II are not interlinked by a consequentialist logic (cf. Chap. 2). Thus, the relationship can be marked by mutual support as well as contestation. Although the discourse dimensions are important for each other, they also stand for themselves.

Gendered discourses in the climate field, being already distilled by other ecofeminist and environmental humanities scholars will be briefly shed light on in the following. Serving the later analysis, the reconstruction will of course be supplemented substantially by findings from the empirical studies of India's climate change discourse as well as that of the EU's. Therefore, working inductively-deductively is imperative (cf. Sect. 2.2.4).

First Dimension: Discourses Shaping Subject Positions

The first dimension of gendered discourses builds manifold subject positions. Pointing to the findings of the previous chapter, these are constructed by dichotomous thinking, which comprises a specific hierarchisation determining the value and power of the subject positions; examples include the subject positions of the group of small-scale farmers in India, the EU Commissioner for Climate Action, or the spokesperson of an environmental organisation. These actors are a substantial part of the discourse and some of them try to influence the second discourse dimension. Whether this influence is successful depends on the hegemonic structure and other factors, which have been discussed above (cf. Sect. 2.2.3). Consequently, the table depicting Plumwood's and Gaard's dichotomies (cf. Sect. 2.3.2) must be completed by the level of valence or in poststructuralist wording hegemony. The additional level shows which aspect of the dualisms to be deconstructed has been socially created as more valuable than the opposed by Western culture and philosophy (Tickner 1993; Hekman 1990, 5) and is therefore part of the master identity, which seems to be hegemonic. Thus, these dichotomies are better illustrated with a horizontal table emphasising the state of domination (cf. Table 2.4).

[21] Without explicitly distinguishing between the two discourse dimensions, Seema Arora-Jonsson (2009, 216) already stressed that national discourses and policies, for example regarding equality, play an important role in the subjectification of women.

Table 2.4 Overview of the dichotomies supplemented by the state of domination (based on Plumwood 1993, 43 and Gaard 1997, 116)

Mind	master	reason	mind, spirit	production	white	heterosexual
Associated with masculinity; dominating the opposite categories						
Associated with femininity; dominated by the opposite categories						
Body	slave	matter	nature	reproduction (nature)	non-white	queer

Gender and climate change are often intuitively associated with women as victims. This subject position (Discourse Dimension I) has already been reconstructed multiple times in the context of different environmentally induced aspects, such as the serious deterioration of their livelihoods (Agarwal 1992, 140; Terry 2009, 7) or disasters (Resurrecci6n 2017, 77; Neumayer and Plümper 2007). The female victim of climate change is discursively embedded in a net of different components such as the historical context and the living conditions in the pre-disaster time (Enarson 1998, 161,162) and of course aspects such as classism, racism or ageism. This analysis shows the different dimensions forming a subject position, even if, for example, it is narrowed solely to femininity. It is therefore necessary to deconstruct those discursive mechanisms when it comes to women as victims of climate change and natural disasters.

From the assumption that women are made responsible in particular for care work within families and communities, scholars showed that a further subject position arises, the female actor in environmental protection (Arora-Jonsson 2011, 747). Although ‘the women’ are thus granted agency, this subject position is not unproblematic, since agency seems to be a causal product of re/productive capacities (Bauhardt 2019a, 25) and the main responsibility for care work, which is usually not only financially devalued, seems thus attributed exclusively to women. Consequently, there is a justified fear that protecting the environment in the framework of everyday life will also be unpaid or poorly paid and therefore feminised (MacGregor 2006).

Looking at the discourses of sustainable development, Foster (2011, 137) shows that women are also mostly produced as victims of environmental deterioration while the solution is seen in their economic success being hence able to adapt to the changing climate (Foster 2017, 219). Within this strand of discourse, a distinction is made between women of the Global North and the Global South, as the analyses of several UN documents reveals that "The universalising tendencies within the discourses of *Agenda 21* construct non-Western women as a homogenised, non-liberated, oppressed, and victimised mass" (Foster 2011, 141; emphasis in original). Pointing to Chowdhry, Foster (2011, 141) underpins the notion within the sustainable development discourse, that writing non-Western women as victims "tends to undermine and reduce the multiplicity of interests and identities of women of the 'developing world'", whereby the Western women are also homogenised. In contrast, women of the Global North are discursively produced as environmentally conscious in a more liberal way, concerning their consumption choices or household-waste management. Arora-Jonsson (2011, 745) aptly describes this disparity associating the two differently characterised groups of women with the words "Virtue and Vulnerability". Thus, women are portrayed in a completely one-dimensional way, either as victims or as virtuous environmentalists. Other subject positions, for example demanding political action or producing climate problems themselves, are thus inconceivable and dilute the view of the complex reality. Similarly, the under-complex images of femininity or masculinity can be used to justify specific political approaches.

What further gendered subject positions have been reconstructed by other scholars in addition to those for women? Admittedly, it is considerably more difficult to identify research in the field of the first dimension of gendered climate discourses besides *feminised* subject positions. Masculine constructed subject positions can be presumed in the decision-making bodies, other spheres of politics or in climate sciences, which are all male-dominated (MacGregor 2014, 625). However, these subject positions are barely named as such, because they are regarded as the unquestioned norm, or universal. In fact, there is a lack of research when it comes to masculinities regarding environmental problems (Hultman 2013, 81; Gaard 2015, 22), even though a growing interest in diverse masculine subject positions is evolving (Pulé and Hultman 2021), demonstrated by Martin Hultman's (2013) study of Arnold Schwarzenegger as an "Environmental Hero". The reconstructed subject position of the actor-turned-politician shows a discursively-produced environmentally-friendly person, promoting both fuel cell and hydrogen technologies and market solutions to fight climate change. Schwarzenegger personifies this image, for instance, through the car he drives, a large hydrogen Hummer SUV (Hultman 2013, 89). Hultman, nevertheless, notes that Schwarzenegger's lifestyle as well as his policies should primarily "deflect criticism in order to perpetuate hegemony, to ensure 'business as usual'" (Hultman 2013, 89). Together with his colleague Jonas Anshelm, Martin Hultman can be identified as one of the few researchers of masculinities in the field of climate policy (Hultman and Anshelm 2017). They have also turned to the second dimension of gendered discourses, which will be examined in more detail below.

In the field of queer research in environment and climate change, it is often stated that the subject positions of queer people are sidelined and discriminated against, due to the fact that "Queer and transgendered persons already live on the margins of most societies (…)" (Gaard 2015, 24). This situation can be compared to the suppression of the subject positions of women of the Global South, following the discursive mechanism of privileging the position of the so-called 'master identity'. Having said this, at the same time, the positions as carers for the environment are not attributed to the same extent; here the findings differ as queered climate subject positions are just not visible at all. Nonetheless, research in this area is still in its infancy, as only a few studies exist (Gaard 1997; Erickson and Mortimer-Sandilands 2010; Gaard 2021) and several people interviewed for this study stated that in-depth research into queer subject positions, in India, for example, is still pending (Member of an Indian national network of climate change organisations 2019).

Second Dimension: Discourses Influencing Perceptions of the Problem of and Responses to Climate Change

The foreign policy actors' identity-building elements are the second dimension of gendered discourses, targeting the second research question of the study. Interestingly, detecting research concerning the second dimension of gendered discourses is more complex than the reconstruction of the first dimension. Even though Rachel Carson identified a discursive pattern of 'masculinised' control over the environment by the chemical industries in the 1960s, it seems that Hultman and Anshelm—of course with reference to Carson—are currently the only ones to turn to gendered discourses in this manner. They found that technical solutions to combat climate change are advocated for via masculinised discourses (Hultman and Anshelm 2017, 26), which are often antagonistic to the feminised discourses. Anshelm and Hultman (2015) distinguish four different discourses of global climate change. From these discourses, they deduced in later projects several masculinities, such as ecomodern, industrial and ecological masculinities (Hultman 2017; Hultman and Anshelm 2017; Hultman and Pulé 2018), which encompass specific ideas how to combat climate change. While ecological masculinity favours sufficiency (Hultman and Anshelm 2017, 25), it cannot assert itself against ecomodern masculinity in its preference for technical solutions in the sense of ecological modernisation, either in the Swedish climate change discourse or globally (Hultman and Anshelm 2017, 32). Roughly speaking, due to discursively reproduced masculinised norms, *mitigation*—understood as an innovative, technologised and marked-based instrument—is favoured instead of sufficiency or the idea to question the current economic order.

Following the antagonistic logic of the ecofeminist approaches elaborated above, femininity is associated with care, which means in relation to the climate catastrophe *adaptation*. Hence, we can understand demands such as adaptation, the precautionary principle or sustainable lifestyles as being associated with care and therefore feminised (Djoudi et al. 2016, 257; Johnsson-Latham 2007, 6). Furthermore, the conclusion closes in that degrowth, understood as saving emissions of greenhouse gases (GHGs) through less consumption, is embedded in a feminised discourse. Even though we can find notions of feminisation or feminised discourse patterns concerning countries suffering from climate change, it is not possible to detect systematised research on the second dimension of gendered discourses focusing on feminised or queered discourses in the sense of gendered identities of states or other entities such as the EU. It has therefore been one goal of the study to implement a methodological-theoretical framework allowing the reconstruction of both dimensions of gendered discourses shaping the gendered character of the actors' identities, in this case India's or the EU's.

2.4 Conducting Ecofeminist Hegemony Analysis

How is ecofeminist hegemony analysis conducted in research practice? Sect. 2.4.1 will first outline the six steps of hegemony analysis developed by Nonhoff (2019, 84) which is combined with ecofeminist approaches. Sect. 2.4.2 will illuminate the data selection process and thus the construction of discourse corpus.

2.4.1 The Six Steps of Hegemony Analysis

A comprehensive ecofeminist hegemony analysis of the Indian and EU identities and the question of to what extent these foreign policy identities are gendered requires a systematic and at the same time open approach. The inductive-deductive study must therefore, on the one hand, be structured to be able to analyse the amount of data with the help of MAXQDA software; on the other hand, it is not standardised, so that additional strategemes can be identified and the process of the analysis can be adapted to gendered patterns concerning both the subject positions as well as the second discourse dimension producing a specific understanding of climate change and necessary answers to it. As already outlined, the chosen discourses for study allow us to follow the procedure proposed by Nonhoff (2006, 2019).

The hegemony analysis is carried out in six steps (Nonhoff 2019, 84), and is partly enhanced by integrating ecofeminist approaches: (1) preparatory work, (2) construction of the discourse corpus, (3) structuration of the analysis, (4) analysis of single texts, (5) analysis beyond the single text/on the discourse level, (6) final checks. Steps 4 and 5 are the heart of the analysis and the point where ecofeminist hegemony analysis leads to the answers of the first and second research question. The analysis of single texts (step 4) is again divided into six stages (Nonhoff 2019, 88), enabling a granular reconstruction of the content as well as the context and the location of where gendered subject positions might appear. The concrete steps of the hegemony analysis are shown in the following Table 2.5 outlining the research practice. Although this table insinuates a strictly regulated procedure, the steps are not necessarily feasible in this very order; moreover, due to the reproductive nature of discourses, it is evident that different steps of the analysis merge into one another or occur several times.

Table 2.5 The six steps for conducting hegemony analysis (based on Nonhoff 2019, 84–99)

(1) Preparatory work	(a) Theoretical and methodological work (b) Preliminary studies
(2) Construction of the discourse corpus	
(3) Structuration of the analysis	(a) On the level of a single text (b) On the level of discourse
(4) Analysis of single texts	(a) Description of the historical context (b) Reconstruction of the location of articulation (c) Short overview of the text contents (d) Reconstruction of the hegemonic core strategemes (e) Reconstruction of other strategemes (f) Other interesting aspects
(5) Analysis beyond the single text / on the discourse level	Reconstruction of the strategemes on the discourse level
(6) Final checks	(a) Reconstruction of a discourse coalition (b) Check of the reconstructed strategemes for their validity (c) Check for possible supplementations of the set of strategemes

(1) The theoretical and methodological groundwork (1a) has been covered in the previous chapters. However, the analytical strategy in particular can be adapted on the basis of the research project and its questions (Nonhoff 2019, 84), as I do, adding ecofeminist thinking and incorporating Butler's approach. Fulfilling step 1b, the development of this research programme included the selective examination of individual texts by both EU and Indian actors.
(2) The second research step is the construction of the discourse corpus, which has to be "strictly guided by the goal to scrutinize and better understand the discursive function of hegemony" (Nonhoff 2019, 85). The selection of the corpus is made in Sect. 2.4.2 based on the central actors of the Indian and EU climate discourses (cf. Sect. 2.4.1) considering the research period from 2010 to 2021. Searching especially for the discursively (re-)produced demands and the strategemes used, different types of articulations are of interest due to the wide discourse understanding outlined above (cf. Sect. 2.2.1). Texts to analyse can therefore be speeches by the various actors, performative actions, such as announcements, documents and reports or adopted plans and laws as well as social practices such as solely combatting climate change in the sense of emergency aid or high-level attendance at climate summits. Limitations of the corpus may be evident due to the number of documents and social practices of the overwhelming climate change discourses of India and the EU. It is impossible to grasp the discourses completely regarding both space and time. From a theoretical point of view I have to concentrate on the central texts and social practices of the discourses, as a literally boundless discourse cannot be conceptualised as a closed corpus (Nonhoff 2006, 251). Furthermore, decades of discourse research have shown that researchers reach a point where they have achieved a certain saturation due to a well understanding of the research subject (Keller 2010).
(3) The structuration of the analysis, which is the third step, takes place on two levels: at the level of single texts and at discourse level (Nonhoff 2019, 87). This is not the core analysis, but the immediate preliminary work, developing a chronology of the discourse contributions to be analysed.
(4) After structuring the text corpus, the analysis of single texts follows. Task 4a is dedicated to the historical context and its description, as discourses are historically contingent (Epstein 2008, 9). In the following step 4b the location of articulation will be scrutinised determining the subject position of the actor. This is where I am able to search for interpellating demands in line with Butler's concept of subjectivation (cf. Sect. 2.2.2) as to whether the actor's position is specifically gendered. Summarising the text contents is the function of 4c. However, rather than give a detailed presentation of the content

prior to the discussion of the results of the ecofeminist hegemony analysis later in the chapters, the content is embedded in the discussion of the results. The consecutive step 4d refers to the reconstruction of the hegemonic core strategemes, which are strategemes I-III. Here, it is specifically the search for the different demands and the strategemes used to install a chain of equivalence. Reconstructing both the followed strategemes and therewith the articulated demands, I will code the relevant text passages with the analysis programme MAXQDA. Taking an explorative approach, I first search for discursive patterns in order to code them, so I mark terms or overarching themes such as 'climate justice', 'economic success' or 'culturally rooted' as well as the various strategemes used by the actors. Most often, specific demands are (re-)articulated following specific strategemes. However, it is also possible that the same demand is related with more than one strategeme as will be discussed in the later analysis. The next research step is 4e, the search for further strategemes existing on the single text level. The last step of the single text analysis 4f addresses other interesting aspects, where typical gendered discourse patterns as well as other interesting patterns such as additional strategemes could be found, which will be identified for certain in step 6 on the discourse level. The latter three steps of 4 are the core of the work, as they are "employed to reconstruct the (offensive-) hegemonic strategy" (Nonhoff 2019, 88) leading us to the determination of the gendered identity of India and the EU in the climate change context.

(5) In step number 5 the analysis shifts to the discourse level (Nonhoff 2019, 92) gaining a complete picture of the followed strategemes of the offensive-hegemonic strategy. The remaining strategemes V–IX tend to take place at the discourse level, which does not mean that individual aspects cannot be attested at the single text level as well. However, to reconstruct, for example, the gradual constitution of greater leeway for interpreting the symbolic equivalent of the universal (strategeme V) in its wide range, the selection of the discourse should be considered as a whole. This is made possible within the study in particular by coding via MAXQDA which offers different tools collating the results and identifying linkages (Herschinger 2011, 56). Strategemes VI and VII would also be examined at the discourse level. As already discussed, I anticipate that the strategemes of the second order will not be detectable in the analysed discourses due to their topicality. Nevertheless, step five of the examination has another relevant function. On the discourse level we are only able to search explicitly for the interplay of opponents (Nonhoff 2019, 92), reacting on countering demands or further statements, which will then be deconstructed.

(6) The final task of the analytical strategy provides various checks (Nonhoff 2019, 99). Some of them can have far-reaching significance for the analysis of the hegemony. First, the possible discourse coalitions consisting of different actors are to be reconstructed (6a). Which subject positions are won for the pursuit of the hegemonic project? Which coalitions work towards the hegemony? In 6b, I turn to examine the reconstructed strategemes for their validity. This critical reflection on one's own results is of enormous importance in reconstructing an offensive-hegemonic strategy and not to overvalue or appropriately contextualise individual—albeit conspicuous—aspects. The final step is the check for possible additional strategemes which have not been part of the investigation so far but which favour the implementation of the discursive formation towards a hegemony. It is an expression of the openness and explorative nature of the analytical scheme. The set of reconstructed strategemes is later expanded, both in the case of India and the EU, as derived in detail in the analysis.

These six steps are in particular the realisation of the poststructuralist hegemony analysis. As shown, the gendered demands and patterns are named and reconstructed as such in parallel. However, the ecofeminist foil is then placed on the analysed discourse in a subsequent step and discussed in a separate chapter (cf. Chaps. 3.3 and 4.3). In fact, the outlined analytical steps will not always be clearly named in the study, as they are often obvious or intertwined. Moreover, the open procedure of the ecofeminist hegemony analysis requires the flexibility to repeat the different steps, for example due to new findings, or to carry them out in a different or interwoven order. Thus, coding via MAXQDA creates a certain simultaneity in the later discussion of the analysis results.

2.4.2 Data Selection and Construction of Discourse Corpus

All in all, this study has analysed a corpus of 482 documents, of which 189 related to the Indian discourse and 293 to the EU discourse. A detailed overview of all analysed documents can be found in the electronic supplementary material. Several considerations guided the data selection and thus the construction of the discourse corpus as the second step of the analytical strategy laid out above.

(1) First, it is necessary to consider which groups of actors of the Indian and EU climate change discourses have had an interest or a say in shaping the

identity to be reconstructed in this study's period of investigation (2010 to 2021). However, as already indicated, it is neither possible nor necessary to analyse the entire discourse in order to be able to address the research questions; accordingly, not all actors and their subject positions are taken into account. Rather they are prioritised based on their contributions' significance. A detailed list of all individual actors whose discursive contributions are critically analysed can be found in the electronic supplementary material. At this point, only a more general overview of actor categories shall be presented.

On the Indian side, we are confronted with a multitude of actors in relation to climate policy. Of course, state and government actors such as the prime minister, the ministry of the environment, diplomats or parties have to be considered. Given that Indian states are required to implement climate policy, their respective climate plans will also be examined, although the focus clearly remains at the Union level. Beyond this category of actors, representatives of economic and development interests, various grassroots organisations and NGOs as well as journalists or scientists can be identified today. Initially, the circle of additional actors—besides politicians and diplomats—contributing to the climate discourse was limited. Only a few, such as The Energy and Resources Institute (TERI) as well as the Centre for Science and Environment (CSE) have already been important institutions advising the government in the 1990s (Jäger 2018, 60). But this circle has expanded considerably so that today, climate experts do also have a say as they are for instance involved in the high-level Prime Minister's Council on Climate Change, which reports directly to the Prime Minister. Furthermore, numerous environmental and climate-focused initiatives and NGOs such as Greenpeace India or Fridays for Future India have become an inherent part of the contemporary Indian climate discourse. Thinking specifically of gender and climate change in India, some organisations focusing primarily on women and gender issues have been working on climate issues for ages and have increasingly tried to extend their influence on the respective climate discourse, The All Indian Women's Conference (AIWC), for example, initiated projects on sustainable energy production and have observer status with UNFCCC (United Nations Framework Convention on Climate Change), which strengthened the gender lens within worldwide climate networks. It is therefore worthwhile to consider their contributions as well.

The actors in the EU climate discourse are broadly similar, although the fact that the EU is not a nation state affects the diversity as well as the composition of the climate discourse actors. The EU legislative process generates numerous actors on the policy side alone: the Commissioner on environment or the reshaped position of the Commissioner for the implementation of the European Green Deal, individual heads of states or governments meeting in the European Council and national environment ministers acting in the Council of the EU—who are only considered in the EU context and not with their articulations at the national level. EU parliamentarians can also hold important subject positions. Needless to say, the EU discourse is characterised additionally by a multitude of lobby actors, trying to make their reading of the problems as well as their solutions the hegemonic ones. Since climate policy is intertwined with so many different policy fields such as energy, fisheries, economics, environmental protection and many more, a clear selection has to be made here identifying the really important actors and contributions to the presumably gendered identity, without overburdening the study. These are to be found in particular in the areas of climate change debates among EU actors discussing the main direction of travel, such as the EU's role at the global high table of climate politics, future EU climate targets and the related measures to reach the EU's mostly ambitious goals. Besides environmental NGOs, such as CAN Europe (Climate Action Network Europe) representing more than 170 European organisations fighting against climate change, these are economic lobbyists. As in the Indian case, scientific and activist figures also play a role in the EU discourse, although they are mostly national actors from individual EU countries. If actors mostly operate on a national level, they will not be taken into account. However, Greta Thunberg for example is well known EU-wide and has managed to take up an important subject position pushing the Fridays for Future's demands in the discourse.

(2) Second, the selection of the data to be examined will be derived from these central actors and their statements as well as from social interactions. Obvious texts include speeches by Indian and EU politicians concerning both occasions national/regional as well as international. Furthermore, some of the climate protection programmes of India and the EU, which passed their parliaments between 2010 and 2021 or were submitted at the level of the United Nations—such as the Nationally Determined Contributions (NDCs) under the Paris Agreement—or more recent updated versions, will be analysed for their inherent demands and followed strategemes. In both cases, these plans and agreements were accompanied by extensive debates, symbolic meetings and various demands from the different politicians and further actors outlined in the latter chapter. Therefore, the public debates like newspaper articles or interviews, as well as reports and statements by NGOs in the periods surrounding the respective negotiations will be examined for their hegemonic

contributions as well as the reactions to them. To this purpose, individual contributions, for example in newspapers, are analysed, although a systematic examination of specific media products will not be undertaken, not least as these already exist. Living in an era of social media, occasional posts by the various central figures are also additional sequences to be evaluated, especially because climate policy posts are increasingly discussed under the scrutiny of a broad public.

Since social actions are also essential for the creation or reproduction of a hegemony, data such as the initiation of a political climate alliance or the consultation of important stakeholders, such as women's groups, are also examined in the analysis—if information is available. This can be scrutinised, among other things, through corresponding reports by NGOs, but also through debates with the opposition in the respective parliament. This chapter does not give a finished answer to the selected and studied sequences. Since in the course of the analysis, additional discourse contributions can occur, for example because various demands are contradicted, the corpus has remained open throughout the research process.

Accordingly, the conceptualisation of the text corpus has also been derived from the following historical overview of the discursive constitution of the climate identity that precedes the period under investigation from 2010 until 2021. The application of ecofeminist hegemony analysis to determine the contribution of gendered patterns to climate policy identities will now begin with the historical overview of the case of India.

References

Agarwal, Bina. 1992. "The Gender and Environment Debate: Lessons from India." *Feminist Studies* 18 (1): 119–58. https://doi.org/10.2307/3178217.

Agarwal, Bina. 1998. "Environmental management, equity and ecofeminism: Debating India's experience." *Journal of Peasant Studies* 25 (4): 55–95. https://doi.org/10.1080/03066159808438684.

Agarwal, Bina. 2000. "Conceptualising environmental collective action: why gender matters." *Cambridge Journal of Economics* 24 (3): 283–310. https://doi.org/10.1093/cje/24.3.283.

Agarwal, Bina. 2010. "Does Women's Proportional Strength Affect their Participation? Governing Local Forests in South Asia." *World Development* 38 (1): 98–112. https://doi.org/10.1016/j.worlddev.2009.04.001.

Agarwal, Bina. 2014. "Food sovereignty, food security and democratic choice: critical contradictions, difficult conciliations." *Journal of Peasant Studies* 41 (6): 1247–68. https://doi.org/10.1080/03066150.2013.876996.

Agarwal, Bina. 2015. "The power of numbers in gender dynamics: illustrations from community forestry groups." *Journal of Peasant Studies* 42 (1): 1–20. https://doi.org/10.1080/03066150.2014.936007.

Agarwal, Bina. 2020. "A tale of two experiments: institutional innovations in women's group farming in India." *Canadian Journal of Development Studies / Revue canadienne d'études du développement* 41 (2): 169–92. https://doi.org/10.1080/02255189.2020.1779673.

Alaimo, Stacy. 2010. *Bodily Natures: Science, Environment, and the Material Self.* Bloomington: Indiana University Press.

Alaimo, Stacy, and Susan J. Hekman, eds. 2008. *Material Feminisms.* Bloomington: Indiana University Press.

Angermuller, Johannes, Martin Nonhoff, Eva Herschinger, Felicitas Macgilchrist, Martin Reisigl, Juliette Wedl, Daniel Wrana, and Alexander Ziem, eds. 2014. *Diskursforschung: Ein interdisziplinäres Handbuch.* Sozialtheorie. Bielefeld: Transcript-Verl.

Anshelm, Jonas, and Martin Hultman. 2015. *Discourses of Global Climate Change: Apocalyptic Framing and Political Antagonisms.* Routledge studies in environmental communication and media. Abingdon, Oxon: Routledge.

Arora-Jonsson, Seema. 2011. "Virtue and vulnerability: Discourses on women, gender and climate change." *Global Environmental Change* 21 (2): 744–51. https://doi.org/10.1016/j.gloenvcha.2011.01.005.

Arora-Jonsson, Seema. 2009. "Discordant Connections: Discourses on Gender and Grassroots Activism in Two Forest Communities in India and Sweden." *Signs: Journal of Women in Culture and Society* 35 (1): 213–40. https://doi.org/10.1086/599259.

Aufhauser, Elisabeth, and Rosa Diketmüller. 2001. "Überbevölkerung Macht Armut – schafft Bevölkerungspolitik Wohlstand?" *Journal für Entwicklungspolitik* 17 (1): 47–67. https://doi.org/10.20446/JEP-2414-3197-17-1-47.

Baier, Andrea. 2008. "Subsistenzansatz: Von der Hausarbeitsdebatte zur ‚Bielefelder Subsistenzperspektive'." In *Handbuch Frauen- und Geschlechterforschung: Theorie, Methoden, Empirie,* edited by Ruth Becker. 2., erw. und aktualisierte Aufl., 75–80. Geschlecht und Gesellschaft 35. Wiesbaden: VS, Verl. für Sozialwiss.

Barad, Karen. 2003. "Posthumanist Performativity: Toward an Understanding of How Matter Comes to Matter." *Signs: Journal of Women in Culture and Society* 28 (3): 801–31. https://doi.org/10.1086/345321.

Barad, Karen. 2007. *Meeting the Universe Halfway: Quantum Physics and the Entanglement of Matter and Meaning.* North Carolina: Duke University Press.

Barca, Stefania. 2020. *Forces of Reproduction:* Cambridge University Press.

Bauhardt, Christine. 2019a. "Nature, Care and Gender: Feminist Dilemmas." In *Feminist Political Ecology and the Economics of Care: In Search of Economic Alternatives,* edited by Christine Bauhardt and Wendy Harcourt, 16–35. Routledge Studies in Ecological Economics Ser. Milton: Routledge.

Bauhardt, Christine. 2019b. "Ökofeminismus und Queer Ecologies: feministische Analyse gesellschaftlicher Naturverhältnisse." In *Handbuch Interdisziplinäre Geschlechterforschung.* Vol. 65, edited by Beate Kortendiek, Birgit Riegraf, and Katja Sabisch, 467–77. Geschlecht und Gesellschaft. Wiesbaden: Springer Fachmedien Wiesbaden.

Bauhardt, Christine, and Wendy Harcourt, eds. 2019. *Feminist Political Ecology and the Economics of Care: In Search of Economic Alternatives.* Routledge Studies in Ecological Economics Ser. Milton: Routledge.

Butler, Judith. 1990. *Gender trouble: Feminism and the subversion of identity.* New York: Routledge.
Butler, Judith. 1997. *The Psychic Life of Power: Theories in Subjection.* Stanford: Standford University Press.
Butler, Judith. 2004. *Undoing Gender.* New York: Routledge.
Cannon, Terry. 2002. "Gender and climate hazards in Bangladesh." *Gender & Development* 10 (2): 45–50. https://doi.org/10.1080/13552070215906.
Carney, Megan. 2014. "The biopolitics of 'food insecurity': towards a critical political ecology of the body in studies of women's transnational migration." *Journal of Political Ecology* 21 (1). https://doi.org/10.2458/v21i1.21123.
Carson, Rachel. 1962. *Silent Spring.* Boston: Houghton Mifflin.
Crenshaw, Kimberlé. "Demarginalizing the Intersection of Race and Sex: A Black Feminist Critique of Antidiscrimination Doctrine, Feminist Theory and Antiracist Politics." *University of Chicago Legal Forum* 1989 (1): 139–97.
Daly, Mary. 1978. *Gyn/Ecology: The Metaethics of Radical Feminism.* Boston: Beacon Press.
Demirović, Alex. 2007. "Hegemonie und die diskursive Konstruktion der Gesellschaft." In *Diskurs radikale Demokratie Hegemonie: Zum politischen Denken von Ernesto Laclau und Chantal Mouffe*, edited by Martin Nonhoff, 55–85. Bielefeld: transcript Verlag.
Djoudi, Houria, Bruno Locatelli, Chloe Vaast, Kiran Asher, Maria Brockhaus, and Bimbika Basnett Sijapati. 2016. "Beyond Dichotomies: Gender and Intersecting Inequalities in Climate Change Studies." *Ambio* 45 (Suppl 3): 248–62. https://doi.org/10.1007/s13280-016-0825-2.
Eaton, Heather, and Lois A. Lorentzen, eds. 2003. *Ecofeminism and Globalization: Exploring Culture, Context, and Religion.* Lanham, Md: Rowman & Littlefield.
Eaubonne, Françoise de. 1974. *Le Féminisme ou la mort.* Femmes en mouvement. Paris: Horay.
Elmhirst, Rebecca, Mia Siscawati, Bimbika S. Basnett, and Dian Ekowati. 2017. "Gender and generation in engagements with oil palm in East Kalimantan, Indonesia: insights from feminist political ecology." *Journal of Peasant Studies* 44 (6): 1135–57. https://doi.org/10.1080/03066150.2017.1337002.
Enarson, E. 1998. "Through Women's Eyes: a Gendered Research Agenda for Disaster Social Science." *Disasters* 22 (2): 157–73. https://doi.org/10.1111/1467-7717.00083.
Epstein, Charlotte. 2008. *The power of words in international relations: Birth of an anti-whaling discourse.* Politics, science, and the environment. Cambridge, Mass. MIT Press.
Epstein, Charlotte. 2010. "Who speaks? Discourse, the subject and the study of identity in international politics." *European Journal of International Relations* 17 (2): 327–50.
Erickson, Bruce, and Catriona Mortimer-Sandilands. 2010. *Queer Ecologies: Sex, Nature, Politics, Desire.* Bloomington, Ind: Indiana University Press. http://www.jstor.org/stable/ https://doi.org/10.2307/j.ctt16gzhnz.
Flavell, Joanna. 2023. *Mainstreaming Gender in Global Climate Governance.* London: Routledge.
Foster, Emma A. 2011. "Sustainable Development: Problematising Normative Constructions of Gender within Global Environmental Governmentality." *Globalizations* 8 (2): 135–49.
Foster, Emma A. 2017. "Gender, Environmental Governmentality, and the Discourses of Sustainable Development." In MacGregor 2017, 216–28.

Gaard, Greta. 1997. "Toward a Queer Ecofeminism." *Hypatia* 12 (1): 114–37. https://doi.org/10.1111/j.1527-2001.1997.tb00174.x.

Gaard, Greta. 2011. "Ecofeminism Revisited: Rejecting Essentialism and Re-Placing Species in a Material Feminist Environmentalism." *Feminist Formations* 23 (2): 26–53. https://doi.org/10.1353/ff.2011.0017.

Gaard, Greta. 2015. "Ecofeminism and climate change." *Women's Studies International Forum* 49 (1): 20–33. https://doi.org/10.1016/j.wsif.2015.02.004.

Gaard, Greta. 2021. "Queering the Climate." In *Men, Masculinities, and Earth: Contending with the (m)Anthropocene*, edited by Paul M. Pulé and Martin Hultman, 515–36. Cham: Springer International Publishing; Palgrave Macmillan.

Gaard, Greta C. 2017. *Critical Ecofeminism.* Ecocritical theory and practice. Lanham: Lexington Books.

Glasze, Georg, and Annika Mattissek, eds. 2009. *Handbuch Diskurs und Raum: Theorien und Methoden für die Humangeographie sowie die sozial- und kulturwissenschaftliche Raumforschung.* Sozialtheorie. Bielefeld: transcript.

Glynos, Jason, and David R. Howarth. 2007. *Logics of Critical Explanation in Social and Political Theory.* Routledge Innovations in Political Theory v.10. New York, NY: Routledge. http://site.ebrary.com/lib/alltitles/docDetail.action?docID=10204765.

Gratz, Mandy, and Anja Zürn. 2020. "Naturverhältnisse feministisch denken: Anforderungen an einen aktuellen Ökofeminismus." *Forum Wissenschaft* 37 (3): 22–25.

Griffin, Susan. 1978. *Woman and Nature: The Roaring Inside Her.* New York: Harper & Row.

Gururani, Shubhra. 2002. "Forests of Pleasure and Pain: Gendered practices of labor and livelihood in the forests of the Kumaon Himalayas, India." *Gender, Place & Culture* 9 (3): 229–43. https://doi.org/10.1080/0966369022000003842.

Hansen, Lene. 2012. "Discourse Analysis, Post-Structuralism, and Foreign Policy." In Smith, Hadfield, and Dunne 2012, 94–109.

Haraway, Donna. 1987. "A manifesto for Cyborgs: Science, technology, and socialist feminism in the 1980s." *Australian Feminist Studies* 2 (4): 1–42. https://doi.org/10.1080/08164649.1987.9961538.

Haraway, Donna J. 2012. *The Companion Species Manifesto: Dogs, People, and Significant Otherness.* 6th print. Paradigm 8. Chicago, Ill. Prickly Paradigm Press.

Harris, Leila M. 2006. "Irrigation, Gender, and Social Geographies of the Changing Waterscapes of Southeastern Anatolia." *Environment and Planning D Society and Space* 24 (2): 187–213. https://doi.org/10.1068/d03k.

Hawkins, Roberta, and Diana Ojeda. 2011. "Gender and environment: Critical tradition and new challenges." *Environment and Planning D Society and Space* 29 (2): 237–53. https://doi.org/10.1068/d16810.

Hekman, Susan J. 1990. *Gender and Knowledge: Elements of a Postmodern Feminism.* Boston: Northeastern Univ. Press.

Herschinger, Eva. 2011. *Constructing Global Enemies: Hegemony and Identity in International Discourses on Terrorism and Drug Prohibition.* London: Routledge.

Herschinger, Eva. 2012. "'Hell Is the Other': Conceptualising Hegemony and Identity through Discourse Theory." *Millennium* 41 (1): 65–90. https://doi.org/10.1177/0305829812449501.

Herschinger, Eva, and Judith Renner. 2014. "Einleitung: Diskursforschung in den Internationalen Beziehungen." In *Diskursforschung in den Internationalen Beziehungen*, edited by Eva Herschinger and Judith Renner, 9–36. Baden-Baden: Nomos.

Holland-Cunz, Barbara. 2014. *Die Natur der Neuzeit: Eine feministische Einführung.* Opladen: Barbara Budrich.

Hooks, Bell. 1988. *Feminist theory from margin to center.* 4. pr. Boston: South End Pr.

Hoppe, Katharina. 2019. "Donna Haraways Gefährt*innen: Zur Ethik und Politik der Verwobenheit von Technologien, Geschlecht und Ökologie." *Feministische Studien* 37 (2): 250–68. https://doi.org/10.1515/fs-2019-0026.

Howarth, David. 2010. "Power, discourse, and policy: articulating a hegemony approach to critical policy studies." *Critical Policy Studies* 3 (3–4): 309–35. https://doi.org/10.1080/19460171003619725.

Howarth, David, Jason Glynos, and Steven Griggs. 2016. "Discourse, explanation and critique." *Critical Policy Studies* 10 (1): 99–104. https://doi.org/10.1080/19460171.2015.1131618.

Hultman, Martin. 2013. "The Making of an Environmental Hero: A History of Ecomodern Masculinity, Fuel Cells and Arnold Schwarzenegger." *Environmental Humanities* 2 (1): 79–99. https://doi.org/10.1215/22011919-3610360.

Hultman, Martin. 2017. "Natures of masculinities: Conecptualising industrial, ecomodern and ecological masculinities." In *Understanding climate change through gender relations*, edited by Susan Buckingham and Virginie Le Masson, 87–103. Routledge studies in hazards, disaster risk and climate change. London: Routledge.

Hultman, Martin, and Jonas Anshelm. 2017. "Masculinites of Global Climate Change: Exploring Ecomodern, Industrial and Ecological Masculinity." In *Climate Change and Gender in Rich Countries: Work, Public Policy and Action*, edited by Marjorie G. Cohen, 19–34. Routledge studies in climate, work and society. London: Routledge.

Hultman, Martin, and Paul M. Pulé. 2018. *Ecological Masculinities: Theoretical Foundations and Practical Guidance.* Milton: Routledge.

Hummel, Diana, and Immanuel Stieß. 2017. "Social ecology: a transdisciplinary approach to gender and environment research." In MacGregor 2017, 186–201.

Inayatullah, Naeem, and David L. Blaney. 2004. *International Relations and the Problem of Difference.* New York: Taylor & Francis.

Jacobs, Thomas. 2018. "The Dislocated Universe of Laclau and Mouffe: An Introduction to Post-Structuralist Discourse Theory." *Critical Review* 30 (3–4): 294–315. https://doi.org/10.1080/08913811.2018.1565731.

Jäger, Manuela. 2018. *Indien in den internationalen Klimaschutzverhandlungen.* Marburg: Metropolis.

Johnsson-Latham, Gerd. 2007. "A study on gender equality as a prerequisite for sustainable development. What we know about the extent to which women globally live in a more sustainable way than men, leave a smaller ecological footprint and cause less climate change.".

Keller, Reiner, ed. 2010. *Handbuch sozialwissenschaftliche Diskursanalyse.* 4. Aufl. Wiesbaden: VS Verlag für Sozialwissenschaften.

Kennedy, Emily H., and Liz Dzialo. 2015. "Locating Gender in Environmental Sociology." *Sociology Compass* 9 (10): 920–29. https://doi.org/10.1111/soc4.12303.

Kronfeld-Goharani, Ulrike. 2015. "Zum Verständnis von Nachhaltigkeit in meeresbezogenen Diskursen." Accessed May 17, 2022. https://publikationen.soziologie.de/index.php/kongressband_2014/article/view/288/pdf_170.

Laclau, Ernesto. 1980. "Populist rupture and discourse." *Screen education* 34 (Spring): 87–93.

Laclau, Ernesto. 1994. "Why Do Empty Signifiers Matter to Politics?" In *The Lesser Evil and the Greater Good: The Theory and Politics of Social Diversity*, edited by Jeffrey Weeks, 167–78. London: Rivers Oram Press.

Laclau, Ernesto. 1996. *Emancipation(S)*. Phronesis. London: Verso.

Laclau, Ernesto. 2005. *On Populist Reason*. London: Verso.

Laclau, Ernesto, and Chantal Mouffe. 1985. *Hegemony and Socialist Strategy: Towards a Radical Democratic Politics*. London: Verso.

MacGregor, Sherilyn. 2006. *Beyond Mothering Earth: Ecological Citizenship and the Politics of Care*. Vancouver: UBC Press.

MacGregor, Sherilyn. 2009. "A Stranger Silence Still: The Need for Feminist Social Research on Climate Change." *The Sociological Review* 57 (2_suppl): 124–40. https://doi.org/10.1111/j.1467-954x.2010.01889.x.

MacGregor, Sherilyn. 2014. "Only Resist: Feminist Ecological Citizenship and the Post-politics of Climate Change." *Hypatia* 29 (3): 617–33. https://doi.org/10.1111/hypa.12065.

MacGregor, Sherilyn, ed. 2017. *Routledge international handbook of gender and environment*. Routledge international handbooks. London: Routledge.

Marttila, Tomas. 2019. "Die kritische Epistemologie der poststrukturalistischen Hegemonietheorie." In *Diskursanalyse und Kritik*. Vol. 6, edited by Antje Langer, Martin Nonhoff, and Martin Reisigl, 149–68. Interdisziplinäre Diskursforschung. Wiesbaden: Springer Fachmedien Wiesbaden.

McDonald, Matt. 2012. "Australia and Global Climate Change." In Smith, Hadfield, and Dunne 2012, 384–99.

Member of an Indian national network of climate change organisations (Interviewee 1). 2019. Interview by A. Zürn. September 19.

Merchant, Carolyn. 1989. *Ecological Revolutions: Nature, Gender, and Science in New England*. Chapel Hill: University of North Carolina Press.

Merchant, Carolyn. 1990. *The Death of Nature: Women, Ecology, and the Scientific Revolution*. [Nachdr.]. San Francisco: Harper.

Merchant, Carolyn. 2005. *Radical ecology: The search for a livable world*. 2nd ed. New York: Routledge.

Merrick, Helen. 2017. "Naturecultures and feminist materialism." In MacGregor 2017, 101–14.

Mies, Maria, Veronika Bennholdt-Thomsen, and Claudia von Werlhof. 1988. *Women: the Last Colony*. London: Zed.

Mies, Maria, and Vandana Shiva. 1993. *Ecofeminism*. London: Zed Press.

Mohanty, Chandra, Ann Russo, and Lourdes Torres. 1991. *Third World Women and the Politics of Feminism*. Bloomington: Indiana University Press.

Molyneux, Maxine, and Deborah L. Steinberg. 1995. "Review: Mies and Shiva's "Ecofeminism": An New Testament?" *Feminist Review* (49): 86–107.

Moraga, Cherríe, and Toni C. Bambara, eds. 1983. *This Bridge Called My Back: Writings by Radical Women of Color.* 2. ed., 7. printing. New York, NY: Kitchen Table.

Müller-Brandeck-Bocquet, Gisela, and Carolin Rüger. 2015. *Die Außenpolitik der EU.* Berlin: De Gruyter Oldenbourg.

Nabers, Dirk. 2015. *A Poststructuralist Discourse Theory of Global Politics.* New York: Palgrave Macmillan. http://gbv.eblib.com/patron/FullRecord.aspx?p=4082533.

Neumayer, Eric, and Thomas Plümper. 2007. "The Gendered Nature of Natural Disasters: The Impact of Catastrophic Events on the Gender Gap in Life Expectancy, 1981–2002." *Annals of the Association of American Geographers* 97 (3): 551–66. https://doi.org/10.1111/j.1467-8306.2007.00563.x.

Nightingale, Andrea. 2006. "The nature of gender: work, gender and environment." *Environment and Planning D: Society and Space* 24 (2): 165–85.

Nightingale, Andrea J. 2011. "Bounding difference: Intersectionality and the material production of gender, caste, class and environment in Nepal." *Geoforum* 42 (2): 153–62. https://doi.org/10.1016/j.geoforum.2010.03.004.

Nightingale, Andrea J. 2017. "Power and politics in climate change adaptation efforts: Struggles over authority and recognition in the context of political instability." *Geoforum* 84 (1): 11–20. https://doi.org/10.1016/j.geoforum.2017.05.011.

Nightingale, Andrea J. 2018. "The socioenvironmental state: Political authority, subjects, and transformative socionatural change in an uncertain world." *Environment and Planning E: Nature and Space* 1 (4): 688–711. https://doi.org/10.1177/2514848618816467.

Nonhoff, Martin. 2006. *Politischer Diskurs und Hegemonie: Das Projekt »Soziale Marktwirtschaft«.* Bielefeld: transcript Verlag.

Nonhoff, Martin. 2010. "Hegemonieanalyse: Theorie, Methode und Forschungspraxis." In *Handbuch sozialwissenschaftliche Diskursanalyse*, edited by Reiner Keller. 4. Aufl., 299–331. Wiesbaden: VS Verlag für Sozialwissenschaften.

Nonhoff, Martin. 2019. "Hegemony Analysis: Theory, Methodology and Research Practice." In *Discourse, Culture and Organization: Inquiries into Relational Structures of Power*, edited by Tomas Marttila, 63–104. Cham: Springer International Publishing.

Ortner, Sherry B. 1972. "Is Female to Male as Nature Is to Culture?" *Feminist Studies* 1 (2): 5. https://doi.org/10.2307/3177638.

Parker, Andrew, Mary Russo, Doris Sommer, and Patricia Yaeger. 1992. *Nationalisms and Sexualities.* Routledge revivals. New York: Routledge.

Plumwood, Val. 1993. *Feminism and the Mastery of Nature.* Reprinted 1997. Transferred to digital print 2003. Feminism for today. London: Routledge.

Prügl, Elisabeth. 2021. "Untenable dichotomies: de-gendering political economy." *Review of International Political Economy* 28 (2): 295–306. https://doi.org/10.1080/09692290.2020.1830834.

Pulé, Paul M., and Martin Hultman, eds. 2021. *Men, Masculinities, and Earth: Contending with the (m)Anthropocene.* Cham: Springer International Publishing; Palgrave Macmillan.

Resurrección, Bernadette P. 2017. "Gender and Environment in the Global South." In MacGregor 2017, 71–85.

Resurrección, Bernadette P., and Rebecca Elmhirst, eds. 2021. *Negotiating gender expertise in environment and development: Voices from feminist political ecology.* Routledge studies in gender and environments. London: Routledge, Taylor et Francis Group.

Reyes, Rosa R. 2002. "Gendering responses to El Niño in rural Peru." *Gender & Development* 10 (2): 60–69. https://doi.org/10.1080/13552070215907.

Rocheleau, Dianne E., ed. 1996. *Feminist Political Ecology: Global Issues and Local Experience.* International studies of women and place. London: Routledge. http://www.loc.gov/catdir/enhancements/fy0649/96000425-d.html.

Rogers, James. 2009. "From 'Civilian Power' to 'Global Power': Explicating the European Union's 'Grand Strategy' Through the Articulation of Discourse Theory." 47 (4): 831–62.

Rorty, Richard. 1975. *The linguistic turn: Recent essays in philosophical method.* Chicago: University of Chicago Press.

Ruether, Rosemary R. 1975. *New woman/new earth: Sexist ideologies and human liberation.* New York: Seabury Press.

Ryberg, Jesper. 1998. "The Argument from Overpopulation: Logical and Ethical Considerations." *Population and Environment* 19 (5): 411–26.

Saeger, Joni. 2017. "Rachel Carsons was right—then, and now." In MacGregor 2017, 27–42.

Salleh, Ariel. 2017. *Ecofeminism as politics: Nature, Marx and the post modern.* 2nd revised edition. London: Zed Books Ltd.

Sandilands, Catriona. 1997. "Mother Earth, the Cyborg, and the Queer: Ecofeminism and (More) Questions of Identity." *NWSA Journal* 9 (3): 18–40.

Sandilands, Catriona. 1999. *The Good-Natured Feminist: Ecofeminism and the Quest for Democracy.* Minneapolis: University of Minnesota Press.

Schieder, Siegfried, and Manuela Spindler. 2010. *Theorien der Internationalen Beziehungen.* 3. Aufl. Opladen: Budrich.

Schroeder, Richard A. 1987. "Gender Vulnerability to Drought: A Case Study of the Hausa Social Environment." Working Paper 58. Accessed March 03, 2021. University of Wisconsin.

Shiva, Vandana. 1988. *Staying Alive: Women, Ecology and Development.* London: Zed.

Shiva, Vandana. 2008. *Soil not oil: Environmental justice in a time of climate crisis.* Brooklyn: South End Press.

Shiva, Vandana. 2013. *Making Peace with the Earth.* London: Pluto Press.

Shiva, Vandana. 2020. *RECLAIMING THE COMMONS: In defense of biodiversity and traditional knowledge:* SYNERGETIC Press.

Shrestha, Gitta, Deepa Joshi, and Floriane Clément. 2019. "Masculinities and hydropower in India: A feminist political ecology perspective." *Int J Commons* 13 (1): 130. https://doi.org/10.18352/ijc.920.

Smith, Steve, Amelia Hadfield, and Timothy Dunne, eds. 2012. *Foreign Policy: Theories, Actors, Cases.* 2. ed. Oxford: Oxford Univ. Press.

Snyder, R. C. 2008. "What Is Third-Wave Feminism? A New Directions Essay." *Signs: Journal of Women in Culture and Society* 34 (1): 175–96. https://doi.org/10.1086/588436.

Tanyag, Maria. 2023. "Feminist Governance and Climate Change." In *Handbook of Feminist Governance*, edited by Marian Sawer, Lee A. Banaszak, Jacqui True, and Johanna Kantola, 262–73. International handbooks on gender. Cheltenham, UK: Edward Elgar Publishing.

Terry, Geraldine. 2009. "No climate justice without gender justice: an overview of the issues." *Gender & Development* 17 (1): 5–18. https://doi.org/10.1080/13552070802696839.

Thompson, Charis. 2006. "Back to Nature?" *Isis* 97 (3): 505–12. https://doi.org/10.1086/508080.

Thompson, Charis, and Sherilyn MacGregor. 2017. "The Death of Nature: Foundations of ecological feminist thought." In MacGregor 2017, 43–53.

Tickner, J. A. 1993. "States and Markets: An Ecofeminist Perspective on International Political Economy." *International Political Science Review* 14 (1): 59–69. https://doi.org/10.1177/019251219301400104.

Torfing, Jacob. 2005. "Discourse Theory, Achievements, Arguments and Challenges." In *Discourse Theory in European Politics: Identity, Policy, and Governance*, edited by David R. Howarth and Jacob Torfing, 1–32. Basingstoke: Palgrave Macmillan.

Truelove, Yaffa. 2011. "(Re-)Conceptualizing water inequality in Delhi, India through a feminist political ecology framework." *Geoforum* 42 (2): 143–52. https://doi.org/10.1016/j.geoforum.2011.01.004.

Valentine, Gill. 2007. "Theorizing and Researching Intersectionality: A Challenge for Feminist Geography*." *The Professional Geographer* 59 (1): 10–21. https://doi.org/10.1111/j.1467-9272.2007.00587.x.

Villa, Paula-Irene. 2012. *Judith Butler: Eine Einführung*. 2., aktualisierte Aufl. Frankfurt, M. Campus-Verl.

3 The Indian Climate Policy Identity and its Gendered Components

After laying the theoretical foundation of the study, the findings of the ecofeminist hegemony analysis will be presented and discussed in detail. The demands articulated and the strategemes employed in order to build specific hegemonic formations will be reconstructed and analysed in the context of ecofeminist thinking.

First comes a brief overview of the Indian climate change discourse up to 2010 (Sect. 3.1). The intention is not to conduct an extensive discourse analysis in advance of the period under study (2010–2021), but rather to present relevant information for later contextualisation with sporadic references to gendered discourse contributions which could become virulent and visible at a later point. Even though the reconstruction of these earlier discourses is not comprehensive, it is still crucial because discourses are historically contingent. Current strands of climate change discourse are informed by knowledge of former key developments that might be gendered. Yet, in the context of the reconstruction of the discourse prior to the research period, the gendered patterns are scarce and, moreover, only implicit and should be understood merely as indications for the subsequent ecofeminist hegemony analysis.

To address global climate policy as a state, national institutions and discursive spaces are necessary in order to deal with political questions of adaptation, mitigation, climate protection as well as its implementation. Therefore, how the climate discourse in India developed and who was initially able to use the discursive spaces for their own demands must first be clarified. Moreover, the question arises as to how India has behaved in the context of global climate discourses,

---

Supplementary Information The online version contains supplementary material available at https://doi.org/10.1007/978-3-658-46774-6_3.

A. Zürn, *Of Climate and Gender*, https://doi.org/10.1007/978-3-658-46774-6_3

and what India's demands and positions have been on the global stage. Since both levels influence each other and set impulses, Sect. 3.1 will first consider the general genesis of the discourse, before Sect. 3.1.1 will scrutinise India's role on the global stage in more detail. Sect. 3.1.2 will then reconstruct the national climate policy discourse and its specific actors as well as political structures. Together, these chapters will set the stage for the results of the ecofeminist hegemony analysis of the years 2010–2021 and their in-depth discussion in Sects. 3.2.1 to 3.2.7.

3.1 The Genesis of India's Climate Change Governance in an International Context

At the first international conference on the protection of the environment in Stockholm in 1972, one of the few participating heads of government was Indira Gandhi, then Prime Minister of India (United Nations 1972, 43). Attending the conference and adopting first environmental laws, such as the "Wild Life (Protection) Act" in 1972 (Government of India 1972) or the 1986 "Environmental Protection Act" as a direct reaction to the catastrophe of Bhopal (Swenden and Saxena 2020, 19) could underline the meaning that India henceforth attached to environmental protection. But the simultaneous emphasis on India's right to develop and the challenge of poverty eradication limited the agenda substantially. India instead adopted the narrative of being the advocate of the Global South, making the industrialised countries responsible for their historical emissions (Michaelowa and Michaelowa 2012, 576). However, this role has been contested in recent years (cf. Sect. 3.1.1). The once seemingly monolithic group of developing countries differentiated itself, not least as the climate catastrophe in certain countries materialised more strongly or more visibly than in other states (Thaker and Leiserowitz 2014, 112). Hence, India finds itself in a constellation of greater complexity. On the one hand, the states particularly affected by climate change are calling emerging powers to take far-reaching mitigation measures (Banerjee 2012, 1778). On the other hand, the Indian Union is demanding space to develop (Plagemann and Prys-Hansen 2020, 284), while relying on fossil energy sources. This apparent fundamental contradiction is also reflected in Indian climate policy at the national level (cf. Sect. 3.1.2). One can observe various developments and actors trying to reconcile these opposites or to advance their own agenda more strongly. However, India often emphasises its long tradition of closeness to nature (Jayaram 2018, 189). Its need for protection is often highlighted by tribal groups and rural citizens (Priyadarshini and Abhilash 2019), but also by the government (Narlikar 2017, 102)—if it fits into the particular interest or hegemonic project.

How India's environmental and climate governance is structured will be looked at in more detail later (cf. Sect. 3.1.2). The Indian Ministry of Environment was introduced in 1980 at the national level—it was one of the first Ministries in the world solely responsible for environmental questions (Ciecierska-Holmes et al. 2020, 4), one of which is how to tackle climate change. For a long time, only a very small group of people were able to take a vital part in the climate discourse of the Indian Union (Thaker and Leiserowitz 2014, 110). Apart from bureaucrats of the Ministry of the Environment and the Ministry of Foreign Affairs, the Centre for Science and Environment (CSE) and The Energy and Resources Institute (TERI) were the only other actors. In particular, the critical report "Global Warming in an Unequal World: A Case of Environmental Colonialism" by the CSE (Agarwal and Narain 1991) paved the way for the Indian negotiating line at international climate conferences (Rajamani 2009, 355), pointing to the historical responsibility of developed countries. At least until 2007 we speak about an almost closed group of actors (Fernandes, Jörgensen, and Narayanan 2020, 169). With the National Action Plan on Climate Change (NAPCC) in 2008 and the Clean Development Mechanism (CDM) of the Kyoto-Protocol, scientists and economists gained a larger role (Sengupta 2018, 220–21), as did further NGOs. Those developments led to a diversification of the climate change discourse concerning both the participating actors and the debated topics (Thaker and Leiserowitz 2014, 110).

The individual subnational states of the Indian Union are also significant to the implementation of the discussed measures and policies (Jörgensen, Mishra, and Sarangi 2015) as they simultaneously face huge challenges, such as the lack of resources (Dubash 2021, 7). At the national level, actors such as ministers, government officials, and political parties are also confronted with various shortcomings: Highly divergent interests and resulting political pressure as well as increasing public awareness that human suffering from climate change is already considerable, as eroding coasts, heat waves or the loss of natural livelihood sources dramatically show.

Thus, the following chapters will address what topics are relevant in the Indian climate change discourse and what demands are articulated. Furthermore, gendered aspects of the discourse will also be addressed. In order to set the context for the later ecofeminist hegemony analysis, I will also reconstruct the time immediately prior to the period of investigation. First, however, India will be examined in the context of international climate policy, as the debates and decisions taken on the global stage have an influence on national climate protection measures

(Atteridge et al. 2012, 71) and the global discourse also informs India's political efforts shaping its identity by differing from the radical other.

3.1.1 India on the Global Stage until 2010

Indira Gandhi set the tone of India's international stance with her speech in 1972, insisting that poverty reduction must have clear priority. The goal was to generate growth by following the Western model (Sengupta 2012, 112), which was based on fossil fuels. Accordingly, *Development first* became the central demand (Fernandes, Jörgensen, and Narayanan 2020, 161), insisting on the Indian paradigm of climate equity (Dubash 2013, 192). Equity meant in this context, that the developed countries caused the partly irreparable consequences of climate change and are therefore responsible for combatting the problem. This paradigm created the central antagonism on the global stage between developing and developed nations: development on the one side and climate protection on the other, framed as incompatible.

India thus became a key actor of the Global South (Saran and Jones 2017, 23), uniting behind the 'polluter pays' principle. In 1990, the Union government hosted the first conference in Delhi to jointly coordinate on environmental issues with allied developing countries. It was basically decided that the obligation of reducing GHGs lies with the developed countries (Saryal 2018, 6) and that the Global South has no need to assume mitigation obligations; also that technical and financial support must be provided by the Global North (Sengupta 2012, 106). Anchoring the CBDR-RC principle (common but differentiated responsibilities and respective capabilities) in the UNFCCC (United Nations Framework Convention on Climate Change) is one success of India "as a deal maker" (Fernandes, Jörgensen, and Narayanan 2020, 163), following the concerns of the G77 and further developing countries in the 1990s.

In contrast, India played only a minor role in the creation of the Intergovernmental Panel on Climate Change (IPCC) in 1988 (Dubash et al. 2018, 401). The IPCC regularly publishes Assessment Reports updating the public and global actors on the current scientific knowledge of climate change, concerning both physical and socioecological research findings. One reason for India's limited involvement may be the fact that Indian climate change research was not well developed at these times (Ramachandran 2012, 69). Today, climate sciences in India are more advanced and in 2002, the then-Director General of the Indian TERI, Rajendra Pachauri, was appointed as the IPCC Chair, which "raised the attention to the topic within the domestic scientific community" (Michaelowa and

Michaelowa 2011, 7–8) not least as the IPCC was awarded the Nobel Peace Prize in 2007.

Reflecting again on the UNFCCC process for negotiating a joint agreement to combat climate change, the Indian Union rejected mitigation pledges due to sovereignty reasons and continued to stress this during the pre- and post-Kyoto processes (Raghunandan 2020, 206), at least until the Copenhagen Summit in 2009. Although in 1992 the Rio Summit took place and tried to soften the dominating antagonism between economic growth and environmental protection by introducing the approach of sustainable development (Chaturvedi and Doyle 2015, 48), India's paradigm didn't shift.

Interestingly, in Rio sustainable development (SD) was conceptually closely linked to the role of women (Foster 2017, 220). Based on the Brundtland Report, SD is defined as "development that meets the needs of the present without compromising the ability of future generations to meet their own needs" (World Commission on Environment and Development 1987, 24). Even though women's groups were involved in Rio and its preparation (Bretherton 1998, 86), the definition of SD does not mention the role of women. Nevertheless, a separate chapter was dedicated to women in the adopted Agenda 21 (United Nations Conference on Environment & Development 1992, 271). In this chapter, particular attention is paid to the vulnerability of women to environmental and climate degradation, producing the role of women as victims (Foster 2017, 221). Further constructed roles shaped in the Rio process included the caring actor for environmental protection and women as workers contributing to development (Bretherton 1998). Emma A. Foster (2017, 222) clearly places this new development in the discourse of environmental protection and development within a neoliberal framework promoting especially the economic success in SD. Whether these explicitly gendered discourse contributions at the global level had an influence on the Indian climate change discourse will be shown in the following chapters. However, these initial impulses trying to reconcile development and climate action within the framework of SD do not seem to have had any influence on India's position at the global level in these times.

An example of the consistency of the Indian stance is the text of the Kyoto Protocol (KP), adopted in 1997, reaffirming the CBDR-RC principle and distinguishing between Annex 1 and non-Annex 1 countries. The latter are not committed to mitigation measures, while industrialised states have to fulfil concrete Greenhouse Gases (GHGs) reduction obligations. India's understanding of equity in the context of climate governance was clearly anchored in the KP with the distinction of the states and the CBDR-RC principle;[1] a lucid success for the country's interests.

Another KP-agreement is the Clean Development Mechanism (CDM). Annex 1 and non-Annex 1 countries cooperate to save GHG emissions in the non-Annex 1 country and credit the saving to the industrialised state of Annex 1. India initially opposed the CDM mechanism (Sengupta 2012, 107), but it was supported by domestic business actors and therefore the Union government agreed to the adoption of the mechanism (Fernandes, Jörgensen, and Narayanan 2020, 163; Thaker and Leiserowitz 2014, 112). In the meantime, India had become one of the largest partners of CDM-projects (Pulver 2012, 256). The advantages of climate protection were recognised and utilised in this specific case. Nevertheless, further Indian commitments concerning mitigation efforts were unthinkable for a long time. Negotiating the roadmap to initiate a successor agreement of Kyoto, India underlined its traditional position at COP13 in Bali 2007 insisting on the exclusive mitigation responsibilities of the developed countries (Saryal 2018, 8). Working intensively with developing countries, attention was paid to both the distinction between Annex 1 and non-Annex 1 countries and the rejection of MRV (measurement, reporting and verification) actions for developing countries (Sengupta 2012, 109). From the Indian side, MRV is opposed particularly for reasons of sovereignty, a sacrosanct value of India, especially because of its colonial past (Lowinger et al. 2021, 43). The closer the world community moved towards Copenhagen, the more signs of a shift in India's rigid stance became visible—or at least first steps in the direction of a broader and more nuanced position at the global high table.

Confronted with international power dynamics and further developing the will to gain an active global role, India felt the need to move closer to the USA, who were highly interested in a viable climate agreement in Copenhagen in 2009 (Raghunandan 2020, 207). While the bloc of states of the Global South was still dissolving, new alliances were forged. The 2009-built collaboration of India, China, Brazil, and South Africa—BASIC—deals exclusively with climate policy and demonstrates the Indian ambition for international recognition, as well as a new self-confidence of these emerging powers. Interestingly, India is credited with playing a central role in the formation of this alliance (Hallding et al. 2013, 612). At the first glance, this cooperation seems surprising, especially because of the rivalry with China. However, BASIC is just one example of several such issue-specific clubs (for example G20, BRICS or IBSA) of which BASIC was

[1] For a detailed discussion of India's understanding of CBDR-RC as a vital part of the climate justice norm contrasted with the EU's understanding, see Lowinger et al. 2021, 51–56.

particularly successful in the past. The fact that these four countries have joined forces can be explained, among other factors, by their incorporation by the major industrialised countries in the debates leading up to Copenhagen—for example in the context of the G8 + 5 summit in Heiligendamm in 2008 or the inclusion of the states in the Major Economies Forum on Energy and Climate (MEF) initiated by the USA (Hallding et al. 2013, 616–617). Building a counterweight to those developed states seems smart, not least because all BASIC countries, including India, wanted to strengthen their global role. Furthermore, the Indian Union recognised that it could no longer avoid an active role in global mitigation efforts, due to pressure from many different sides. In parallel to the industrialised countries, the LDCs (Least Developed Countries) and especially the island states insisted on greater effort from emerging countries such as India (Raghunandan 2020, 207–208).

To fulfil this ambitious new role, India made its first commitments in the run-up to Copenhagen, which seems to contradict the previous Indian stance of equity and justice. Then-Prime Minister Manmohan Singh undertook the official recognition of the two-degree target in early July 2009 at the MEF summit in l'Aquila (Singh 2009), which was a momentous step at that time and can be seen as the first ever binding climate statement of India. A rather vaguely formulated project of "per capita plus", which later led to the concept of "equal access to sustainable development" at COP16 in Cancún was presented by then-Environment Minister Jairam Ramesh (Atteridge 2013, 61) in an attempt to balance traditional and new Indian approaches to climate change (Atteridge et al. 2012, 71). Partly due to pressure from first announcements of China, India has also pledged in Copenhagen to reduce its emissions intensity by 20–25% by 2020 (compared to 2005), which is however not that ambitious and could be reached by "business-as-usual strategies" on the part of the ministries (Betz 2012, 17). Nonetheless these are historical first commitments. The deviation from the traditional Indian path was highly controversial on the domestic front and was hotly debated from every angle, politicians, NGOs and further civil society actors as well as receiving heavy media coverage (Vihma 2011, 82). Due to its active role and commitments undertaken, India slowly but surely became a crucial actor who was no longer 'solely' the naysayer.[2] A first step in this direction was Singh's promise back in

[2] In the context of its international politics and diplomacy, India has a reputation of not being able to say "yes" (Cohen 2001, 66), Michaelova and Michaelova (2011) have dealt with this analysis in relation to India in the international climate negotiations. Timo Lowinger (2019) traced the causes of this finding and dismantled historical and structural domination reasons, arguing that Indian behaviour is often reduced to the formula "The India that can't say yes" (Cohen 2001, 66) without factoring in the underlying root causes.

2007 that India will never reach the per capita emissions of developed countries (Atteridge et al. 2012, 70), also known as the "Singh Convergence Principle" (Dröge and Wagner 2015, 3). The subsequent drive to slightly change India's narrative in Copenhagen and thereafter (Betz 2012, 9) can be seen as a turning point.

But overall, the Copenhagen Accord (CA) must be deemed a failure in terms of global climate policy and the urgent need of massive mitigation efforts (Müller-Brandeck-Bocquet, Pietzko, and Zürn 2018, 13). In contrast, India can broadly chalk up the outcome of Copenhagen as a success, even though the Global North only pledged superficial measures. On the one hand, India's role was massively strengthened; the final agreement was negotiated by the USA and BASIC, a group brought together by India (Qi 2011, 308), while excluding the hitherto mostly involved EU (Falkner, Stephan, and Vogler 2010, 257). On the other hand, the securing in the CA of the right of developing countries to an equal access to the atmosphere, and thus the right to growth, bears India's hallmark (Fujiwara 2010, 12). In addition, the new global climate governance approach of bottom-up action was introduced in Copenhagen (Falkner, Stephan, and Vogler 2010, 259). This new approach is in India's interest, as bottom-up climate action pledges allow for a uniting of the positions both of rejecting any legally binding mitigation targets and of embracing the new self-confidence as a proactively engaging emerging power. Since the bottom-up-initiative can also be traced back to BASIC, the crucial Indian role in Copenhagen becomes clear.

Keeping the huge shift in India's until then stable climate stance in mind, the phase after Copenhagen with the follow-up negotiations from Cancún in 2010 onwards become highly interesting. Therefore, it is important to examine in depth the Indian climate discourse regarding its hegemonic strategy and the strategemes that were employed, as they might have changed fundamentally during and after Copenhagen. The reconstruction leads to an Indian climate identity which might be gendered, helping to vindicate the specific Indian climate change policy or specific measures. In order to deconstruct the discourse, we first need an insight into the national climate change discourse up till 2010 to become familiar with the roots of the Indian climate governance as well as important structures and protagonists trying to shape the Indian climate discourse.

3.1.2 Indian Climate Change Governance on a National Level

"Climate change was for a long time only a foreign policy issue in India" (Isaksen and Stokke 2014, 113), not least due to the already-delineated position that developed countries alone must combat climate change, whereas the Union government followed the 'development first' credo even before the liberalisation of its markets in 1991. India's domestic climate protection efforts are more diverse today than its initially restrictive stance would suggest. As already shown, various measures were taken to protect the environment beginning in the 1970s. Although their effectiveness and lack of support will be discussed in the following chapters, it is already evident that industrial development projects can produce conflicts with nature protection. Hence, the Indian society looks back on manifold social practices of fighting for environmental protection for more than five decades (Lele 2012, 214). Social movements and NGOs working on biodiversity, chemical pollution, and environmental degradation due to development projects such as dams or the protection of forests from industrial exploitation, are just a few examples (Andharia and Sengupta 1998, 429–31). The groups were united in their diversity concerning both their different strategies (Andharia and Sengupta 1998, 435) as well as various organisational forms (Andharia and Sengupta 1998, 426). Their common ground included both the factoring in of social aspects of environmental protection such as specific vulnerability of the poor (Lele 2012, 211)—especially of women—being highly concentrated at the local level of environmental or developmental threats (Aamodt and Stensdal 2017, 121), and the negation of any commitment to mitigation measures with the outlined argument of global justice.

This position was largely influenced by the 1991 CSE (Centre for Science and Environment) Report "Global Warming in an Unequal World" (Agarwal and Narain 1991) addressing a World Resources Institute (WRI) report which calculated probably for the first time the global shares of GHGs (Nachiappan 2019, 559). According to the CSE, methane, for example, which is produced inter alia in the context of rice production, was calculated disproportionately by the WRI compared to the luxury emissions of developed countries (Agarwal and Narain 1991, 2). This painted a distorted picture of the emissions as well as the resulting responsibilities to the detriment of the Global South—the authors call it "A case of environmental colonialism" (Agarwal and Narain 1991, 1). In the eyes of the CSE, it is nevertheless the rich North's historical obligation to combat global warming (Agarwal and Narain 1991, 16). Concretising its approach, this line of argument was rapidly taken up by the Indian government (Vihma 2011,

78), although the MoEF (Ministry of Environment and Forests) was attacked by the CSE for not taking any action against the WRI study (Agarwal and Narain 1991, 9). Moreover, the CSE demanded that the government take serious engagement to regenerate its environment and improve energy efficiency (Agarwal and Narain 1991, 19). It seems striking that the CSE held one of the central positions of consultation to government officials despite in some cases being highly critical. As one of the few non-governmental institutions in India's exclusive round of climate change discourse actors it was—besides the other central think tank, TERI—the CSE that provided decision-makers with scientific climate change expertise. These think tanks had played an important role in the Indian climate change policy, serving as thought leaders and mediators between politics and the public, since policy research had been established at universities only after the environmental discourse was already underway (Ganguly 2020, 75). However, the role of these think tanks is not static and has changed in the last decade, as the later analysis will show.

Along with a few think tanks, a number of NGOs are also important to generating the historical context of the study at hand. One of them is the internationally well-known Chipko Movement, founded in 1973, opposing the industrial use of forests especially in the Himalaya region (Bahuguna 1980, 2). Chipko later became a nationwide campaign as the struggle of forest resources were discussed in different spheres such as the technical, economical or ecological (Shiva and Bandyopadyay 1986, 14). The questions of who the forest belongs to and who is allowed to harvest its goods—the tribal inhabitants ensuring their livelihoods or the industry striving to achieve development goals of the state—were thus prominently raised (Lele 2012, 211). Women had a special role within the movement, not least as they are the first to suffer from lack of natural resources needed for fodder or as fuel for their daily living (Bahuguna 1980, 13). Ecofeminist literature in particular has dealt with this grassroots organisation (Mies and Shiva 1993; Warren 1988). Attesting that these works placed women together with nature and the belief in their purity, these texts are discussed as controversial (Sandilands 1999, 55). However, what lasts to this day when it comes to the Chipko movement is the semantic construction of the 'natural' protection of forests and environment by women.

Even if it's uncertain whether the trees were literally hugged by the women (Mitra 1993) it remains clear that especially in the 1970s their engagement in the form of demonstrations, occupations or other peaceful actions—in the Gandhian tradition of "satyagrahas"—was used to protect the forests (Shiva and Bandyopadyay 1986, 10–11). As a reaction, the government in 1980 placed a moratorium on tree-cutting at the foot of the Himalayas for 15 years until nature had recovered.[3] Apart from this success, the forest-based movement, comprising many additional groups besides Chipko, was also able to forestall the controversial Forest Bill of 1982 (Andharia and Sengupta 1998, 429). Chipko is today known as one of the pioneers of significant environmental protection organisations in India. While the forest movement has been successful in parts, taking stock of the environmental movement is difficult, not least because the multifaceted landscape of diverse environmental NGOs has continued to engage only locally or at incident-level (Andharia and Sengupta 1998). Although the movement achieved an increase in awareness of environmental issues—especially due to the catastrophe in Bhopal 1984 (Ciecierska-Holmes et al. 2020, 4)—climate change continued to play only a minor role. This changed at the turn of the century (Aamodt and Stensdal 2017, 121) and at least from 2007, it can be attested that the civil societies' climate change action expanded heavily (Swarnakar 2019, 256).

International pressure as well as rapidly developing global networks of climate-change NGOs which on the one hand were already present in India and on the other hand integrated existing Indian NGOs, led to enhanced engagement with the climate change questions of the Indian environmental civil society (Ylä-Anttila and Swarnakar 2017, 274), looking beyond single environmental protection issues. Interestingly, expanding the focus and therewith actively engaging in the climate change discourse, for instance presenting their struggles and interests at the COPs (Ylä-Anttila and Swarnakar 2017, 277), changed the NGOs' approach from generally fighting poverty in harmony with nature, to searching for justice (Ganguly 2020, 76). However, different NGOs concentrate on justice in different ways. While the CSE is still committed to international justice, Greenpeace India (GPI) triggered a broad debate on domestic social justice in 2007. The report "Hiding Behind the Poor" shows that people's emissions vary greatly depending on their income classes (Greenpeace India 2007, 10). According to the report, rich Indians seem to be "hiding behind the poor", as poverty remains widespread, keeping the emissions-intensive lifestyles of the top income classes out of reach of the 800 million people of the poor who are at the same time much more vulnerable to climate change effects (Greenpeace India 2007, 4). GPI derives from the study results the demand to decarbonise India's development,

[3] The effectiveness of the law is evaluated differently today. On the one hand, it must be stated that fires were started in order to be able to use some trees, as green felling was forbidden (Chandran 2012, 13). On the other hand in some Indian regions the people were able to achieve similar successes (Schreurs and Papadakis 2009, 128) as various laws were passed based on the moratorium leading to today's more closely interwoven network of connected forest areas.

insisting that it does not mean, that "India should not develop or the wealthy should stop consuming" (Greenpeace India 2007, 2).

These findings and political claims of GPI and later even more so the planned NAPCC broadened the debates. The GPI and the later CSE study led to a multifaceted vibrant Indian discourse, concerning both domestic as well as international policies for fighting climate change (Chakravarty and Ramana 2012, 218). The CSE challenged the GPI narrative, showing that emissions in the industrialised North continue to be the real problem, as they claim that the "richest Indians emit less than poorest Americans" (Centre for Science and Environment 2009). Indeed, it is certainly no coincidence that the CSE factsheet answering GPI's report was published in 2009. It was the year of the Copenhagen Summit and the broad political and media debate about India's role and responsibility in global climate change policy. India faced international pressure inter alia because Western states argued that consumption will increase in emerging countries in the future. Therefore, the GPI report fed the industrialised state's storyline (Chakravarty and Ramana 2012, 226). During these times, the domestic Indian debate was dominated by the questions of what justice means and what significance the first major climate policies of India—with the adoption of the NAPCC in 2008—have for the people and their abilities to develop. Reconstructing the genesis and content of the NAPCC is a particularly important element laying the groundwork for the following ecofeminist hegemony analysis.

Representing a milestone of Indian domestic climate change policy, the eight primarily technical missions of the NAPCC such as the "National Mission for Enhanced Energy Efficiency" or the "National Mission on Sustain Habitat" (Government of India 2008) visualise the focus on co-benefits, the new narrative of the plan's policies (Pillai and Dubash 2021, 103). Besides the clear mitigation focus, the NAPCC also deals with adaptation action pointing to the vulnerability of women and the need for adaptation programmes to improve and secure the livelihoods they care for (Government of India 2008, 14). The adoption of the NAPCC had already been announced in the 11th Five-Year Plan of the Indian Government, where the dichotomy of climate protection and development (Planning Commission Government of India 2008, 22) was dissolved with the demand for "inclusive growth", as already stated in the title of the plan. The NAPCC's narrative of co-benefits is also implicitly reflected in the Five-Year Plan, published shortly before the NAPCC. Referencing the co-benefits of the promotion of renewable energies in order to guarantee energy security as well as climate protection (Planning Commission Government of India 2008, 23), the Five-Year Plan put the narrative of co-benefits into the framework of inclusive growth, which should therefore be understood in this context as sustainable development. Co-benefits are elaborated in the NAPCC as "measures that promote our development objectives while also yielding co-benefits for addressing climate change effectively" (Government of India 2008, 2). The new approach of co-benefits combines the previously antagonistic demands of development first and CBDR-RC (Dubash 2021, 9).

Within the Prime Minister's Council on Climate Change (PMCCC), which was established in 2007 to develop the NAPCC, it was the group of the representatives of TERI, CSE and the MoEF that shaped the new narrative of co-benefits and tried to show that climate change measures can be an integral part of other policy goals (Aamodt and Stensdal 2017, 121); it seems that the traditional actors again shaped the discourse. However, this impression is deceptive, and the discourse was henceforth influenced by many different actors, especially through the drafting of the NAPCC. The Council consisted of 26 people, including different ministers, prominent representatives of leading NGOs, and retired experts (Dubash and Joseph 2016, 47). The newly-established Council was considered to play an important role in the reconfiguration of India's climate policy because, as the name suggests, it is directly attached to the Prime Minister, who also held the chair. However, after the drafting of the plan, which was finalised by a small group "from within the Council—the principal scientific advisor, former secretary to the then Union Ministry of Environment and Forests and the director-general of TERI" (Sangomla 2021), the PMCCC met less frequently authorising the implementation of the plans until the Council was effectively dead by 2011 (Pillai and Dubash 2021, 104). This notwithstanding, the importance of the NAPCC can hardly be overestimated. It initiated a broad debate (Vihma 2011, 77), materialised the new narrative of co-benefits and at least the Solar Mission and the National Mission on Energy Efficiency took off to be implemented relatively quickly (Never 2012, 371). Furthermore, the NAPCC sent a clear message to the international partners: Count India in.

The implementation of the NAPCC was accompanied by another initiative of the then Prime Minister Singh who appointed a Special Envoy on Climate Change (Pillai and Dubash 2021, 99). This position was realised to guarantee the central coordination of the implementation of the NAPCC and was also to lead the negotiating team at climate conferences (Pillai and Dubash 2021, 103). The Special Envoy was located directly in the Prime Minister's Office, as was the PMCCC. Although the position of the Special Envoy did not last long due to responsibility conflicts, and the NAPCC has only a non-binding character (Pillai and Dubash 2021, 102–3), these initiatives show the new importance that was henceforth attached to climate policy at the national level. In addition, the hitherto exclusive group of actors in the climate discourse expanded significantly.

The shift in India's climate policy was driven by different factors. The already-shown international pressure was highly influencing (cf. Sect. 3.1.1) but national considerations such as energy security (Vihma 2011, 76) and economic interests as well as the increasing knowledge of future environmental threats due to climate change were also driving forces. The most important person embodying the shift in India's climate change track was the Environment Minister Jairam Ramesh, appointed in 2009 (Never 2012, 380). Although he was only in office for two years, in retrospect, Ramesh enhanced the significance of climate change as a domestic political issue, not least due to his affinity for the media (Michaelowa and Michaelowa 2011, 17). Because of Ramesh's controversial standpoints, Indian media coverage of climate change—with respect to both the enforced domestic mitigation measures and the international dynamics—increased enormously in the lead-up to Copenhagen (Michaelowa and Michaelowa 2012, 578–587). Even though Ramesh's personality was an essential aspect of reshaping India's climate discourse, the shift was also in Singh's particular interest. Ramesh described the instructions he received from Singh to change India's international perception; in the field of climate policy, this meant that India should become part of the solution to the problem even though it did not cause it (Ramesh 2015, Chap. 10).

In fact, Ramesh's initiatives, subsumed under the term "per capita plus"[4], and his political impetus—as revealed in a leaked letter addressed to PM Singh—to abandon the Kyoto Protocol and unbundle from G77 (Sethi 2009) have been strongly criticised from many sides (Rastogi 2011, 135). For example, criticism came from different politicians in the context of the until then very rare climate policy debates in parliament taking place before and after the summit in Copenhagen (Vihma 2011, 82). On the one hand, these high-profile debates underlined the importance of climate change, while on the other hand the government was subject to allegations of terminating the informal domestic consensus on "development first". However, the tabled initiatives also provoked debates within the government (Sengupta 2012, 110). Interestingly, most of the NGOs excoriated the government's new course in climate policy (Vihma 2011, 84) as they did not

[4] In an Interview with Padmaparna Ghosh (2009), Ramesh discussed various measures that fall under his rather vague concept of per capita plus. Among them are domestic mitigation efforts in the industrial, transport and energy sectors. Ramesh's main argument was that India itself will be a major victim of climate change and therefore finding good regulations is in its own interest. Furthermore, Vihma (2011, 76) mentions India's new negotiating approach of listening more and rejecting less as part of the per capita plus approach.

want India to give up its opposition to the industrialised countries and the specific understanding of climate justice. Given that Ramesh was called the "NGO Minister" because of his greater cooperation with civil society instead of business actors (Never 2012, 380), this finding may be surprising. However, not all NGOs shared this criticism; a few were growth-critical and did not condemn India's initiative to be open to climate change commitments. Further actors opposing Ramesh included a part of the existing negotiating team (Rastogi 2011, 135–36), and the Special Envoy Shyam Saran (Vihma 2011, 84) who served prior as the Foreign Secretary of India and therefore had an excellent network.

The opposition from so many sides also explains why the negotiating position in Copenhagen was less proactive than it initially seemed it could be. Nevertheless, the main commitments made by Singh and Ramesh, such as reducing energy intensity and not exceeding the per capita volume of industrialised countries, held. Furthermore, they are part of Ramesh's legacy showing that India is willing to be part of a global solution, easing the discourse on climate justice and international equity and promoting climate policy initiatives on a domestic level (Never 2015, 98). The idea of initiating individual State Action Plans on Climate Change (SAPCCs) ensuring that all Indian states have the best solutions for their needs was also developed by Ramesh (Dubash and Joseph 2016, 49). Some of the SAPCCs will be examined in more detail later (cf. 3.2.5), as they contain specifically gendered discourse elements.

Ramesh also set up an expert group on low-carbon growth, which submitted a report to inform the influential Five-Year Plan by the Planning Commission. Aditya Valiathan Pillai and Navroz K. Dubash (2021, 103) note, however, that these and other projects, such as the newly installed climate science network, connecting the fragmented Indian researchers working on climate change questions, did not really endure Ramesh's tenure. Rather, it seems that the dynamic period of the new institutionalised groups in the field of climate policy took its toll, with overlaps leading to unproductive conflicts. For example, it is assumed that Ramesh with his advisers and the low-carbon group has outstripped the PMCCC (Never 2015, 101).

Later, Ramesh became the Minister for Rural Development in July 2011 and entered the rank of a full cabinet minister. Although this ministry is known to be more prestigious than the MoEF it may be expected that he acted in too polarising a manner as environmental minister and was therefore forced to change his post. The period of Ramesh's tenure from Copenhagen to Cancún in late 2010, which has not yet been explored, will be looked at in more detail in the following analysis (cf. Sect. 3.2.1). Reflecting on the period of Copenhagen, it seems that the time was not yet ripe for Ramesh's proactive approach or for his initiatives to change the institutional structures of Indian climate change policy. Nevertheless, Singh and Ramesh raised important stimuli during this lively period of domestic climate change discourse. The fact that India is already suffering from climate change and will be an even greater victim in the future is a case in point. But the most important legacy of Ramesh and Singh is probably the creation of the new narrative of co-benefits which was, together with the first international commitments, a turning point. Whether the anchoring of this new narrative is relatively stable or rather fragile, and whether or for what purpose this discourse strand is gendered, the following analysis will show; there, the reconstruction of Indian climate policy, both domestic and international, is wrapped up in the background of the ecofeminist analysis of the 2010–2021 Indian climate discourse.

To sum up the comprehensive reconstruction of the preceding period, the current chapter has shown that gender was sometimes discussed in specific contexts such as the NGOs' struggle for development opportunities in harmony with nature. Overall, however, if gender references were made, they were only implicit. For the later analysis it should be kept in mind that the topics closely linked to climate change, such as development, co-benefits and justice, can certainly show gendered patterns, but again that these appear only in a strongly implicit or non-visible way. The part-absence of the gender topic so far is therefore not necessarily due to the fact that the discourse was not gendered until 2009. It was probably gender-blind so that gender relations or other aspects were not addressed. What this means exactly will be illuminated later in the discussion of the findings of the subsequent analysis (cf. Sects. 3.1.1 and 3.1.2).

3.2 India's Gendered Climate Change Discourse: (Co-) Benefitting from Sustainable Development and a Significant Role

Following the reconstruction of the main aspects of the Indian climate change discourse until 2009, the results of the ecofeminist hegemony analysis of the research period from 2010–2021 will be delineated and discussed in the upcoming chapters. The reconstruction of the single texts of the corpus, concerning the historical context, the location of the articulation, and the content (step 4 (a)–(c), cf. Table 2.5 above) will be succinctly put; in fact, the focus lies on the identification of the strategemes (step 4 (d)-(e)), and the rearticulated demands analysis in the context of the research questions.

The text corpus with 189 texts (cf. annex the electronic supplementary material) was analysed using the qualitative analysis software MAXQDA. Different strategemes and demands, as well as subject positions reconstructed in the single texts have been coded, identifying frequent and salient discourse patterns leading to a hegemony. Operationalising the ecofeminist hegemony analysis, the strategemes were coded and the diverse demands were separately marked. As discussed in the theoretical framework, the character of the demands can differ, as they are cumulative, subsuming or encompassing (cf. Sect. 2.2.1). In order to reduce complexity, I will only point to a demand's character if it is likely that it is the encompassing one which might prove to be the empty signifier. Of course, demands such as *climate justice* can be articulated following different strategemes, for example to separate the discursive space (strategeme II) or when different demands are equivalised (strategeme I) building a chain of equivalence. As already shown, after identifying particularly the core strategemes on the single text level, the strategemes that occur at the discourse level will be reconstructed, and the results analysed in a subsequent step (step 5). The combined findings entail the full deconstruction of the strategemes shaping the offensive-hegemonic strategy, which leads to the gendered Indian identity in the context of climate change.

The distinction made in Sect. 2.3.3 between gendered discourses in the first and second dimension is operationalised as follows. In the single text analysis, the gendered subject positions can be identified, whereas the gendered discourse patterns in the second dimension, which influence the possible solutions of the climate change problem, can especially be reconstructed on the discourse level of the analysis, exploring whether the specific reading is shared by the decisive discourse participants. In concrete terms, this means reconstructing the dichotomies introduced in the ecofeminist chapter, which sort the world along gender stereotypes.

Although it is primarily the core actors in climate policy—already identified in the previous chapters—who shape the Indian climate discourse, the actors contributing to the discourse, and therefore those analysed, will change in the course of the research period. The main reason is the change of the Union government in 2014 from the Congress-led coalition, which had been in power for a long time up to that point, to the coalition led by the Bharatiya Janata Party (BJP). Consequently, the group of important actors varies over time and furthermore expands considerably. In the analysis of gender structures, even though they haven't featured largely in previous chapters, the demands of gender organisations such as the All Indian Women's Conference (AIWC) concerning Indian climate change policies, are also of interest, even if they can only be taken up sporadically due to a lack of documents.

These and further circumstances, such as elections or events on the global climate stage structure the chapters into different time periods and sections. Depending on how virulent the climate discourse was in the respective time span and affected by important stimuli from outside, the number of analysed discourse contributions differs. After reconstructing the discourse on single text as well as at discourse level, step 6 (c) is followed in Sect. 3.2.8 discussing whether the hegemonic protagonists of the gendered Indian climate change discourse use additional strategemes which have not yet been identified to make their hegemonic project a success. The first dimension of gendered discourses regarding gendered subject positions as well as the second dimension dealing with the question on how to answer the consequences of climate change will be discussed in the chapters reconstructing and analysing the discourse in time spans (cf. Sects. 3.2.1–3.2.7). Both dimensions will be examined through the gender lens of ecofeminism in the final section (cf. Chap. 3.3–3.3.2). To do so, it is necessary to build on the insights gained in the former chapters, as the gendered dichotomies structuring the climate change discourse are mostly only implicitly detectable. Beginning with the focus on Ramesh's term (2009–2011) and the remaining period of the congress-led coalition which ended in 2014, the main findings of the in-depth ecofeminist hegemony analysis will be presented and discussed. In this context, I will work out how the already reconstructed shift in India's climate policy stand has been consolidated—or perhaps contested—as groundwork for the subsequent terms of different governments trying to shape the Indian climate change discourse.

3.2.1 The Different Facets of Sustainable Development

In 2010 in the run-up to Cancún, Jairam Ramesh still acted in the spirit of Manmohan Singh's mandate to "give India a positive and constructive image in global negotiations" (Ramesh 2015, The Running Thread I).[5] Ramesh stressed the importance of Indian action to play an important global role in his keynote address at the Asian Development Bank Conference in 2010 with the words: "As responsible global citizens of a responsible power, we should be concerned and

[5] Most of the speeches by Ramesh analysed in this project were published in his book "Green Signals", which at the time of writing was solely available as an e-book version without page numbers. For better traceability, the book chapter is given in the text and the occasion and time of the analysed speech or document in the footnote. The section of text reproduced here is the prologue of the book formulated by Ramesh.

we need to act. An increase in our international role comes with certain responsibilities" (Ramesh 2015, Chap. 3). Under these responsibilities, Indian mitigation measures can be understood, as Ramesh shows the connection between the particular vulnerability of the poor and the need for adaptation measures. Ramesh articulates an equivalence of the resulting demands *poverty eradication* and *Indian adaptation measures* and points additionally to *Indian mitigation actions*, as he argues that all those aspects are needed to improve both poverty reduction and also environmental protection (Ramesh 2015, Chap. 3). Equivalising those demands, he follows core strategeme I (cf. Sect. 2.2.4) of the offensive-hegemonic strategy (re-)producing the Indian chain of equivalence. Simultaneously, the use of strategeme VII (Occasional but deliberate breach of the antagonistic line of division) can be reconstructed. The demand *mitigation action* as part of "the radical other's" chain of equivalence, which is most likely one of the industrialised states, was picked and integrated into the Indian chain of equivalence. Thus, the outreach and success of the hegemonic project can be increased through strategeme VII (Nonhoff 2006, 234). The benefit of moving this demand across the boundary of the two antagonistic chains of equivalence will be made clear later, when the relevant hegemonic project of the time has been elaborated.

Ramesh's reasoning served the purpose of justifying, especially domestically, the abandonment of the policy of strictly rejecting (voluntary) climate commitments in Copenhagen in late 2009. The analysed discursive contributions manifested the demand that India should play a *stronger international role* in climate change politics. It was anchored successively as a new nodal point within the Indian climate change discourse, mainly by the Congress-led government of Singh and particularly by Environmental Minister Ramesh. Indeed, India's own international role was also of interest to the government before 2010. Yet, the ecofeminist hegemony analysis shows that this demand gained a new quality as it put other vibrant demands such as *mitigation action solely by developed countries* (Singh 2010; Ramesh 2015, Chap. 3)[6] into perspective following the fundamental shift of Copenhagen opposing the former image of the Indian 'naysayer' (cf. Sect. 3.1.1).

As the demand *strong role of India in international climate negotiations* brackets together other demands, such as *Indian mitigation measures* or *progress of Indian climate change sciences*, it might be the encompassing demand we are searching

[6] The text analysed here was his keynote address at the Asian Development Bank Conference in 2010 with the title "The Environments of the Poor: Making Sustainable Development Inclusive".

for. Pushing this demand, Ramesh (2015, Chap. 5)[7] argued that India benefits in international negotiations from a better domestic scientific knowledge basis and autonomy from merely "borrowed" climate data and models. In his COP16-speech in Cancún, the then environment minister (Ramesh 2010, 3) pointed to the newly established Indian Network for Comprehensive Climate Change Assessment (INCCA). The INCCA was already providing various expert reports on climate issues in India at that time (Ministry of Environment and Forests 2011, 265) clearly focusing on technical aspects of climate change instead of the consequences for the society and its structures. One of INCCA's first published reports "Climate Change and India: A 4×4 Assessment. A Sectoral and Regional Analysis for 2030s" focuses on agriculture, forests, human health and water (INCCA 2010). Strikingly, only in the area of human health are the socio-political consequences of climate change considered in more detail. The social practices—both the initiation of the INCCA and its reports—are discursive contributions in the sense of the demand *progress of Indian climate change sciences* stabilising the demand *strong role of India in international climate negotiations* as a nodal point. Visually spoken, *Indian mitigation measures* as well as *progress of Indian climate change sciences* are demands of the spider's web, leading to the nodal point of a strong Indian role.

Ramesh's high priorities of Indian scientific progress in climate change and the implementation of a network of knowledge production (Fernandes, Jörgensen, and Narayanan 2020, 169) also became clear in his speech at the Conference of the International Association for the Study of the Commons, entitled "Managing the Commons: From Academics to the Real World" (Ramesh 2015, Chap. 4). Reaffirming his position, he drew the attention to the importance of science and emphasised that the key scientific findings are not sufficiently taken into account, especially in the context of climate negotiations (Ramesh 2015, Chap. 4). By recognising scientists as important stakeholders, seeking their advice and engaging with them on an equal footing, he repeatedly installed them as important subject positions in the Indian chain of equivalence. Accordingly, strategeme VI "(Re-)articulation of subject positions for politico-societal forces" has been pursued by Ramesh. Whether and how this serves his potential hegemonic project of a strong Indian role will be shown in the course of the analysis.

The demand for progress in Indian climate sciences can be inter alia traced back to an error by the IPCC that received worldwide attention. The fourth Assessment Report in 2007 insinuated that the glaciers in the Himalayas will have

---

[7] This statement is taken from the introduction to the fifth chapter of Ramesh's book and not part of a speech.

melted completely by 2035 (Intergovernmental Panel on Climate Change 2007, 493). Due to multiple complaints, Ramesh initiated the preparation of a status report. This was written by the well-known Indian geoscientist V.K. Raina who revealed the IPCC's error in his late 2009 report (Raina 2009). Rajendra Pachauri, the then head of the IPCC initially rejected Raina's report in an Indian news-show contemptuously as "voodoo science" (Pachauri 2009, 4:27–4:34). After a later review the IPCC had to admit the mistake made in its 2007 Assessment Report in 2010 (The Guardian 2010). Yet, this IPCC error should not be overestimated, as the conclusions drawn from the various calculations are still valid. Rather, what it shows is the growing importance of scientists in the Indian climate discourse and their increased recognition by politicians as well as by the media, where previously they had not belonged to the small circle shaping the climate discourse in India (cf. Sect. 3.1.2).

Even though the demand *progress of Indian climate change sciences* and the new subject positions of scientists within the Indian discourse were not contested by other actors, with the exception of Pachauri[8], the subsuming demand that India should play a *stronger role* internationally was challenged by different stakeholders, pointing to another central demand: *climate justice* (CJ). The CSE insisted on *climate justice,* engaging with other actors like policy makers (Centre for Science and Environment 2011a, 23). As the think tank traditionally demands *CJ* in its various forms, such as equity in burden-sharing (Narain 2010a, 2010b) meaning that solely the North is responsible for mitigation action, the CSE contested the demand for a *stronger Indian role* since Ramesh linked it with *Indian mitigation measures.* Following strategeme VI ((Re-)Articulation of subject positions for politico-societal forces) the CSE tried to include further subject positions in the ongoing discourse, enhancing its standpoint which can be understood as a counter-hegemonic project opposing the government's interest in a stronger role. Journalists had been trained in the run-up to international climate negotiations since 2009 by the CSE (Centre for Science and Environment 2010, 50) to increase the hitherto marginal media coverage of global climate politics in India (Boykoff 2010, 21). Furthermore, the CSE financially facilitated the reporting of journalists directly from the COPs since 2009 and also briefed them there (Centre for Science and Environment 2010, 50). Although this does not mean that the journalists worked on behalf of the CSE or exclusively published its view, they had special access to the CSE experts. Consequently, the CSE's

[8] The outlined example of the error in the IPCC report does not mark a radical opposition of Pachauri to his colleagues of Indian climate sciences, rather Pachauri explicitly welcomed the initiative for stronger networking of Indian climate scientists at the Sustainable Development Summit 2010 (Pachauri 2010, 4).

position, which contradicts Ramesh's particular interest of a *strong Indian role*, has been disseminated in the Indian media. Interestingly, one of the supported journalists reporting from Cancún in 2010 criticised the government for "further dilut[ing] its stance on climate change negotiations" (Sethi 2010). He pointed to the dichotomy of emission reduction and economic growth in the same way the CSE regularly used to. Underlining the need for climate justice to increase the living standard of the poor, which in the CSE's view automatically means rejecting binding emission measures, was for a long time the defining component of Indian environmentalists (cf. Sect. 3.1.2). The critical position of most Indian journalists is confirmed by Anu Jogesh (2012) via his quantitative study of Indian media coverage of climate policy. He concludes that in the period from September 2009 to March 2010, 55% of opinion pieces studied in the Indian English-speaking print media criticised the government's new position of being open for emission targets and only 27% of the columns supported the stance of those in power (Jogesh 2012, 281).

The hegemony analysis shows so far that although the new nodal point of a strong India in climate negotiations was introduced volubly by powerful government actors, it was also challenged and questioned by well-established stakeholders such as the CSE. Since the progress of Indian climate science was not opposed by the CSE, the demand for a stronger role was not attacked in all its components, or in other words, the demand has not been completely invalidated as a nodal point. While the counter hegemonic project pushed by the CSE must be considered not entirely successful, the demand *strong role of India in international climate negotiations* did not generate the necessary support to serve as an encompassing demand able to gain the status of an empty signifier representing all demands of the chain of equivalence and therefore the discursively constructed identity. Nevertheless, Ramesh and Singh were successful in establishing the demand *strong Indian role* with all its masculinised facets of strongness and economic success as a new central reference point of the Indian climate change discourse.

Returning to the prevailing chain of equivalence during Ramesh's term, further nodal points can be identified, which are in some cases more broadly supported. Besides *energy security, environmental protection* and *co-benefits*, this was especially *sustainable development (SD)*. Different actors, such as TERI, Greenpeace India or the Federation of Indian Chambers of Commerce and Industry (FICCI) and especially the government, repeatedly demanded that SD features its importance for India's wellbeing concerning both the economy and the ecology.

SD as a concept has been virulent since Rio 1992 (cf. Sect. 3.1.1), gaining momentum when the idea of the Sustainable Development Goals (SDGs) was discussed at the Rio + 20 Summit in 2012 as a follow-up to the Millennium Development Goals (MDGs). Developing a more holistic approach, the SDGs combine sustainability with development and an inclusion of climate aspects addressing all countries of the global community. SD as a concept has also been discussed in India for many years. The best example is the annual SD Summit in Delhi initiated and organised since 2001 by TERI and paying special attention to the situation in developing parts of the world (The Energy and Resources Institute 2022). With renowned international guests from politics, economics, research and civil society and many further participants, the summit enjoys a strong reputation, which is why the speeches given there by central actors of the Indian climate discourse are also part of the analysis.

In his 2010 speech Singh followed strategeme II dividing the discursive space antagonistically when he called for *sustainable growth* in India, distinguishing the Indian Union from the 'radical other' (cf. Sect. 2.2.3), the industrialised countries. However, the specific meaning of SD seems to be left open: "We recognise that we have to adopt a different model of growth to that followed by the industrialized countries. But a lot of effort is needed to operationalise the meaning and content of sustainable development." (Singh 2010, 2). Singh split the discursive space at further points in this speech, for example by emphasising the *historical responsibility of industrialised states* to combat the climate catastrophe or demanding *financial support* from them (Singh 2010), insinuating that India stands as part of a poor South on the one side and the rich North on the other. Furthermore, he highlighted the Indian *co-benefits* of *mitigation measures* as they go hand-in-hand with better health conditions or a cleaner environment (Singh 2010, 2). Therewith, Singh reproduced *co-benefits* as an additional nodal point of the then Indian chain of equivalence, which remains in the tradition of the narrative enshrined in the NAPCC. Singh's 2010 speech accurately represents the discourse of the time as it was shaped by the government. He equivalised several nodal points such as *co-benefits*, *poverty eradication* and *a strong Indian role* using strategeme I. Presenting SD as another equivalised nodal point he followed once again strategeme I. Arguing that "In a poor country like India, where hope and deprivation co-exist, sustainable development requires that the needs of the present are given at least as much attention as the needs of the future" (Singh 2010, 2), Singh tries also to construct SD as a common societal reference point which most of the people can identify with, following strategeme III 'representation' (cf. Sect. 2.2.4).

Since the three core strategemes are followed in this speech, it must be assumed that Singh is pursuing his hegemonic project to install SD as the empty signifier representing India's identity. This impression must first be compared with that of other central actors, so that it can be analysed later as to whether this hegemonic project also finds the necessary dissemination in the discourse—beyond relevant actors such as the government or NGOs.

What is certain so far, is that SD serves also as a specific point of reference in Singh's reasoning since several nodal points are linked discursively to it. This already known pattern of justification follows Ramesh defending his actions at the 2010 COP16 to India's members of parliament (MPs). They heavily criticised Ramesh for his compromising formulation at the summit in Cancún alluding that "all countries must take on binding commitments in an appropriate legal form" (Ramesh 2015, Chap. 10)[9], which was domestically understood as another unilateral affront to abandon the decades-old consensus to reject any binding measures for reasons of sovereignty and historical responsibility. In his justification to the MPs, Ramesh rebuffed the accusation that any liabilities were incurred, using the nodal point SD in particular to show the added value of his concession at the global level. In doing so, he followed strategeme I discursively producing various equivalences, linking SD and *growth* as well as *a strong Indian global role*, *climate justice* and the demand of *poverty eradication* (Ramesh 2015, Chap. 10). Using strategeme II the discursive space was divided between India and developed countries raising SD to distinguish India from the CO_2-intensive growth history of developed countries. The latter shows that only SD seems to be able to contradict the antagonistic demands of the chain of equivalence of the "radical other", such as the growth model of the Global North. This ability of the encompassing demand *sustainable development* is an important indication that it serves as the empty signifier of the Indian chain of equivalence (Nonhoff 2019, 91).

When Ramesh followed strategeme I equivalising many different demands culminating in SD, *climate justice* in particular was repeatedly connected with SD emphasising that "Largely due to India's efforts, references to 'equity' and 'equitable access to sustainable development' were included" due to India's initiative in Cancún (Ramesh 2015, Chap. 10). Thus, the criticism by various actors, such as the CSE, which particularly contests the demand of a *strong India* pointing to the paradigm of *climate justice*, is invalidated by Ramesh linking *climate justice* with SD. The impression is created that if SD is achieved, *climate justice* serves as one mosaic stone of the empty signifier.

Strategeme I worked in many directions: The approach of otherwise fundamentally critical voices such as Greenpeace India (GPI) connected well with SD.

[9] This and the following passages are taken from the letter sent to all MPs after the Cancún Climate Change Conference in Chap. 10.

Without literally framing their activities with references to SD, GPI campaigned for renewable energy (Greenpeace India 2012, 13) and initiated the famous Delhi-project "Switch On The Sun" pointing to the importance of renewable energy sources to be used in parallel both to meet the people's needs and to combat climate change (Greenpeace India 2014, 46). The project has received wide attention from the media and politicians of different parties such as the Congress and the BJP (Greenpeace India 2014, 48). GPI's subject position was therefore integrated in the chain of equivalence due to their engagement in the renewable energy sector. Other stakeholders are more concrete in their commitment to SD such as different business actors. While fearing strict regulations (Carbon Disclosure Project 2011, 11), they have enjoyed participating in CDM projects in the past and therefore have a clear interest in intensifying this engagement regarding SD for economic reasons (Federation of Indian Chambers of Commerce and Industry 2012, 22). As economic success is culturally linked with masculinity (Magnusdottir and Kronsell 2015), we can speak about a masculinised demand which is followed here by FICCI and CDP.

In its annual report 2012, the CSE also dealt with SD, and particularly urged India to make its growth paradigm inclusive and sustainable, hence it follows a mixed understanding of SD, containing both masculinised and feminised patterns. The CSE announced that it would help to develop approaches in response to the question of how to achieve sustainable growth (Centre for Science and Environment 2012a, 5). Interestingly, the CSE's discursive contribution as well as a statement of the largely conservative FICCI shows that the Indian concept of SD is open for very different interpretations and understandings, hence it is not possible to define whether it is a masculinised or feminised demand at this point of the analysis. Recognising leeway, FICCI stressed that "countries should be given freedom to develop their own sustainable development indicators/criteria" (Federation of Indian Chambers of Commerce and Industry 2012, 6) when it comes to concrete SD-projects.

Neither did the Congress-led government develop its own definition of SD, as the WSDS speech of Singh already showed. Instead, SD is so vaguely specified that Ramesh (2015, Chap. 4) even situates the redefinition of GDP namely in terms of his idea of "Green Domestic Product" in the framework of SD within a speech to the Finnish Parliament in 2012[10], and on further occasions (Ramesh

[10] The title of the speech, held on 11.04.2012 was "Making Sustainable Prosperity Happen NOW".

2015, Chap. 7)[11] without concretely outlining what it means to him. Therefore, it can be assumed that SD is open for different interpretations and priorities—this corresponds to the use of strategeme V. However, in its 12th Five-Year Plan the Indian Government pointed to the well-known but also highly unspecific Brundtland definition of SD (cf. Sect. 3.1.1). Insisting on the tripartite understanding of SD in the context of agriculture declaring that the economic, ecological as well as social factors of SD are equally important (Planning Commission Government of India 2013b, 115), the 12th Five-Year Plan follows the UN's understanding of SD in the framework of the global SDGs (United Nations General Assembly 2015, 1). In fact, sustainability in different contexts is an omnipresent reference point of the Five-Year Plan, titled "Faster, More Inclusive and Sustainable Growth" (Planning Commission Government of India 2013b). Its fourth chapter "Sustainable Development" (Planning Commission Government of India 2013b, 112) particularly considers economic growth and therewith related low carbon strategies needed to improve different economic sectors such as transport, housing, cement production and further technical facets to generate SD. The SD chapter concentrates mainly on the economic aspects of SD, while the ecological and social facets play only a minor role and are dealt with in later Five-Year Plan chapters without a specific SD focus. Accordingly, a masculinised understanding of SD can be assumed, at least on the part of the government, which primarily seems to underline the growth aspect.

Nevertheless, in the third volume of the Five-Year Plan, a chapter is dedicated to "Women's Agency and Child Rights" dealing in a short paragraph with women's agency in climate change (Planning Commission Government of India 2013a, 171). Women are constructed as subject positions as the government promises to take their needs in the areas of mitigation and adaptation into account, especially within the framework of the assessments of the future State Action Plans on Climate Change (SAPCCs). Referring specifically to poverty reduction and development programmes in which gender and women's aspects will henceforth also be addressed (Planning Commission Government of India 2013a, 171–72), a gendered climate discourse of the first dimension can be identified. The later analysis of the SAPCCs will show for which subject positions the women are recognised (cf. Sect. 3.2.5) and whether the assumption is confirmed that women's subject positions are primarily narrowed to the realm of care and vulnerability against the background that women are constructed as a homogeneous group.

---

[11] Ramesh hold the address 'The Way to Green GDP' at the India Today Conclave on 18.03.2011.

Regarding the SD chapter of the Five-Year Plan, a notable example of the technical focus are the recommendations of the Expert Group on Low Carbon Strategies for Inclusive Growth which was appointed by the Planning Commission. Twelve focus areas were identified, only one of them addresses the ecological dimension: Improving the stock of forest and tree cover (Planning Commission Government of India 2013b, 118). The other focus areas are dedicated to economic facets, such as energy production, technical improvements in the steel or cement industry, and transport (Planning Commission Government of India 2013b, 118). Overall, it can be suspected that it is primarily masculinised economic interests that are linked to SD by the then Indian government and that technical progress and its co-benefits stay in the foreground while the protection of the environment or social aspects of sustainable development are not that high on Singh's and Ramesh's priority list. As an economist who has spent decades of his political career working on India's economic growth and administration (Ramesh 2015, The Running Thread I), this is perhaps not surprising in Ramesh's case. But his far-reaching calls for rethinking GDP indicate a process of change and a proactive and innovative way of engaging with climate change issues.

The structure of the Five-Year plan and the way SD is discussed in further discursive contributions such as Singh's WSDS speeches embody the offensive-hegemonic strategy of the Indian government to anchor SD in the Indian climate discourse. The encompassing demand SD is undoubtedly an important reference point going hand-in-hand with other demands of the Indian chain of equivalence. Furthermore, SD is presented as the solution—or at least a part of it—to manifold challenges of the climate catastrophe. Indeed, in the eyes of the government SD is proposed as the answer to demands for energy security (Singh 2010, 2011), as well as the challenges of widespread poverty (Singh 2011), to issues of unsustainable land management (Ministry of Environment and Forests 2012, 285) or to the problem of environmental toxins such as increasing smog in Indian cities. Equivalising the resulting nodal points such as *poverty eradication*, *energy security* or the *strong role of India* with *SD*, the government places the encompassing demand SD as its hegemonic project at the centre of the climate discourse. Even though the government's particular interest of a strong global role is also vehemently pursued, the hegemony analysis shows that only SD has found the necessary broad support from other actors whose subject positions are an integral part of the Indian chain of equivalence.

Establishing the Indian Network for Comprehensive Climate Change Assessment (INCCA) and demanding progress in Indian climate sciences, Ramesh followed strategeme VI "(Re-)Articulation of subject positions for politico-societal forces" and offered Indian climate scientists new highly credible subject positions generating support for his hegemonic project. Strategeme VI was also used in further cases, when GPI was recognised by the ruling party and further politicians for its engagement for solar energy in Delhi holding therefore an important subject position. Likewise, TERI's subject position was recognised when the cooperation with the think tank was repeatedly emphasised with a specific focus on the masculine connotated aspects of technical (MacGregor 2014, 626) and economic (Magnusdottir and Kronsell 2015) added value of SD. Furthermore, additional NGOs such as the CSE or GPI were recognised for being part of the group supporting some aspects of SD as well as the business community, as already discussed above.

Besides strategeme VI, strategeme V "Gradual constitution of greater leeway for interpreting the symbolic equivalent of the universal" was also followed in the framework of the offensive-hegemonic strategy, as the various outlined actors share numerous interpretations of sustainable development. In the period of Ramesh's tenure studied here, strategeme VII "Occasional but deliberate breach of the antagonistic line of division" can also be reconstructed. The demand for Indian mitigation measures, which are mostly associated with a stronger Indian role and therewith equivalised with SD, are one such breach of the antagonistic line. In fact, mitigation measures are traditionally demanded from the antagonistic players, the developed states. The partial incorporation of this demand—weakened as the mitigation goals are non-binding and self-determined—strengthens the empty signifier SD.

It has thus been reconstructed that various nodal points in India's chain of equivalence have been equivalised during Ramesh's tenure, such as *strong Indian role*, *energy security*, *poverty eradication*, *co-benefits*, *environmental protection* and *sustainable development*. This list is not exhaustive, but the hegemony analysis has proven that SD was pushed as a hegemonic project and—at least temporarily—can be attested to have been achieved, given that a hegemony was established by winning a critical mass of actors as a discourse coalition to support this identity. While the core strategemes I-III have been reconstructed regularly in the central speeches of Ramesh and Singh and contributions of further actors, strategemes V-VII, which are the supplementary hegemonic strategemes extending the outreach of the hegemonic formation, can also be proven. Instead, strategeme IV "the super-differential demarcation of the Indian climate change discourse's limits" is not traceable. Since super-differential boundaries are only drawn if necessary (Nonhoff 2006, 343), this strategeme does not appear to be needed, since the boundary of the Indian climate change discourse is not in danger of becoming fragile at this point. Moreover, the discourse arena is not challenged in its existence by another discourse arena, so it is also not necessary to stabilise its discursive boundaries as a reaction to any attack. Attesting that the core strategemes as well as the supplementary hegemonic strategemes are reconstructed focusing on SD, SD represents the empty signifier at that time of the Indian climate change discourse. The fact that the understanding of SD is divergent to the different actors and therefore left open for their interpretation, speaks also for SD as the ruling empty signifier of the Indian chain of equivalence. Accordingly, SD is an encompassing demand that combines many different demands in order to compensate for the lack of the common good representing the (unachievable) universal of a society (cf. Sect. 2.2.3). However, it lies in the nature of an empty signifier that it is unclear what exactly is understood to fall under SD and whether it serves especially to vindicate different aspects of growth insisting on co-benefits. Nonetheless, the strong emphasis on masculinised demands can be noted, the meaning of which will be taken up again later (cf. Sect. 3.3.1). Against the background that in 2011 the head of the environment ministry changed, it will also be interesting to see if this empty signifier is stabilised or if the hegemonic project changes.

3.2.2 Back to the Roots? Climate Justice in the Indian Discourse

After the new Environment Minister Jayanthi Natarajan (Congress) replaced Ramesh in July 2011, she took a dramatically different line of reasoning with regard to Indian climate policy from that of her predecessor. At COP17 in Durban, Natarajan appealed above all to the demand *climate justice* (CJ). Equivalising the demand *climate justice* with other former Indian certainties, such as the demand to *fight poverty first,* not to break the *"firewall of CBDR"* and to create space for *basic development* (Natarajan 2011), she followed strategeme I. Using strategeme II, Natarajan rejected Indian binding commitments, delimiting India from the developed countries and therewith dividing the discursive space. At the same time, Natarajan tried to negate the Indian image of a naysayer pointing to the Indian energy intensity actions announced in Copenhagen (Natarajan 2011) which could mean that she goes along with the demand of a *strong Indian role.* However, in Natarajan's COP17 speech examined here, this appears more as one argument among many, but not a specific concern of the environment minister. Whether this nodal point remains a central demand of the government in Natarajan's turn is hence to be found out. Although Natarajan was neither able to sustain the image of a pro-active India in Durban (Never 2015, 98) nor to prevail in maintaining the strict firewall (Sengupta 2019, 126), these discourse contributions are crucial as they contribute to changing India's chain of equivalence regardless of what was actually decided at the COP. Insisting sharply on *climate justice*, the arrangement of the chain of equivalence has changed, as the already existing nodal points have been given a new weight due to the re-organisation of the nodal point's outreach through different actors and their support as well. Therefore, whether CJ was supported by influential subject positions must also be reconstructed. If this is the case, it can be assumed that Natarajan tried to push the hegemonic articulation CJ in the sense of a hegemonic project.

Other speeches of Indian representatives in the UNFCCC framework followed patterns similar to those of Natarajan's COP17 speech using strategemes I and II specifically insisting on climate justice. At the meeting of the Ad Hoc Working Group on the Durban Platform for Enhanced Action, India pointed again to domestic mitigation actions and linked them with the central demands of *substantial mitigation action by the developed countries and technological transfers* by them to support the developing state's fight against climate change (India 2012, 36). In particular, demanding *mitigation action by the North* is combined with demanding *climate justice* (India 2012, 37). Reasoning this way, India again divides the discursive space following strategeme II. Equivalising notably the demands for *mitigation action by the north* but also "appropriate" *mitigation action by India* and *climate justice* (India 2012, 36), strategeme I is used. Adding the word "appropriate" to mitigation actions by developing countries and pointing to their ability to increase their efforts if the developed countries stick to more ambitious mitigation measures (India 2012, 37) shows the discursive linkage of Indian mitigation action with specific aspects of climate justice when strategeme I is used. Interestingly, the third core strategeme "representation" was followed neither in this contribution nor in Natarajan's COP17 speech.

In Warsaw 2013 at COP19, the ductus of Natarajan's speech changed. Attesting that India is a victim of climate change and noticing an "ambition gap" (Natarajan 2013, 2) on the behalf of the developed countries, she demands of the Global North *more action concerning both mitigation efforts* and *support for developing countries*. Also, pointing *to poverty eradication* and *energy security*, she again used strategeme I equivalising different demands. Furthermore, strategeme II was followed constructing the developed world on the one side and the developing world on the other. Contrary to earlier contributions, Natarajan underlines the importance of *sustainable development* and combines the demand with *climate justice* mentioning that "supplementary actions for increasing ambition at the international level must fully respect the national priorities of sustainable development and be based on the availability of means of implementation in accordance with the principles the Convention (sic!)" (Natarajan 2013, 3). As CJ does include loss and damage (L&D) (Huang 2017, 40) which was pushed as a major theme of the developing states in Warsaw it would have been obvious to give a speech focused on L&D. The form and future importance of L&D was the subject of heated debates in Warsaw, with developed states trying to avoid any right for reparations after extreme weather events and also wanting to prevent the establishment of L&D as a third pillar alongside mitigation and adaptation under the UNFCCC (Vanhala and Hestbaek 2016, 119). Interestingly, the final documents of Warsaw also show a gender-sensitive view of L&D as the very brief and technical decision of the implementation of the Warsaw mechanism on L&D includes the call for "Collection, sharing, management and use of relevant data and information, including gender-disaggregated data" (UNFCCC 2014, 6). Whether this is reflected in India's climate policy or its demands at the global level will be shown by the analysis of the SAPCCs in the course of the study (cf. Sect. 3.2.5), and the actors' statements will also be examined for these indications identifying possible gendered discourse patterns.

Besides the striking fact that Natarajan did not address these gender aspects in her speech, yet another important observation can be made: Although India strongly demanded L&D (Natarajan 2013, 4), significant changes in her general reasoning can be attested to. These—at first glance paradoxical—changes are: At the 2011 COP the then environmental minister linked the highly important demand *poverty eradication* with *climate justice*, and two years later she linked the very same demand with *SD* in Warsaw. Indeed, due to the L&D debate it would have seemed more useful to stick with the former reasoning focusing on *climate justice*, as L&D is often perceived as one mosaic stone of CJ. Instead, Natarajan now used *SD* representing her different demands. This supposed triviality therefore offers serious indications that Natarajan, as an important actor of the Indian discourse, made a strategic change focusing more on SD for growth and therefore masculinised reasons.

Between these two COPs came COP18 in Doha in 2012. Various strands of international climate policy were negotiated there, as usual at the annual summits. What was new was the UNFCCC's intense engagement with gender in the context of international climate policy. In fact, in Marrakesh in 2001 gender equality was also a topic of the COP. However, "the progress on implementing (...) has been slow and women continue to be under-represented on the delegations of many countries, particularly those most vulnerable to climate impact" (Gupta 2015, 409). Within the framework of the so-called "Doha Miracle" (Nagel 2016, 201), after lengthy preceding discussions, decisions were taken to increase the number of women in the different bodies for reasons of gender balance and to invite the delegations to aim for better parity in the future (UNFCCC 2012, 48). These decisions are an opportunity to take a brief look at gender relations within the Indian COP-delegations to gain an impression of its probably gendered subject positions. Over the years, a cautiously positive development can be observed: The proportion of women was 15% in Doha and increased to 19% in Paris 2015. Although a positive trend towards increased participation by women, especially compared to the early 2000s, can be attested to, the proportion remains relatively low. According to various studies, only when participation of women reaches 30% can the involvement of their different views be secured (Ergas and York 2012, 968; Burk and Hartmann 2017, 143). It must therefore be assumed that India's climate policy at the international level follows a male gaze, even though Natarajan, the chairperson of the delegation at that time, was a female minister. Indeed, the mere increase in participation says nothing about the quality of involvement. This aspect will be taken up again later.

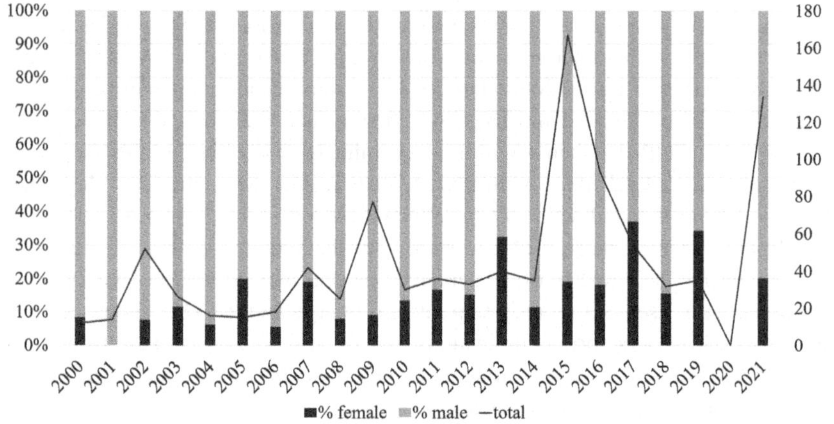

Fig. 3.1 Members of the COP delegations by sex, 2000–2021. (Own illustration with data from UNFCCC 2000, 2001, 2002, 2003, 2004, 2005, 2006, 2007, 2008, 2009, 2010, 2011, 2012, 2013, 2014, 2015, 2016, 2017, 2018, 2019, 2021b)

The central statement in Doha was made by the representative of the Indian Ministry of Environment, also equivalising *energy security* and *climate justice* closely with *SD* (Mehrishi 2012, 4). However, only a general analysis of the discourse will show whether SD is once again made the central point of reference or whether CJ is perhaps upgraded. Therefore, we should consider how other important actors in India's climate discourse engaged during Natarajan's tenure, especially concerning her focus on CJ.

Climate justice was supported by politicians from both major parties, the oppositional BJP (Modi 2011, 12) as well as the Congress and many further stakeholders such as the Indian forerunner of justice in global climate negotiations, the CSE. Following strategeme VI, (re-)articulation of subject positions for politico-societal forces, the MoEF under Natarajan cooperated with the CSE at COP17 in Durban as well as at COP18 in Doha in the framework of co-hosted side events. In Durban the joint session "The imperative of equity for an effective climate agreement" was chaired at high level by the environment minister herself (Centre for Science and Environment 2011b) insisting on the—in Natarajan's eyes—cornerstones of every climate agreement: Equity and historical responsibility (Centre for Science and Environment 2012b). The 2012 joint side event also put equity at the heart of the debate (Centre for Science and Environment 2012b), but unfortunately the concrete text of the contributions there cannot be reproduced and therefore it is not possible to explore whether CJ was linked to SD or other nodal points. In fact, MoEF's collaboration with civil society actors at international venues was not new. Nevertheless, these events and their thematic focus are a contribution to the discourse not least as they have the potential to influence the negotiations (Schroeder and Lovell 2012) and in this specific case lend particular emphasis to the demand of *equity* which is closely linked with the demand CJ. Especially against the background that Natarajan's predecessor did not attach similar importance to equity questions, these co-hosted initiatives embody the employed strategeme VI to increase the outreach of the demand CJ.

Yet, *equity* has always been part of India's climate discourse, pushed by both governmental and non-governmental actors. Due to India's history as a colonised country, sovereignty and equity are of enormous importance for its foreign policy (Michael 2013, 32–33) as well as its climate policy (Pathak and Parris 2021). Even though the shift in Copenhagen had redefined this historical certainty in the context of Indian climate policy, which forms a part of India's foreign policy, with Natarajan the demand for equity reached a new importance.

As already shown, the CSE supported Natarajan's course emphasising the importance of equity. In the climate change chapter of its 2013/2014 annual report, the CSE pursues two strategemes: Strategeme I is followed when various demands are equivalised, such as *mitigation action by developed countries*, *renewable energy* and *co-benefits* (Centre for Science and Environment 2014, 18–19), centring *equity* and *climate justice* (Centre for Science and Environment 2014, 18,19,21) inter alia in the form of CBDR-RC (Centre for Science and Environment 2014, 18). Furthermore, the discursive space is also divided here (strategeme II) by demanding aid from the North because the South, in this case India, is particularly vulnerable to the consequences of climate change, as the CSE declares (Centre for Science and Environment 2014, 19). The authors of the report also state their own recognition as an important subject position within the Indian climate discourse and attribute the agenda-setting of the past regarding equity to themselves (Centre for Science and Environment 2014, 18–19). History proves the CSE right in this respect (cf. Sect. 3.1.2). And while its importance certainly declined during Ramesh's tenure, Natarajan seems to have found allies in the CSE. How this position developed and what significance the CSE and its stance have for the Indian climate discourse with the BJP-led government will be discussed in the next chapter.

Returning to Natarajan's impact, the media landscape must also be considered. As was the case with her predecessor, the journalists adhered to the traditional reasoning reproducing inter alia *growth* and *climate justice* as important nodal points (Jogesh 2019, 318, 320). In a survey of English-speaking Indian print media from 2010–2017, Jogesh (2019) reconstructed that the discussion of climate justice essentially comprises five categories. Besides the category of financial and technical transfer to support Indian climate action, CBDR was also one of the categories, as was the criticism of industrialised countries and the demand that they get more involved. The other two categories are, on the one hand, that India should only take action if it is in line with its own interests and development context and, on the other hand, that India should take mitigation action (Jogesh 2019, 319). During Natarajan's tenure, the discussion of financial and technological transfer decreased, but so did the reporting in the CBDR category, whereas the criticism of developed countries and the need for Indian mitigation efforts were slightly increasingly discussed (Jogesh 2019, 319). It can be assumed that the journalists took into account that the Indian foreign policy entered into a time of pragmatism (Mohan 2017, 22) that Natarajan was seen as an ally who made CBDR a central demand anyway. What is certain, however is that the demand for increased *engagement of the developed countries* by pointing out their failures flanked the Indian narrative especially as the developing states tried to make loss and damage the focus of COP19 in 2013.

Sethi, an Indian journalist who was again supported by CSE to cover the 2013 COP19 in Warsaw (Centre for Science and Environment 2014, 42),[12] used the

[12] Before joining The Times of India in 2007 to cover development, agriculture, environment and climate change, Sethi worked for the CSE journal Down to Earth as Senior Assistant Editor.

second strategeme to differentiate between the vulnerable South and the responsible North stating that the developed countries "stayed from acting climate responsible" (Sethi 2013b). Even though this is just one example of the wide support of Natarajan's particular interest of climate justice by the Indian media landscape, it must be noted that the different facets of climate justice hardly envisioned the future of India in climate change scenarios, but the contributions were rather a defensive struggle against ambitious Indian mitigation measures. Since the assumption was that mitigation measures could hinder India's growth, a masculinised colouring of the discourse can be discerned at this point—for economic success and growth in a dichotomously structured society are assigned to the dominant, masculinised sphere (cf. Sect. 2.3.3).

Still, *co-benefits* or *energy security* were also reproduced as demands and put on an equivalent footing with CJ and other demands. Strikingly, not all demands can be represented by CJ; for example, Indian mitigation action cannot be represented by CJ if it is not only a co-benefit of other aspects such as the promotion of renewable energies to guarantee energy security. The hegemony analysis has shown that at this point in time, the demand for *climate justice* is based on the understanding that India has equal rights to pollution in terms of catch-up development, which does not necessarily include investing solely in renewable energy to cover India's energy needs. Since the nodal point *climate justice* lacks the ability to represent all the equivalenced nodal points, it cannot serve as an empty signifier; the demand *climate justice* lacks an identificatory nucleus around which a hegemonic project can comprehensively develop (Nonhoff 2006, 309). In sum, from a theoretical viewpoint strategemes I and II are traceable, while strategeme III "representation" is absent.

A further aspect of the reason that the hegemonic articulation *climate justice*—which can be understood as the particular interest of Natarajan—does not succeed could certainly be the hegemonic project of Singh and other actors, *sustainable development*, which he still continued to pursue. Singh underlines this demand again at the WSDS 2013, equivalising SD with *environmental protection* (Singh 2013, 1) or *international cooperation* to unlock the Indian renewable energy potential (Singh 2013, 3). Interestingly, Singh also equivalises *CBDR* and *equity* with *SD* in the context of global climate negotiations (Singh 2013, 2). Given that Singh gave CJ as a demand in all his former WSDS speeches some space, especially in 2012 (Singh 2012), it is remarkable that only SD is able to create representation of the Indian climate change policy. Uniting all the demands listed—of course in all their contradictions—Singh returns to SD again and again. It would be a one-dimensional argument simply to say that because the head of government continues to pursue SD, it still serves as the empty signifier, especially concerning the fact that his speeches analysed here were mostly part of a SD summit. On the one hand, however, it must be noted that MoEF had lost strength and attention under Natarajan (Sethi 2013a), so the Prime Minister's statements and other contributions such as the Five-Year Plan are important and prestigious sources of the Indian climate change discourse. On the other hand, the circle of actors in India's climate discourse had widened by the time of the implementation of the NAPCC, and continued to significantly broaden during Ramesh's term. Accordingly, Natarajan's allies such as the press and the CSE are important, but not decisive. New actors such as industry associations, scientists and other NGOs besides the well-known CSE and TERI are increasingly important—and also recognised—subject positions. Thus, new discourse coalitions were formed.

Regarding the tenure of Ramesh, by making climate scientists important actors and as economic actors broadly supported SD, the hegemonic project SD underwent an enormous expansion. Although the former 'natural' partner CSE criticised this stance, the demand SD was supported by other civil society actors. Only the media were strongly critical until the end. But even though politicians from different parties perpetuated the supposed dichotomy of growth and sustainability, Singh and Ramesh managed to build a sufficiently broad coalition.

Natarajan was also able to rely on some actors, but against the already well-entrenched narrative of SD suggesting economic growth in a sustainable variant, this coalition had a hard time as various actors continued to favour the co-benefits closely associated with SD. These actors, such as business associations but also regions and NGOs, already noticed the first successes of the development of masculinised aspects such as renewable energies.[13] The Singh administration paved the way for these as well as their co-benefits such as technical progress especially by new policies and laws (Raina and Sinha 2019, 334–35). Therefore, *climate justice* no longer had the sole authority as key reference point as it had in earlier decades.

3.2.3 Everything New with Modi?

In the 2014 election campaign, climate policy was, as in former elections, not a defining issue (Pillai and Dubash 2021, 105). But since the parties campaigned for votes on the questions of growth, poverty reduction and India's standing in the

[13] For a detailed explanation of why renewable energies as part of the technological fixes is a masculinised idea to combat climate change, see MacGregor 2010, 231.

world, climate policy was definitely a subset in some of the pre-election debates. It is therefore worthwhile to analyse the manifestos of the major Indian parties—Congress and BJP—with a focus on their environmental or climate change chapters.

The Congress Manifesto (Indian National Congress 2014) demands *growth* (16, 49), *environmental protection* (17, 40) and substantially higher use of *renewable energy* (41) announcing that they will continue the prestigious project of the NAPCC. Interestingly, the election programme, which is mainly an action plan for the upcoming term, emphasises the demands for the *protection of livelihoods* (Indian National Congress 2014, 41) from different consequences of climate change, as well as *environmental protection* in general (Indian National Congress 2014, 17, 40, 41). Both demands only played a minor role in former discussions und can be associated with the feminised sphere, as the private sphere in which livelihood is localised, and nature are linked to femininity as different ecofeminists worked out referring back to the dichotomous structures shaping society and its norms (cf. Sect. 2.3.3). In fact, yet again, *sustainable development* can be identified as the central point of reference. Showing a variety of possibilities for representation, SD is primarily linked to *environmental protection* (Indian National Congress 2014, 40) and a *strong global role* (Indian National Congress 2014, 47) but only peripherally to *growth.* Although growth is the dominant theme of the manifesto, the environment and climate chapter (Indian National Congress 2014, 40–41) gives more space to genuine environmental protection issues and stands out due to the described equivalences, while the empty signifier *sustainable development* persists. Both the first and the third strategeme can be reconstructed. In contrast, strategeme II which separates the discursive space, cannot be traced. Since the manifesto is a document addressed specifically to potential voters, this aspect of clearly distinguishing oneself from the outside was probably not that necessary. However, it seems that the Congress party still follows its hegemonic SD project with a new focus on environmental and therewith also livelihood protection. With regard to the gendered structure of the demands SD, it is therefore no longer possible to speak unambiguously of a primarily masculinised pattern.

The BJP's manifesto is also primarily dedicated to poverty reduction and growth, stating that the fight against poverty would become a national priority (BJP 2014: 15). Accordingly, the climate section "Flora, Fauna and Environment—Safeguarding Our Tomorrow" (Bharatiya Janata Party 2014, 36) prioritises the demands for *cleaner production* and *cleaner fuels* which are an important part of future growth. Later adding *pollution control* and other more human-oriented demands such as *social forestry*, the BJP electoral-programme also draws on *sustainable development* (Bharatiya Janata Party 2014, 36). Whether or not this demand turns out to be a central reference point or rather a potential empty signifier under the Modi-government, is still to be determined and will be discussed in Sect. 3.2.6. Strikingly, it seems that the Hindutva ideology does not play any role in the context of BJP's suggested climate change policy. This will change tremendously in the following years.

The election programmes show that both parties follow a neoliberal understanding of growth and only consider climate aspects in their respective chapters. Interestingly, in contrast to the BJP the Congress prioritises social aspects of climate politics, such as clean rivers, the consequences of pollution for women due to their social roles or the needs of forest-dwellers (Indian National Congress 2014, 40–41). Of course, this says little about the actual policies of the Congress party, which has so far in reality tended to put these needs on the backburner.[14] Nevertheless, the INC's chain of equivalence differ from that of the BJP, as the increase of social demands, which are associated with the sphere of femininity, articulated by the INC can be attested. The BJP, however, makes social aspects of climate change only a subordinate issue to economic concerns, even in the environment-dedicated chapter of the BJP manifesto. Hence, the BJP colours its climate policy in masculinised demands. It will be discovered as to whether or not this impression is confirmed during the BJP's first term in office.

After the surprisingly clear success of the BJP, Modi was sworn in as Indian prime minister in late May 2014 and Prakash Javadekar (BJP) took over the Ministry of Environment and Forests, which was changed in its nomenclature and is now called "Ministry of Environment, Forest and Climate Change" (MoEFCC). Whether upgrading the topic of climate change by making it the title of the ministry would significantly change the discourse is not foreseeable, but this contribution to the discourse is able to shape the chain of equivalence. At first glance, the chain of equivalence of the new BJP government did not change significantly compared to the last years of the Singh government. In the first speech to the UN General Assembly, for example, Modi split the discursive space (strategeme II) when he demanded *CBDR* and the technical and financial support from developed states in order to fight climate change (Modi 2014, 55). Following strategeme I *CBDR*, *climate justice* and *SD* were equivalised. Furthermore, *SD*

[14] The Congress-led coalition has been criticised for not sufficiently addressing the social impacts and factors of climate change in its climate policy. It is also often criticised that the MoEF has taken too business-friendly decisions in clearance processes for the environmental impact of industrial projects. However, the introduction of the National Green Tribunal in 2010, for example, has ensured that obstacles raised by third parties, such as tribal communities, in conflicts between environmental protection and economic projects are specifically strengthened in the form of specialised courts.

is used to point to the importance of *renewable energy* (Modi 2014, 55) and the Indian successes in this field which were mostly those of the previous government at that time.

However, criticism of the newly appointed Minister was voiced early, as the government announced that clearance procedures to safeguard environmental interests in economic projects would be carried out more quickly than previously (Gogoi 2014a, 119). Gaurav Gogoi, a Congress Member of Parliament drew a delicate comparison between the current and previous governments:

> *"In the past, we have often seen the Ministry of Environment stand by its ecological objectives while it negotiated with the Ministry of Commerce and Industry. Today under the new BJP government both the ministries seems to be speaking in unison." (Gogoi* 2014a, *119)*

The sharing of this impression in the public debate led to fears that environmental protection would be neglected completely in favour of economic growth (King 2014). Therefore, it seems that the temporary equivalence of growth and sustainability was in fact annulled. Whether this finding is embedded in the discourse can only be determined following the deconstruction of further contributions.

The government was also criticised for reconstituting the Prime Minister's Council on Climate Change (PMCCC). At this point it no longer included any representatives of civil society or non-governmental think tanks, as Rachendra Pachauri (TERI) had quit the post which was then left vacant (Sangomla 2021) and Ajay Mathur (TERI) had become director general of the International Solar Alliance. Jyoti Parekh and Sunita Narain, both NGO chairs, were not reappointed as members of the PMCCC (Government of India 2014). Although the PMCCC did not meet again following its first consultations in the reshaped group, Modi apparently had no particular interest in winning over those who were not reappointed, as well as their organisations, as subject positions shaping the Indian climate discourse or even supporting his hegemonic project. Of course, it is also possible that these actors did not become part of the PMCCC due to their political positions or for other reasons. The fact that the CSE was not part of the Council, even though it is the best-known non-parliamentary representative of the Indian equity-debate in climate change, is at first glance quite surprising, as the CSE appears to be a 'natural' partner when it comes to climate justice. Accordingly, it is worthwhile to monitor the demand for climate justice with the new administration, as the importance of that nodal point in the Indian chain of equivalence may be shifting. While the PMCCC is conceived as an actor working on the strategic lines of the Indian climate change policy, the Executive Committee on

Climate Change (ECCC) is the coordinating and purely bureaucratic committee which had already met in 2013, most likely on the initiative of the MoEF (Pillai and Dubash 2021, 104–5). The ECCC also met under the new government, but at very irregular intervals. Therefore, it must be assumed that neither Council played a significant role in India's climate discourse during Javadekar's term. However, this does not in fact mean that individual actors of the respective body cannot hold a significant subject position.

Following the change of government, it is particularly interesting to scrutinise the parliamentary debates as most of the members of parliament received new subject positions and therefore a new hegemonic constellation was formed. The budget discussion of the MoEFCC provides insights into the slightly changing reasoning of the stakeholders, not the least of which is that the budget of Javadekar's ministry was nearly halved (Shrivastava 2014). The discussion of the government's policies deconstructs the possibly declining stability of specific nodal points of the discourse and their function. In addition to the observation from the ranks of the Congress party that the government is providing too little funding for climate change policy (Nag 2014, 206), the significantly accelerated clearance procedures are the main subject of the debate. The MoEFCC is responsible for the clearance procedures in which the environmental compatibility of industrial projects is verified and, if necessary, approved or has restrictions imposed. These clearance procedures are a concise example of the conflict between the demands for *growth* and *environmental protection.* In the past, this conflict has often been equivalised by the empty signifier *SD*. In this parliamentary debate the government is highly criticised for its clearance policy (Narasimham 2014, 159, 160) and accused of not balancing these interests adequately. Although the previous government was also charged, for example by NGOs, of favouring industrial projects over environmental protection, the conflict now seems to go much further, not only because the decline in environmental standards has been openly acknowledged by the government, but also because low environmental standards are also feared in other areas as well. Supriya Sule (Congress) for example pointed out the need for action in the area of (illegal) sand mining, which is a serious problem in India (Sule 2014, 181). Furthermore, the member of the Congress party made it clear, that "development and industry have to go hand in hand" (Sule 2014, 180). While Sule equivalises these demands, the BJP is accused of failing to balance these likewise important aspects.

However, since it can be expected that the hegemonic power in the Indian climate discourse shifted in favour of the BJP after the elections, these accusations of the opposition parties should be checked against the discourse contributions of the members of the government. *Growth*, a highly masculinised demand and a central reference point in the chain of equivalence pushed by the BJP, can be confirmed, for example, on the basis of one of Javadekar's speeches in the parliament. While fielding various questions from the Marxist Member of Parliament M.B. Rajesh, Javadekar insisted that "We need to grow. We have the challenge of poverty; there is also the challenge of energy access not being available to all the citizens, which we need to provide" (Lok Sabha 2014, 16). The environment minister again pointed to growth in his speech, when he said that "our growth will not be impacted in any way" equivalising the demand for *growth* with the demand for *energy efficiency*, he added that this "will be our contribution to the world environment" (Lok Sabha 2014, 16). Even though he also emphasised that climate actions include welcome co-benefits such as energy efficiency and therewith better air quality (Lok Sabha 2014, 16), this parliamentary debate shows that Javadekar particularly insists on growth. Equivalising *growth* with *energy efficiency*, *poverty eradication* and *co-benefits* attests the use of strategeme I. The discursive space is split in the sense of strategeme II when Javadekar insists that India does not have any binding emission cuts, pointing to the Kyoto Protocol (Lok Sabha 2014, 16) and therewith dividing the space between the Annex 1 and non-Annex 1 states. The third strategeme, represented by an empty signifier, is not followed here. Nonetheless, *growth* should be considered as a possible representative of the hegemonic project of the new government; however, this impression should be reconsidered in the framework of a larger sample of the discourse.

In the run-up to COP20 in Lima, Modi made a memorable appearance on the Indian Teacher's Day. He answered a student's question of how best to deal with the climate catastrophe by pointing out that it was not the climate that had changed, but people and their way of life (Mehra 2014). On the one hand, the approach of climate denial can be read out of this answer. This would fit with the observations of a parliamentarian who accuses the Prime Minister of denying climate change in the framework of different speeches (Gogoi 2014b, 111). On the other hand, this statement can be seen as the beginning of the individualisation of responsibility in dealing with climate change impacts, serving as a justification in the case where the state does not engage with climate protection or adaptation in an appropriate manner. According to ecofeminist criticism, the privatisation of responsibility inevitably leads to more work for the care takers, who in most cases in Indian society are women. In addition, the divide between the public and private spheres is being pushed forward in the framework of a gendered logic (Prügl 2021, 300); shifting the climate catastrophe into the private sphere and thus removing it from the public sphere means to depoliticise the discourse and the question how to answer the climate crisis. The significance of these aspects shaping the Indian identity will be shown in the later discussion (cf. Sects. 3.3.1 and 3.3.2).

A call for the introduction of an international yoga day in Modi's first speech at the United Nations General Assembly (UNGA) is one mosaic stone in the picture of individualisation in dealing with responsibility through the re-traditionalisation of lifestyles, pointing to ancient Hindu traditions. In Modi's view, yoga acts not only as health protection but also as climate protection through the body's connection with nature:

> *"[Yoga] is not about exercise but to discover the sense of oneness with yourself, the world and the nature. By changing our lifestyle and creating consciousness, it can help us deal with climate change." (Modi* 2014, *56)*

In later contributions it becomes apparent how yoga as Hindu tradition is meant to suggest a low-carbon lifestyle, which in turn at least relativises a much-needed collective Indian effort in fighting the climate catastrophe.

Modi's speech is also important for the study at hand in some further respects, not least because the Prime Minister addressed in his speech to the UNGA both the Indian society and the international community—among whom are allies—but also "the radical other" from whom he distances India to form its climate change identity. Moreover, it's possible to trace the offensive-hegemonic strategy of the new Indian government. Modi follows all three core strategemes. Various demands are equalised (strategeme I), especially *growth* with *poverty eradication* and *renewable energy* (Modi 2014, 55). It also becomes clear how Modi divides the discursive space, because the demand for *climate justice* in the form of *CBDR* is emphasised (Modi 2014, 55). Furthermore, the responsibility of the developed states to support the developing states such as India with financial and technical resources in their fight against climate change is explained (Modi 2014, 55), splitting the discursive space. Representation can also be reconstructed in the internationally-recognised speech. Modi repeatedly focuses the chain of equivalence on *growth* (Modi 2014, 48, 55) emphasising that India sees itself as part of the developing states (Modi 2014, 48).

Strikingly, however, Modi opens a new line of reasoning stressing that "we need to change our lifestyles. Energy not consumed is the cleanest energy. This will provide a new direction to the economy, while causing no economic harm" (Modi 2014, 55). This sentence could be understood as a demand for public degrowth, which is highly feminised, but Modi adds: "For us in India, respect for nature is an integral part of spiritualism. We treat nature's bounties as sacred"

(Modi 2014, 55). With this addition, it becomes clear that Modi is again concerned with an individualisation of responsibility for climate protection, because if living in harmony with nature is part of Indian spirituality, the spiritual people by default already care for the environment. Proceeding with this thought the government insinuates that structural changes are not needed—especially as the still spiritually-rooted Indians are environmental-friendly by culture. Linking the individualisation of responsibility to protect the environment with spirituality with this intensity is new in Indian climate discourse and must be considered more closely in the light of the discussion of the contributions to the Paris Agreement, which started with the Lima Conference in late 2014.

At the COP20 in Lima, the Minister of Environment delivered his speech giving further insights into the changing hegemonic project. Javadekar first greets all women in the context of the Gender Day of the COP, who in his opinion play "such a central role to preserving nature" (Javadekar 2014b, 1). Interestingly, he addresses women only as carers of the environment, not in other roles, such as political actors. Interpellating women solely in their role of protecting nature, it seems that if we can speak of a specifically-addressed group of women here, they are only respected by Javadekar in this specific subject position. As shown in Sect. 2.2.2, groups that are only recognised for certain roles as subject of the discourse can initially only act from this position to shape the discourse. Whether the women addressed here are also granted other subject positions in the chain of equivalence or whether they can shape the discourse out of the assigned position will be shown later, especially in the analysis of the State Action Plans on Climate Change (cf. Sect. 3.2.5).

Environment Minister Javadekar reproduces familiar patterns in his Lima speech. Following the first strategeme, Javadekar equivalises the demand *growth* with *inclusive development*, emphasising particularly the fight against climate change (Javadekar 2014b, 1). Here, *growth* is again a central demand. Distancing himself from the wording of former environmental minister Jairam Ramesh, Javadekar accentuates in two phrases that he expects "equitable carbon space to achieve sustainable development and eradicate poverty" for developing countries (Javadekar 2014b, 3, 4). Ramesh (Ramesh 2015, Chap. 4), instead, welcomed his negotiating success in 2010, when he, together with colleagues, was able to replace the old phrase "equitable access to carbon space" with "equitable access to sustainable development". Presumably, the new government tried to soften the wording here and to expand from the right to equitable access to SD, which clearly and unambiguously identifies development as sustainable development, with a preceding step—growth. Javadekar's new 'old' wording implies that development requires carbon space first for reasons of catch-up development based on fossil fuels. Although Javadekar's wording was not adopted in the Lima document, the attempt to reverse supposedly fixed formulations shows that the understanding of development is up for negotiation and is highly contested. Furthermore, it is evident that he tries to re-masculinise the connotation of development underlining the economic growth aspect and fossil fuels.

Equivalising climate justice, explicitly *CBDR,* with necessary *technical and financial aid* by developed countries (Javadekar 2014b, 4), Javadekar uses familiar arguments. As these demands also divide the discursive space and create an 'us' of the developing countries and a 'them' of the developed countries, strategeme II can be clearly identified. In Lima, the states negotiated the character of the future Intended Nationally Determined Contributions (INDCs) which had to be submitted in the run-up to COP21 in Paris. Javadekar clearly underlined the demand for *sovereignty* in the framework of the INDCs: "We firmly believe that the INDCs are to be 'nationally determined'" (Javadekar 2014b, 4). Adding that "We do not see any role for any ex-ante review in this process" (Javadekar 2014b, 4), the demand for Indian sovereignty here is twofold: the goals must be self-determined and a review mechanism in advance of the process of goal-setting is strictly rejected. Regarding the ex-ante review, the Government's stand has been highly criticised by the CSE (Centre for Science and Environment 2015a, 35) suggesting a fear that if there is no comparable and transparent ground, equity and CBDR will not be viable in Paris (The Economic Times 2014) since transparency and comparable factors are missing. The discussion on possible ex-ante reviews shows that the different interests of the CSE and the government lead to the impression that previously common demands, namely *equity* and *sovereignty,* as well as *CBDR* are now played off against each other. On the one hand, the demand for *sovereignty* is used to prevent ex-ante reviews, and on the other hand, the demand for *equity* is used to support them. Until now, both demands have been part of the Indian chain of equivalence. Therefore, the reconstruction of the empty signifier is very important to check whether it features space for both readings to form the common identity of India in the climate context.

Whereas in Lima, addressing the international community in particular, Javadekar referred to sustainable development and the necessary "carbon space" (Javadekar 2014b, 3, 4), in his statement on the Lima negotiations in the Indian Parliament he rather created the picture that India has stood up for its rights to development and growth as it is a developing country (Javadekar 2014a, 62, 63). Strategeme III, "representation" is used in both cases: in Lima to paint the picture of India as a country fighting for growth pointing to its right for inclusive and sustainable development referencing to different forms of climate justice; in

Delhi to show that India fights for growth and is therefore a part of the developing world. It cannot be conclusively clarified which demand represented the hegemonic project of the government at that time. What is clear, however, is that the hegemonic project has changed—at least in its meaning—as SD in the sense of the Modi government focuses more exclusively on growth, and representing India as a developing country would serve the same interest.

In the MoEFCC's annual report SD is downgraded with respect to the chapters structure. The 13th chapter of the former years was titled "Sustainable Development and Climate Change" and was now exclusively dedicated to "Climate Change" (Ministry of Environment, Forests and Climate Change 2015, 234) and the new Chap. 14 covers "International Cooperation and Sustainable Development". Here—at least semantically—SD is separated from climate change. It seems that when it comes to climate change SD is only considered as a subset of international cooperation and no longer as a central reference point. Later, SD was even erased from Chap. 14 of the 2018/19 report and no longer part of any chapter title (Ministry of Environment, Forest and Climate Change 2019). This is another indication evidencing the declining standing of the former empty signifier SD.

The MoEFCC report, like most of the other discourse contributions in 2014 and at the beginning of 2015, was developed in the light of the upcoming COP21 in Paris 2015. In the short phase of the new government on the road to Paris, it can be said that the demand for rapid economic growth had taken on a central role in India's climate discourse—which is quite different from the formulation of SD. Furthermore, different aspects of climate justice, such as sovereignty or the CBDR principle gained higher importance also with an economy-oriented connotation, as a broad analysis of the Indian print media in the period around the Paris negotiations found (Swarnakar, Shukla, and Broadbent 2022). Interestingly, Modi pushed for India to be regarded as a society traditionally linked to nature and therefore caring for it, as portrayed by the practice of yoga. The Indian government got into position vis-à-vis Paris in two respects: in its claim to economic growth without conditions and in its claim to be a culturally-rooted ally of the earth. This has also changed the gendered structure of the discourse. Masculinised demands are gaining in importance, while attempts are being made to transfer the climate problem into the private sphere, depoliticising the answers needed to fight the climate crisis and assigning them to the sphere to which care and therewith women are allocated.

Whether these patterns and demands would be discursively reproduced and stabilised in Paris, and how the Indian climate change discourse developed in preparation for the important COP21 stabilising the Indian chain of equivalence, will be analysed in the next chapter. In addition to political and governmental actors, NGOs and other civil society organisations as well as the media are also of interest, since the Paris COP was an event that attracted a lot of attention and was associated with many different expectations and hopes which were discussed in manifold fora.

3.2.4 India in Paris: A Spiritually-Grounded Developing Country?

On the road to Paris, India's climate discourse was characterised by simultaneities: while the states of the Indian Union were drafting and adopting their SAPCCs, international partners submitted their INDCs. India proposed its own INDCs to the UNFCCC on the prestigious Indian national day—October 2nd, Gandhi's birthday (Press Information Bureau 2015)—symbolically referring back to its cultural roots. Keeping in mind that the USA and China had already committed to the pledging of substantial targets in a bilateral declaration in November 2014 (Office of the Press Secretary 2014), India faced international pressure to submit ambitious climate goals due to its will to play an important active global role (Saryal 2018, 13) and due to its rivalry with China. Even though the Indian INDCs were appreciated as "fair" (Centre for Science and Environment 2015b) and "ambitious" by Indian civil society (Giese 2017, 65), the goals pledged were not sufficient compared to India's "fair share".[15] The Climate Action Tracker attributes India only "medium" INDCs, as they are "not consistent with limiting warming to below 2°C, unless other countries make much deeper reductions and comparably greater effort" (Climate Action Tracker 2015a). In fact, none of the submitted INDCs of the largest emitters US, China, EU or Russia were consistent with the goal to limit global warming below the threshold of 2°C, and all were rated only as "medium" or in the case of Russia "inadequate" (Climate Action Tracker 2015b, 6). However, this is the evaluation of the goals only in terms of their physical significance. Although the assessment of the targets is also the subject of the Indian climate discourse, the analysis focuses on the hegemonic strategy and its use of gendered patterns justifying the Indian targets in line with the Indian climate identity, instead of the physical feasibility of the goals.[16]

[15] "Fair share" is defined by the Climate Action Tracker as "a country's total contribution [that] would [be] need[ed] ... to make a fair contribution to implementing the Paris agreement" (Climate Action Tracker 2022).

[16] For a detailed discussion of India's pledged goals, see Raghunandan 2020, 211–18 or Mohan and Wehnert 2019.

The document on the submission of India's INDCs represents the discourse of the time as it was shaped not solely, but especially by the government. Besides the core strategemes, further strategemes will be reconstructed identifying the offensive-hegemonic strategy of the first Modi government, beginning with the single-text level.

India's INDCs

India's presentation of the goals differs from that of other key actors. The goals are set out in a detailed text document dealing extensively with their justification and the country's self-positioning in the global climate discourse. In contrast, the EU's goals are reduced to the key measurable targets on only a few pages (cf. Sect. 4.2.4; Latvia and the European Commission 2015), as are those of the USA (United States of America 2016). India, however, presents in its very first words the cultural and historical closeness of Indian society to nature:

> *"India has a long history and tradition of harmonious co-existence between man and nature. Human beings here have regarded fauna and flora as part of their family. This is part of our heritage and manifest in our lifestyle and traditional practices. We represent a culture that calls our planet Mother Earth." (Government of India* 2015, *1)*

The opening words of the Indian INDCs show that the demand for *culturally-rooted closeness to nature* reproduced here is permeated by gendered patterns. The narrative of "Mother Earth" embodies the femininity and the associated need for protection of nature. Furthermore, linked value systems of patriarchal societies "continue to reinforce imagery of Mother Nature and Mother Earth that serves to maintain the subjugation of women while legitimizing the exploitation and depletion of natural resources" (Jaremko Bromwich 2020, 53). Nevertheless, the harmonious co-existence of "man" and "nature" indicates that there is a hierarchy between the concepts. This dichotomous juxtaposition gives rise to the so-called master identity, the masculinised figure who, among other things, can utilise nature by means of technology (cf. Sect. 2.3.3). The master identity embodies the "norm" of the human being, while the woman represents the deviation from it. Reducing the complexity, the first sentence juxtaposes man and nature, although co-existence is emphasised. But co-existence is neither the same as a unity of man and nature nor does it mean that man is an integral part of nature. Even though the second sentence insinuates such closeness, it needs to be deconstructed and mirrored with the outlined ecofeminist thoughts. Instead of referring to "man", "human beings" are spoken of, so the circle is expanded to include further genders, even if they are not explicitly addressed. Additionally, the family-narrative is further widened, as it is emphasised that these human beings regard flora and fauna as part of their family (Government of India 2015, 1). Yet, including genders other than solely men, the family narrative of the second sentence does not negate the dichotomisation of the first sentence. Rather, the family-narrative shows that close ties with nature are especially thinkable if further genders, for example women, are included, as they are traditionally strongly linked with both the family life and "Mother Earth". These reconstructed dichotomies are important gendered discourse patterns shaping the Indian identity, which will be discussed in more detail in Sect. 3.3.

The nodal point *culturally-rooted closeness to nature* is reproduced at several additional places throughout the Indian INDCs (Government of India 2015, 1, 2,29). Again directing to the Indian practice of Yoga, the linked "moderation and sustainable lifestyle" is highlighted (Government of India 2015, 1). Following strategeme I the Indian Government equivalised different demands such as *growth, development, renewable energy, energy security, acknowledgement as a vulnerable country, poverty eradication, technical and financial support by developed countries* as well as *culturally-rooted closeness to nature. Climate justice (CJ)* is another central demand of India's INDCs which is equivalised with the relatively new demand for a *low carbon lifestyle by developed countries* (Government of India 2015, 3, 4) and the demand for *technical and financial support by them* (Government of India 2015, 33). Therefore, the simultaneous use of strategeme II can be demonstrated. Through the multiple articulation of *CJ* (Government of India 2015, 3, 4, 33) and various associated demands, for example the *historical responsibility by the north* (Government of India 2015, 2) or a *low-carbon lifestyle by developed countries* and that they should provide both *technical and financial support* (Government of India 2015, 2,3,29,30) the discursive space is clearly divided into two parts. Additionally visualising the antagonistic division of the discursive space, *CBDR* is indirectly demanded pointing to the principles of the convention (Government of India 2015, 33). Strikingly, the reference to the demand *development* in the sense of economic growth runs through the text of the Indian INDCs. Furthermore, the emphasis that India is a developing country and should not be restricted but rather supported in its growth can be found on several pages (Government of India 2015, 3,4). It is therefore apparent that the government discursively constructs the narrative of a developing country as the representative in the chain of equivalence through formulations such as "Climate change is a major challenge for developing countries like India that face large scale climate variability and are exposed to enhanced risks from climate change" (Government of India 2015, 4) or in a more concrete example by discussing the

Indian development statistics: "India is a developing country with a per capita GDP (nominal) of around USD 1408 per annum" (Government of India 2015, 5). As the demand *development* goes hand-in-hand with all other equivalenced nodal points, it is the encompassing demand representing the chain of equivalence at this point in time. Whether the encompassing demand *development* is sufficiently stable enough to serve as the empty signifier has to be elaborated in the context of the other discourse contributions in and around Paris.

However, further strategemes can be identified at the individual text level of the INDCs. Returning to the relatively new demand for a *low carbon lifestyle by developed countries*, the genuine Indian demand to maintain a sustainable lifestyle for cultural reasons is modified here for the developed countries, which are indisputably partly represented by the opposite chain of equivalence. It can therefore be assumed that the Indian government is trying to export the nodal point *low carbon lifestyle* of its own chain of equivalence to the opposite chain, following strategeme VII (Occasional but deliberated breach of the antagonistic line of division). Thereby the Indian government tries to show the developed countries that they have a huge future task while it pretends that the Indian society would already manage to live a sustainable life due to its cultural traditions. In this way, the keenly-desired view of India as a developing state which is to some extent culturally-rooted with nature can be strengthened.[17] At the same time, without this referencing to the traditional Indian cultural roots, the attempt to export the demand for a sustainable lifestyle crossing the antagonistic line of division would be without any argumentative basis or normative power. Therefore, demanding a sustainable lifestyle is not 'just' a demand which would automatically fit in both chains. As the developed states are not known for invoking their historical closeness to nature, especially since mastering the earth is a core idea of the enlightenment (Tickner 1993, 60–63), the attempt to export the claim seems to be both challenging and promising: It would entail overcoming centuries-old Western tenets, but this at the same time would mean filling a void more and more people are lamenting.

Another important discourse-shaping contribution was Modi's speech to open the Indian pavilion in Paris. There, he reproduced already-known demands; in addition to the recognition of *India as being culturally-rooted close to nature* (Modi 2015d, 263, 265), *growth* (Modi 2015d, 260, 262, 263), the positioning of India as a *developing country* (Modi 2015d, 262), *climate justice* (Modi 2015d, 261, 265) and the demand for *support from the developed countries* (Modi 2015d, 262)

[17] Whether the Indian government succeeds in anchoring this demand in the opposite chain of equivalence is a question beyond the scope of this study and has to remain open here.

are equivalised, once more following strategeme I. Splitting the discursive space, the Prime Minister also uses strategeme II, pointing to the historical responsibility of the North and the simultaneous vulnerability of India:

> *"Climate change is a major global challenge. But, climate change is not of our making. It is the result of global warming that came from the prosperity and progress of an industrial age powered by fossil fuel. But, we in India face its consequences today." (Modi* 2015d, *260–61)*

In addition, Modi follows up on the already-reconstructed strategeme VII. In the final sentences of his speech, he addresses the need to disseminate the culturally anchored Indian sustainable lifestyle to other countries referring to India's closeness to nature by pointing to a famous Sanskrit prayer:

> *"In conclusion, therefore, let me return to the theme that defines us in India – the spirit of partnership, the belief in oneness with the whole that is Nature. To the people of India and our friends in the world, I call you to live with the commitment of Lokah Samstah Sukhinau Bhavantu: The desire for well-being should include our planet, our Nature, all countries and the entire humanity. If our thinking is right, we will form a true global partnership of capabilities and needs that leads us to a low carbon age." (Modi* 2015d, *265)*

The chant "Lokah Samstah Sukhinau Bhavantu" means "May All Beings Everywhere Be Happy" and is to be understood as a wish pointing to the future. Interestingly, the Sanskrit prayer has become very common in worldwide yoga practice since 20th century yoga teachers repurposed traditional Hindu prayers as chants in their yoga classes (Fröhlich 2022). Promoting a sustainable lifestyle in other countries is therefore a demand which concerns the future of the entities represented by the opposite chain of equivalence.

However, Modi insinuates again that India has already entered sustainable lifestyle routines referring to the Indian yoga practice as well as Sanskrit prayers and the centuries-old Hindu traditions. Keeping in mind that the outlined Sanskrit prayer was rediscovered and put into a new context, the climate catastrophe, Modi's recourse back to the Indian yoga culture should be understood as a rhetorical rather than historical reference. This well-known phenomenon is described as invention of tradition for political reasons (Fröhlich 2022).

Besides following strategeme VII, the demand that the developed states decarbonise their lifestyles can also be reconstructed in the framework of strategeme II. Splitting the discursive space in the INDCs it is stated that:

> *"... an approach should be anchored in the vision inspired by Mahatma Gandhi's famous exhortation; 'Earth has enough resources to meet people's needs, but will never have enough to satisfy people's greed'. We must promote sustainable production processes and also sustainable lifestyles across the globe." (Government of India* 2015, *3; emphasis in original)*

The demand for worldwide sustainable lifestyles can be found in Modi's COP21 speech, too—albeit less vigorously articulated (Modi 2015e, 3). The foundation for the new tone of reasoning promoting sustainable lifestyles had already been laid out a few months earlier in the Prime Minister's speech to the UNGA on the occasion of the adoption of the SDGs (Modi 2015b). It was again mentioned in Modi's opinion article in the Financial Times published with the beginning of COP21 (Modi 2015c). Addressing the USA, the CSE also formulated this demand, pre-empting the Paris Summit (Narain and Bhushan 2015, 7). The CSE dedicated a study to the—in its eyes—highly insufficient INDCs of the USA (Centre for Science and Environment 2016, 21) and called for a significant reduction of luxury emissions of US society—which account for one of the highest per capita emissions in the world (Narain and Bhushan 2015, 3). Unlike the CSE, the government equivalises this argumentative point with the emphasis of the cultural Indian connection with nature, as already pointed out. The first goal of the list of INDCs shows this combination: "To put forward and further propagate a healthy and *sustainable way of living* based on traditions and *values of conservation and moderation*" (Government of India 2015, 29; emphasis in original).

Further pledged Indian goals are additional carbon sinks, adaptation and technical progress as well as mobilising (international) funds while focusing on energy. Increasing the renewables and striving for energy efficiency are supposed to provide energy security (Government of India 2015, 29). Therewith, the masculinised demands *energy security* and *renewable energy* (MacGregor 2010, 231) are well anchored in the Indian chain of equivalence. These nodal points were reproduced and supported by various stakeholders such as the CSE, emphasising that its proposal on "Global Renewable Energy and Energy Access Transformation" found great resonance in a joint side event at COP21 with the Indian Ministry of New and Renewable Energy as well as in the framework of the Indian INDCs (Centre for Science and Environment 2016, 23). It is not possible to track exactly the influence of the CSE approach on the Indian INDCs with regard to renewable energies, but it is clear here that strategeme VI was followed to win additional subject positions for the demands *energy security* and *renewable energies*.

The launch of the International Solar Alliance (ISA), which India initiated in the run-up to Paris (International Solar Alliance 2022a) further stabilises these nodal points. Today, the Alliance is headquartered in New Delhi and former TERI Director General Ajay Mathur serves as the current ISA Director General (International Solar Alliance 2022b), underlining India's strong interest in renewable energy and the assertation of playing a leading role. Launching the Solar Alliance, Modi used the reception presenting his second book on renewable energy (Jayaram 2021a, 100) "Convenient Action: Continuity for Change" (Modi 2015a) again discursively reproducing and stabilising the demands *renewable energy* and *energy security*. Even though *energy security* is a focus of the Modi government even beyond the Paris negotiations, fulfilling the demand can also cause environmental conflicts and the destruction of specific habitats. As already pointed out, opposition politicians have accused the Modi government of quickly turning the MoEFCC into a "clearance house"—a diagnosis echoed by various NGOs. Here the antagonism between environmental protection and energy security, present in all economies, becomes obvious. While the previous government tried to compensate for this with lengthy discussions and clearance procedures pointing to the need of Sustainable Development, the BJP government opted for quicker and in almost all cases economy-friendly decisions.[18] This underlines that all concerns are subordinated to that of development in the sense of economic growth, while development, as already shown, emerges as representative of the Indian chain of equivalence as an empty signifier—now with a more explicit emphasis on the masculinised demand of economic growth than when compared to the former government.

To anchor an empty signifier, a discourse coalition of different subject positions is needed, and this is indeed generated in the case of *development*, as contributions to be discussed later will show. However, there was also some contradiction: In the case of a planned coal mine in the Mahan woods, Greenpeace India (GPI) took a stand against the mining project together with the local residents, as their livelihood was jeopardised (Greenpeace India 2016, 19). The protest was thwarted using questionable methods. GPI faced many incidents leading to the interim closure of its bank accounts and the arrest of one of

[18] Accusing the MoEF for only clearing the cases in favour of the economy was already raised with the change in the MoEF from Ramesh to Natarajan (Sethi 2013a). In fact, in 2019 almost 100% of the industrial projects were approved (Ganguly 2020, 72), while the preceding Congress-led government cleared only 80% of the applications (Rathee 2019). These are just some cases of social practices showing the Modi government's will to strive for fast economic growth.

its employees (Subramanian 2015).[19] A leaked report of the Indian Intelligence Bureau even insinuates that the interventions of growth-critical NGOs weaken the annual economic growth by 2–3 percent (Intelligence Bureau 2014, 1). The government's actions must then also be factored into this logic. Sunayana Ganguly (2020, 72) concludes that "The state has turned deeply suspicious of environmentalists and indigenous people, advocating for their rights, especially in arenas where their advocacy threatens state access to resources like minerals or energy". Nikit Sud (2022) recognises the same patterns as in Gujarat when Modi led the government there. Critics who opposed economic projects for environmental reasons were branded as "anti-development" or "anti-national" (Sud 2022, 118). Furthermore, in several background discussions conducted for this story, people who wish to remain anonymous confirmed that once Modi came to power as Prime Minister, NGOs came under pressure, especially in the environmental and climate sectors, for example by not having their licenses renewed under the Foreign Contribution Regulation Act[20] and through a diffuse fear of repression. The glimpse into the state of environmental/developmental NGOs as well as the specific case of GPI shows that the discursive hegemony of the Indian government was used in favour of energy security, thus pushing for rapid economic growth while trying to negate GPI's subject position and invalidating its feminised demands for *degrowth* and *sustainable development*.

The resulting chain of equivalence was also reproduced by Environment Minister Javadekar in his speech to Parliament following the Paris Summit. He pointed out that the "development imperative[s] of India and other developing countries" (Javadekar 2015) were defended and that the right to development was anchored in the Paris Agreement. Showing that the protection of the environment and the interests of the most vulnerable had also been brought in line with the development imperative, Javadekar followed strategeme I. As in Modi's Paris speeches, Javadekar also reproduced the demand *CJ* as well as the demand for *financial and technical support by developed states* (Javadekar 2015). Both demands are vocalised to split the discursive space following the second strategeme, as already showed. Pointing to Modi's release of the MoEFCC book "PARAMPARA"—the title meaning "tradition" (The Economic Times

[19] For a more detailed elaboration of the Government's stranglehold of GPI as one example of the behaviour towards growth-critical actors since the arrival of the Modi government, see Lowinger et al. 2021, 47–49.

[20] In 1967 the Indian parliament adopted the Foreign Contribution Regulation Act regulating that receiving funds from abroad is forbidden as organisation unless it is officially registered in this framework (Ministry of Law and Justice 1976). The Act aims to impede external political interference.

2015), accompanied by the subtitle "India's Culture of Climate Friendly Sustainable Practices" (Ministry of Environment, Forest and Climate Change 2015)—in the framework of his public appearance in the Indian pavilion in Paris, Javadekar reproduces the government's demand that India be recognised as a traditionally environmental-friendly culture. In fact, the book as such is a social practice stabilising this nodal point as well as Modi's tweet when he left for Paris announcing that he is "showcasing India's harmony with nature" (The Times of India 2015) through the book launch. As this was Modi's first substantive tweet after proclaiming that he is on his way to COP21, the tweet underscores the importance of the narrative of India as culturally-rooted close to nature.

Overall, Javadekar's statement shows that he portrays India as part of the developing world and its interests, in particular concerning *CJ* and *economic growth* tying in with Modi's narrative of India's closeness to nature. Although Javadekar does not make this demand his central discursive contribution, highlighting these cultural roots supports the holistic appearance of India and relieves possible criticism that the climate engagement might be too one-sidedly focused on growth.

In a nutshell: India's *culturally-rooted closeness to nature* serves as a demand vindicating the need for *development* without too ambitious mitigation measures, which seem already to be an inherent part of India's enduring culture. Amrita Narlikar (2017, 105) describes this as an attempt to improve the acceptance in Indian society of the internationally-promised mitigation measures of the INDCs.

Besides Javadekar's Times of India statement pointing to the government's focus in the year up to Paris: "India has been stressing on two important concepts—climate justice and sustainable lifestyles" (Times of India 2015), further central demands from India have also been reconstructed. The Indian climate discourse vis-à-vis Paris has very much focused on the demand *development* equivalised inter alia with *rapid growth* and the *Indian culturally-rooted closeness to nature*. Accordingly, it can be confirmed on the discourse level that the Indian government pursued India as a developing country as a hegemonic project trying to establish it as the empty signifier of that time. This is supported by the fact that all three core strategemes have already been demonstrated in relation to India's recognition as a developing country. To consolidate the finding that the hegemonic project of the Modi government is to perceive India as a developing state, strategeme VII has already been reconstructed. An attempt should be made to identify strategemes IV–VI in particular, to learn whether this reading of the Indian climate identity is supported and stabilised at least in the short term.

Strategeme V, gradual constitution of greater leeway for interpreting the symbolic equivalent of the universal, can be reconstructed at various points. As already shown, the CSE associates development above all with a major step towards greater implementation of renewable energies, securing energy access for the poor (Centre for Science and Environment 2016, 23; Centre for Science and Environment, Friends of the Earth International, and What next Forum 2015, 2,3). The All India Women's Conference (AIWC), which has been focusing on the education and training of women for decades, has linked these concerns with projects for sustainable energy production calling for greater attention to both women's empowerment and sustainable energy production in the context of India's future development. A central statement of AIWC's president Veena Kohli, presented at a Paris side event: "We strongly advocate shifting over to newer non conventional methods of Low Carbon Development to reduce pollution and mitigate climate change" (Raj 2015, 13) makes abundantly clear that development means particularly low carbon development for the UNFCCC observer AIWC (All India Women's Conference 2013). It cannot be clarified whether the AIWC was offered a subject position in the sense of strategeme VI, and whether therefore the NGO with its interpretation of development was actually included in the chain of equivalence. Nevertheless, it is important to note that AIWC, advocating for women, worked on the Indian development narrative of the time and wanted to shape it in a feminised way instead of the government's masculinised understanding of development.

As the Modi government was relatively new to office when formulating its INDCs and promoting them in Paris, the question is left open as to whether the government was also able, after Paris, to include further subject positions for politico-societal forces in the Indian chain of equivalence, increasing the outreach of the encompassing demand *development* and anchoring it over a greater time period than just a year. However, since the Indian government succeeded in Paris in pushing the demand for a sustainable lifestyle breaching the antagonistic line of division (strategeme VII), it can be assumed that the hegemonic project of the Modi government would receive support in some places beyond Paris, for example from socio-political actors.

Strategeme IV, the super-differential demarcation of the Indian climate change discourse's limits, can once again not be reconstructed. The same reason as in the period of the Singh government (cf. Sect. 3.2.1) can be assumed here: The boundaries of the Indian climate change discourse were not attacked at that time and therefore this strategeme was not used as the different actors seemed to agree on the topics discussed in the context of this discourse. The boundaries of the discourse may be challenged in the further course of the Modi government so that strategeme IV could be followed accordingly. This question will be discussed in Sect. 3.2.7.

3.2.5 The State Action Plans: Winning Women for Work

But first, special attention must be paid to the level of the more than 30 Indian states and Union territories, as a significant part of the Indian climate change policy has to be implemented there. Via State Action Plans on Climate Change, the states were obliged to conceive policies implementing the National Action Plan of 2008. Of course, not all SAPCCs can be analysed in detail, but the research focus on gendered discourse patterns and the support for Modi's hegemonic project *development* or for the predecessor government of Singh who anchored *sustainable development* as the empty signifier will be briefly reconstructed.

Most of the Indian SAPCCs were adopted between 2011 and 2014[21] and were some of the first instances of the states discussing the consequences of climate change for the respective state in a holistic manner (Gogoi 2019, 375). More than one third of the analysed SAPCCs stem from 2014, coinciding with the period of upheaval due to the new government in Delhi readjusting the national climate policy on the road to Paris. Still, most of the states had had a long process of discussion (Gogoi 2019, 370), so the plans were developed especially while the former Singh government was in power. The SAPCCs are created to implement the NAPCC's missions. Vineet Kumar (2018, 39) attests that they are not concrete plans, rather a "wish list", lacking specific financial details. Nevertheless, the SAPCCs provide a picture of which discursively gendered patterns can be found 'on the ground' especially concerning discourse dimension I (cf. Sect. 2.3.3), that is gendered subject positions, as the SAPCCs address, for example, adaptation and daily life routines. These subject positions are built through interpellating demands when hegemonic actors such as state governments demand that specific groups or persons fulfil particular roles or tasks (cf. Sect. 2.2.2). Besides dimension I of gendered discourses, I will also tentatively search for gendered discourses of dimension II in the SAPCCs, without reconstructing all probably-used strategemes. Keeping in mind that the 36 Indian states and union territories are governed by many different party coalitions, it is not the goal to detect this part of the Indian climate discourse or analyse all SAPCCs in detail.[22]

---

[21] In 2018 the Indian states and union territories were asked by the MoEFCC to revise their plans to be able to reach the common post-2020 targets (Ministry of Environment, Forest and Climate Change 2019, 209). In the case of Odisha, the revised plan of 2018 is integrated into the analysis. However, at the time of the study until late 2021, the other states and union territories had not adopted revised plans. Some of these are currently under discussion, such as in Kerala (Sudhi 2021) or in Bihar (Pathak 2021). It must be assumed that inter alia due to the Corona pandemic the revision process in most states has been delayed.

[22] For an analysis of the Indian SAPCCs see Kumar 2018.

In general, the plans are much more oriented towards adaptation than mitigation (Dubash and Jogesh 2014, 6). Although the SAPCCs are not to be understood as genuine implementation plans, the focus on adaptation elucidates that they are designed especially for the local needs. Most of the plans emphasise that greater adaptation efforts are required, pointing inter alia to women as they are particularly suffering from the negative effects of climate change (Government of Bihar 2015, 27; Government of Punjab 2014, 7; Government of Chhattisgarh 2014, 31). As a result, most of the SAPCCs use gendered discourse patterns showing for which roles women are recognised. The discursive production of gendered roles and therewith gendered subject positions can be perfectly reconstructed when women's challenges due to climate change are outlined in the plans. For example, when it comes to water supply (Government of Arunachal Pradesh 2011, xxiii; Government of Odisha 2018, 160), agriculture (Government of Kerala 2014, 59; Government of Nagaland 2012, 2) or forestry (Government of Uttarakhand 2014, 23), the impacts of climate change on socially-constructed roles are thematised. The Government of Arunachal Pradesh points to the need of "gender sensitive adaptation options like effective strategies for ensuring water supply and quality, and reducing the burden on women caused by water collection" (Government of Arunachal Pradesh 2011, xxiii) suggesting inter alia an "awareness campaign involving all water users especially women, and through mass media in rural and urban sectors" to fight the problem of appropriate water supply. This is just one example, but it shows that women's subject positions are socially deduced from the fields of the work and responsibilities they are assigned to. This is the case for the majority of the plans and goes hand-in-hand with the production of women as a monolithic group of care givers. Even though further diversity dimensions such as class or age are mentioned in some cases, their intersection is not explicitly thematised (Singh, Solomon, and Rao 2021, 964)

Therefore, women's subject positions, discursively produced in most of the Indian SAPCCs, are twofold:

(1) On the one hand, women are discursively written as victims of the climate catastrophe. In 26 out of 33 State Action Plans examined, references to women as victims can be identified. This is often linked to the demand that women as well as children and the elderly should be better protected.
(2) On the other hand, their roles are shaped as carers of the environment due to their socially constructed roles and accountabilities.

While the first role is constructed as an almost passive figure, the latter appears ambivalent. In most cases, the position of the female carer for the environment is solely recognised for her (primarily unpaid) workforce in different contexts such as agriculture, and not equipped with an active political role. Singh, Solomon, and Rao (2021, 965) share these findings confirming that only in a few cases are women conceptualised as active agents of change. In particular, only five SAPCCs (Bihar, Chhattisgarh, Gujarat, Madhya Pradesh, Uttarakhand) demand women's empowerment in climate change discussions or integration into decision-making processes, thereby providing women with an explicitly active political part with a specific say (Government of Bihar 2015, 53; Government of Chhattisgarh 2014, 17, 43; Government of Gujarat 2014, 191; Government of Madhya Pradesh 2014, 143; Government of Uttarakhand 2014, 76). Hence, women are thought of as a probably pro-active group with subject positions shaping the climate discourse, foremost when it comes to adaptation reducing their vulnerability.

However, in the former theoretical discussion it has become clear that the subjects can only shape the discourse from the positions for which they are recognised. Thus, there is the potential for women to bring specific demands into the framework of their subject positions. These could be demands for greater inclusion of women into climate change debates and adaptation processes or demands to take more account of existing traditional knowledge in Indian climate policy. The Centre for Science and Environment (CSE) criticises the fact that some SAPCCs do not sufficiently cover the diverse adaptation knowledge of women, while—astonishingly—most State Action Plans do not factor it in at all (Kumar 2018, 24). For the period up to the Paris Agreement studied so far, I can solely consider subject positions in the Indian chain of equivalence stemming from the assigned roles of women in the SAPCCs. Of course, this does not mean that other subject positions, such as those of politicians or various NGOs, are not gendered. Rather, these positions are just not obviously gendered in such a way that interpellating demands clearly targeting a gender dimension of the respective subject position could be reconstructed in the analysis.

How did gender perspectives enter the SAPCCs? The involvement of various national and international institutions such as TERI or the German GIZ in the drafting (Raghunandan 2020, 217) could be one reason for successfully anchoring a gender perspective in the plans without diffusing to the national discourse level. Furthermore, different civil society groups such as women's organisations were also partly involved. However, one interviewee pointed out that these workshops sometimes only exerted limited influence, and that NGOs were therefore seemingly only involved pro forma (Member of an Indian Development and Climate NGO 2019). As gender awareness might not have been that rooted prior to the implementation of the SAPCCs, and as influence by further actors was limited, the resulting gendered subject positions are often traditional ones. These are victim of climate change or carer for the environment—lacking transformative agency as the AIWC already pointed out in 2013 in the context of a hearing on the implementation of the Doha miracle (All India Women's Conference 2013, 1). Women's roles with an agency are only mentioned in the five SAPCCs mentioned above. Hence, it can be assumed that the subject positions for women created in the overwhelming majority of the SAPCCs serve the implementation of adaptation work, but not the inclusion of gender perspectives into the multifaceted Indian climate change discourse. In fact, we should rather speak of women's perspectives when it comes to gender in the SAPCCs, since as other researchers have also noted (Singh, Solomon, and Rao 2021, 967) only women—most often discursively produced as a homogenous group—are addressed. Men are implicitly set as the norm, and the LGBTIQ community is neither addressed nor has subject positions created. This again results in the reproduction of essentialist pictures of women through the SAPCCs: Women are conceived as close to nature, attributing care work on nature solely to them, most obviously concerning concrete adaptation measures or programmes. Although the social construction of these roles is partly addressed in the plans (Government of Gujarat 2014, 169; Government of Odisha 2018, 160; Government of Tamil Nadu 2014, 19–20), this knowledge is not factored in when conceptualising climate change measures. Bearing in mind that in some of the SAPCCs women's subject positions are constructed with agency, these are merely linked to adaptation and not with the all aspects of climate policy.

Deconstructing the second dimension of gendered discourses in the SAPCCs, we could assume that the plans are gendered, focusing on adaptation which is most often linked to femininity (Djoudi et al. 2016, 257), not least due to the outlined constructed subject positions of women linked to adaptation measures. The interplay of the two discourse dimensions theoretically derived in Sect. 2.3.3 becomes graspable here. The extent to which an adaptation focus contributes to gendering the Indian climate discourse is discussed in Sect. 3.3.2. First, though, the post-Paris climate discourse will be further pursued.

3.2.6 India as a Developing Country Striving for a Significant International Role

Although it has already been demonstrated that the first dimension of gendered climate discourses can only be found sporadically in the Indian climate discourse, as in the SAPCCs, the debate about the Lima Work Programme on Gender at the UNFCCC level also generated some insights into women's subject positions in

Indian climate discourse in 2016. The Indian government's idea of the position of women in its climate policy can be reconstructed in the paper contributed to the Subsidiary Body for Implementation (SBI) of the Lima Work Programme on Gender.

In this paper, the Indian government refers first to the mostly traditional positions of women and "cultural barriers [which] may prevent women from asserting their demands effectively" (Government of India 2016, 17). Even though an awareness of the social construction of these roles is evident, the more traditional positioning of women is implicitly adopted. In outlining India's priorities, "new, affordable, cleaner and innovative energy to women, especially rural women" are mentioned, as well as health, forestry and other livelihood aspects and DRR (disaster risk reduction) (Government of India 2016, 17) which are all classical spheres of action to which women are assigned conceptualising their roles as care-givers in a wider understanding.

Hence, it seems that the Indian government tries to follow its hegemonic project of *development* instrumentalising the already existing roles of women in the discourse as subject positions of the chain of equivalence supporting the encompassing demand for development. The outlined demands, such as *energy security, renewable energies, and environmental protection* (Government of India 2016, 17), are demands which are raised here in relation to the complex of gender and climate change. However, these are also the demands that are regularly woven into the chain of equivalence by the Modi government securing the representation of India through the empty signifier *development*.

Astonishingly, the Indian government's approach in this statement goes beyond the reproduction of already familiar patterns of argumentation. For instance, the promotion of increased inclusion of women in decision-making processes is pointed out (Government of India 2016, 18). Fig. 3.1 had already shown the need for action, as before 2016 the share of women in Indian COP delegations only once exceeded 30%, the marker which is necessary for the involvement of women's perspectives in the delegation. However, it is conspicuous that specific reference is made to the fields of activity traditionally attributed to women. Hence, it can be concluded that the discourse producing subject positions at this point is gendered, especially regarding women in the Indian climate discourse. They are discursively produced as protectors of the environment especially linked with adaptation skills, or as workers contributing to the state's development (Government of India 2016, 17, 18). Interestingly, the Government of India argues similarly in its paper to the SBI on the establishment of the Gender Action Plan which was adopted in 2018 under the Lima Work Programme on Gender. While it emphasises that women should be more involved in the various aspects of climate change, for example in sustainable forest management, there is again a clear focus on adaptation and development, as the following section shows:

> *"A number of initiatives, systematically integrating gender considerations, are being implemented in the country to strengthen specialised skills and capacities of women in different spheres of livelihood systems." (Government of India* 2018, *2)*

The government concludes by emphasising that women already constitute a considerable part of Indian delegations in the climate context, both nationally and internationally (Government of India 2018, 3). Yet, in the following COP24 in Katowice, the share of women in the delegation was only 16% and thus 15% points less than in the previous year (cf. Sect. 3.2.2). Regarding both Indian submissions to the SBI, women can initially only gain a voice from the positions as carer or developer. Whether and how the discourse at this time is gendered in the second discourse dimension is shown by deconstructing further discourse contributions.

As the post-Paris period on the international stage was dominated by technical implementation questions, the same was true of India's post-Paris climate discourse. Remembering the submission of the INDCs in 2015, the 2nd of October was again chosen by the Indian government to ratify the Paris Agreement, which then entered into force in November 2016. Besides implementing Paris, India was keen at the 2016 COP in Marrakesh to force the developed countries into transposing another KP period (Ministry of Environment, Forest and Climate Change 2017, 180). India's repeated call for the developed states to initiate further substantive mitigation measures, especially before 2020, followed the logic of the second strategeme antagonistically dividing the discursive space producing a 'we' and 'them'. This is a familiar pattern of the Indian climate discourse, as is the reference to India's *culturally-rooted closeness to nature.* Not least through Modi, for example on the occasion of the 40th anniversary of the Society for the Promotion of Indian Classical Music And Culture Amongst Youth 2017 (Modi 2017), but also through his environment minister Harsh Vardhan (BJP) at COP23 in Bonn 2017, the narrative of India's *culturally-rooted closeness to nature* was again cultivated and also linked to the representation of India as a developing country (Vardhan 2017). Accordingly, it can be attested to that for the years 2016 and 2017 the strategemes used and the hegemonic project strived for by the government actors remained largely the same as vis-à-vis Paris. Yet, the enthusiastically-pursued establishment of the International Solar Alliance (ISA) as well as its first assembly with 25 heads of state attending in Delhi in

2018 (Press Information Bureau 2018) show that India's chain of equivalence is increasingly focused on *renewable energy*, gaining not only *energy security* but also a *more significant role in the international climate arena* equivalising these demands (Jayaram 2021b, 211).

It is against this backdrop that India's reaction to Trump's announcement withdrawing from the Paris Agreement in June 2017 must be regarded. Although there was no specific speech by the Indian government countering Trump's sermon at the White House's Rose Garden, various statements at other occasions indicate that the Indian government was keen to continue playing a cogent role as a climate policy actor. In fact, there was no significant suggestion that India should follow Trump leaving the Paris Agreement. Even though Modi refrained from direct criticism of the US president, he did not tire of emphasising that India's traditional values mean that it is still addressing the pressing need of global climate protection, especially because of its responsibility towards future generations. At the St. Petersburg International Economic Forum Modi pointed out: "India is a responsible nation with regard to climate change… Exploitation of nature is not acceptable to us" (Vishnoi and Roy Chaudhury 2017). In addition to the well-known narrative that India is *close to nature for cultural reasons, India's responsibility* in general as well as *in relation to future generations* are further demands emphasised in this case. All three demands are equivalised following the first strategeme. Anjali Jaiswal from the Indian National Resources Defence Council goes even further, attesting India to be displaying a "bold leadership", indicating that "India is doing several hard things at once, struggling to power its fast-growing economy, lifting millions out of poverty and building a clean energy economy" (The Wire 2017). Underscoring the demand for India to play a significant role in the international climate discourse, against the background of the US exit, Jaiswal holds a subject position backing the Indian chain of equivalence.

Even though none of the Indian officials tried to publicly demand the role of the single forerunner in climate change governance, a more significant role would have been useful regarding the upcoming discussion on the rulebook for the implementation of the Paris Agreement which found its peak in Poland 2018. In Katowice at COP23, environment minister Vardhan repeatedly referred in his speech to India's strong sense of responsibility. In addition to the responsibility towards future generations, Vardhan gave assurances at several points that India would continue to play a substantial role in the joint fight against climate change (Vardhan 2018, 4, 5). Following strategeme I Vardhan equivalised different demands such as the *historical responsibility of the North*, pointing to low Indian per-capita emissions, *climate justice* as well as *Indian mitigation action* emphasising already-reached emission intensity goals (Vardhan 2018, 3). Equivalising the demand for *financial and technological support by the developed countries* with *multilateralism* (Vardhan 2018, 4) the Indian government again follows the first strategeme. While multilateralism was in previous speeches merely paraphrased, for example with the expression of joint action, it seems that this demand is gaining the position of a nodal point, as Modi for example also highlighted the need for multilateralism after the US withdrawal.

The second core strategeme, the antagonistic division of the discursive space, can be reconstructed when Vardhan indirectly distinguishes India's chain of equivalence from the other, in this case the US, emphasising: "we all agree that Paris Agreement is non-negotiable" (Vardhan 2018, 2). As in manifold other Indian discourse contributions, the demand for *climate justice* in its different variations such as *CBDR*, *equity* or directly called *climate justice* (Vardhan 2018, 3) is articulated dividing the antagonistic space. Searching for the third strategeme, the empty signifier representing the Indian identity, is elusive, as it is less obvious than in other government contributions. Pointing at different positions in the speech on *development* and linking it to India's *culturally-rooted closeness to nature*, this speech has at first glance a similar thrust to the government's articulations of previous years. However, it is noticeable that the chain of equivalence no longer runs uncontested and exclusively towards *development*, but that the demand for India to *play a significant role in global climate policy* seems to be more important than in previous years and especially in comparison to Paris.

Vardhan (2018, 4) referred to the prestigious Champions of the Earth Award, which Modi received from the United Nations Environment Programme (UNEP) for his "bold environmental leadership on the global stage" concerning both the announced elimination of single-use plastic in India and his engagement with ISA (United Nations Environment Programme 2022). Thus, the demand for a *significant Indian role* on the global climate stage is given additional attentiveness. Although Modi was castigated by the influential CSE-initiated magazine Down to Earth who argued that the award was inadequate as "his regime's environmental track record indicates anything but that" (Down to Earth 2018) and other environmentalists were also critical due to different legislative changes to the detriment of the environment and climate (Sarkar 2018), the award as such and the associated prestige reproduced and strengthened the nodal point of India as an important climate actor. Nevertheless, various passages in the Indian government's three-year plan support the idea that environmental regulations should be watered down if they "adversely impact growth without protecting the environment" (NITI Aayog 2017, 6) and forest rights of indigenous people should even be partially abolished (NITI Aayog 2017, 169).

Likewise, the debate regarding the systematic persecution of growth-critical NGOs and their framing as "'Foreign Agents' or threats to development" (Matejova, Parker, and Dauvergne 2018, 151) and dramatically increased deaths of environmental activists during Modi's first term in office challenges India's image as a valuable climate actor. The following Fig. 3.2 shows that in 2018, a sorry peak was reached with 23 environmental activists murdered in India, which became the third most dangerous country for land and environmental defenders (Global Witness 2019, 8).

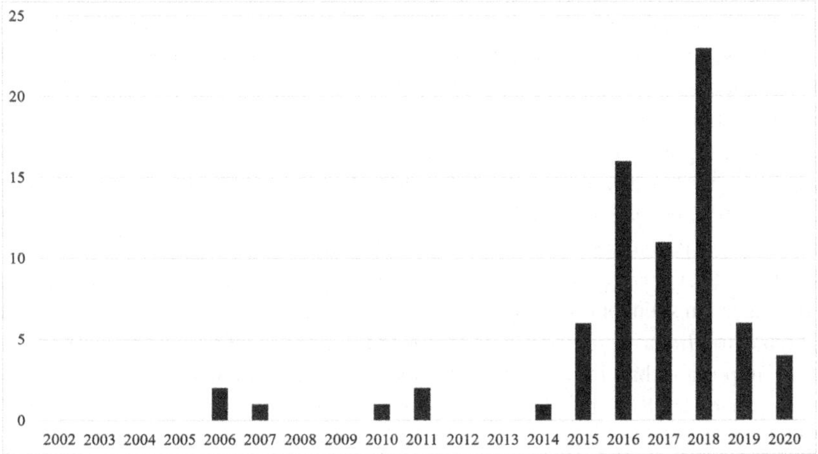

Fig. 3.2 Killings of climate and environmental activists in India, 2002–2020. (Own illustration with data from Global Witness 2012, 2013, 2015, 2016, 2017, 2018, 2019, 2020, 2021)

The exact causes of death are often unclear. Environmentalists report that they are more and more under pressure, reflecting "a climate of fear" (Saldanha 2018) as "India has seen killings spike against a backdrop of heavy-handed policing and the repression of peaceful protests and civic activism" (Global Witness 2017, 6).

These are all indications that the Indian government is unlikely to succeed in achieving the hegemonic project of being recognised as a *significant climate actor with cultural roots*, caring for the environment. The antagonistic juxtaposition of environmental protection, embodied by activists on the one side, and development, for example through mining or the construction of dams on the other side, are often played off by the Indian government referencing Indian history (Matejova, Parker, and Dauvergne 2018, 152). In this logic, environmental activists are portrayed as "foreign agents" and "unpatriotic" (Matejova, Parker, and Dauvergne 2018, 152, 153) so that most parts of the society are often with the government striving for economic growth on the detriment of the environment and climate. To find out whether the Indian government can pursue the project of being recognised as a *significant climate actor* against the outlined contradictions, it must be evaluated in the following whether the government succeeds in creating the necessary subject positions in the chain of equivalence supporting its hegemonic project.

These inconsistencies are only one aspect of India's climate discourse, and the prestige, be it through the award or simply the emphasis that India will by no means follow Trump's example and remain loyal to the PA, is still weighty. What reasons could have led the Indian government to pushing and expanding the demand for recognition as a *significant climate actor*? On the one hand, the emphasis on a significiant Indian role could represent an effort to close the gap created by the USA withdrawing from the PA and to prevent other states from following Trump's step. On the other hand, it seems that the demand recognising India as a significant climate actor has been pushed by the Modi government for longer than 2017, keeping in mind that initiating ISA started before Paris. Playing a significant international role can be understood as having masculine connotations. Not least as this empty signifier is charged by masculinised demands such as economic success, technical innovation and energy security. This suggests that insofar as this hegemonic project can assert itself as hegemony, we must speak of a masculinised identity.

Furthermore, federal elections in India were due in 2019. Hence, it could have been a tactical consideration of the BJP demanding a *significant international role*. To verify this inference the election programme will be analysed to see whether a significant Indian role is also an important concern of the BJP there. A shift in the hegemonic project from recognising India as a developing state with specific cultural roots to seeing India as a significant climate actor which is also culturally-anchored can be presumed on the basis of the developments in the speeches as well as social interactions of the Indian government in the period after Paris. The fact that the Modi government has strongly pushed *development* as the empty signifier striving for economic growth even before the BJP won the elections in 2014 and beyond Paris while "the common masses have hardly seen their lives improve from an economic standpoint, and the rich have benefited immensely from his politics" (Jaffrelot 2021, 153) may have caused or accelerated the process of striving for a new empty signifier, the narrative of *India as a significant climate change actor culturally-rooted to nature.* Of course, the period of the 2019 election campaign must first be considered when searching for a possibly changing hegemonic BJP-government project and reconstructing further followed strategemes that anchor such a project.

3.2.7 The 2019 Elections and Beyond: Striving for a Culturally-Underpinned Significant Role

The starting position in the 2019 election campaign was of course different from that of 2014. As the ruling party, the BJP changed the appearance of its election manifesto to more of a task-list for the upcoming term. Furthermore, instead of questions of development and poverty reduction, the debates addressed in particular security. One event Modi knew how to leverage for his campaign, as Christian Wagner (2019) rightly predicted, was the suicide attack which killed 40 people in the Indian part of Kashmir just two months before the first Indian states went to the ballot boxes. The shift from growth and poverty eradication to "an ethno-nationalist security-dominated repertoire" (Jaffrelot and Verniers 2020, 156) as the election campaign focus in 2019 is clearly visible. Therefore, it is not surprising that "Nation first" is the initial chapter of the BJP-manifesto (Bharatiya Janata Party 2019, 11). In the introduction, as well as in Modi's letter to the electorate, it becomes obvious that besides security, a strong international role or recognition as such seems to be an important campaign goal of the BJP, as Modi insists in his letter to the voters:

> *"Today, India stands tall at the international level. The world recognizes India's immense potential, which is clearly reflected in a record rise in foreign investment. India's stand on subjects like terrorism, climate change and prevention of money laundering are setting the global narrative." (Bharatiya Janata Party* 2019, *4)*

In fact, by ascribing to India a narrative-setting power, Modi goes even further. Since climate change is explicitly listed here as an example of a strong Indian role, it is worthwhile to investigate whether or not the climate part of the manifesto strengthens the *significant role* demand, and which strategemes are used.

The second (and last time) climate change is mentioned in the manifesto, *mitigation measures in India* are initially equivalised with the *significant role* of the subcontinent, stating that "India has become a global champion in addressing the issues of climate change through effective and feasible interventions" (Bharatiya

Janata Party 2019, 22); in a second step, the role of "global champion" is exemplified, when the party equivalises the significant Indian role with the enormous expansion of renewable energies, which the Modi government credits to itself (Bharatiya Janata Party 2019, 22). Indeed, the solar energy capacity in India more than quadrupled from 2015 to 2018 (Statista 2022), but it should not be forgotten that the Congress-initiated NAPCC laid the groundwork for this success. In the context of this brief reference to climate change, the BJP also emphasises the importance of the International Solar Alliance (ISA), invites other countries to join the international organisation and states that India will make its renewable energy efforts a "popular movement" (Bharatiya Janata Party 2019, 22). At this point, the representation of the demand for *renewable energy* by the encompassing demand for a significant Indian international role becomes obvious, as the vindication of the successful role is attributed domestically to the increase of renewables and internationally the initiation of the ISA.

In the later chapter "Forest and Environment", climate change is not mentioned again. Instead, praising the now more effective and much faster clearance procedures leads to the insinuation that these in particular are the reason for reforestation successes, as the forest cover of India was extended by around 9000 square kilometres (Bharatiya Janata Party 2019, 26). It is interesting to note that the dichotomy between the clearance process—which since the advent of the BJP government is basically synonymous with the success of industrial projects (cf. Sect. 3.2.4)—and the suffering nature is not addressed, but rather it is suggested that the clearance process leads to afforestation. This equivalency of the demands *growth* and *environmental protection* completely obscures the fact that the Modi government generated the increase in forest cover foremost through monocultures and insufficiently dense tree cover (Saxena 2019, 33). Accordingly, we can trace here the power of discourse to present obvious dichotomies as supposedly compatible in the sense of an equivalence.

Although in the BJP's 2014 election manifesto the environment chapter also addressed issues of climate change such as mitigation, a clear separation of environment and climate change now appears to have been made, despite no chapter having been explicitly dedicated to climate change. The manifesto does not explicitly refer to the pattern that India is *culturally-rooted close to nature.* In the introduction to the manifesto, ecology and cultural roots are linked: "The cultural, ecological and social capital of a country are just as important as its economic capital" (Bharatiya Janata Party 2019, 10), but not equivalised as in earlier contributions such as Modi speeches. Although yoga and many other aspects of the Hindu tradition play an important role, no connection with nature or climate is made. However, this may be due to the manifesto's character as a universal, and at the same time compact, political plan. Whether the demand *culturally-rooted closeness to nature* loses significance overall and what this could mean for the Indian identity remains to be analysed.

The manifesto of the Congress Party dedicates two chapters to specific climate change aspects and was billed as "its greenest manifesto to date" (Sauer 2019). Chapter 49 "Environment And Climate Change" (Indian National Congress 2019, 51–52) and Chap. 50 "Climate Resilience and Disaster Management" (Indian National Congress 2019, 52) both deal with various climate change topics. The already well-known demands of the Congress Party, such as *climate justice*, *poverty eradication*, *environmental protection* and *adaptation* or *sustainable forestry* are rearticulated. Interestingly, mitigation measures are hardly mentioned, although it is emphasised that Indian companies producing clean technology should be supported with fiscal instruments (Indian National Congress 2019, 52). In addition to the focus on *technological progress*, which is a clearly masculinised demand, it is also mentioned that local communities must be included in the various measures and that their rights must be especially protected (Indian National Congress 2019, 51), which is a demand of the feminised sphere. Demanding to protect tribals' rights can be seen as an attack on the BJP, with the radical clearance procedures which regularly their rights. The programme highlights the target of a *strong Indian role* promising that "Congress (…) will place India at the forefront of the battle against global warming and for the protection of the environment" (Indian National Congress 2019, 51). Although Congress insists on a strong Indian role in climate change policy, as does the BJP, this demand does not seem to be the central one. Instead, the programme comes back again and again to demands concerning the implementation of measures protecting the environment with a social impetus and with risk reduction instruments. Since the Congress Party lost a lot of votes in the 2014 elections, especially among the poorer classes (Rodrigues 2014, 15), it was probably intended to address them, which furthermore explains the focus on adaptation.

Although worthy attempts can be identified in their manifestos (Pandey 2019), the efforts of both parties to address the various facets of climate change were ultimately judged by observers to be inadequate (Singh Panwar and Unni 2019). NGOs for example noticed that when it came to climate change problems, the BJP was unable to deliver during its term in office (Tripathi 2019; Sauer 2019). It seems that this is one reason why the party barely addressed climate change in its manifesto. However, even though they are an important discourse contribution indicating future priorities, the election programmes did not spark subsequent broad debates. As the INC focused mainly on the issue of development and poverty eradication while the BJP focused on security, the share of climate policy issues was front-loaded by the INC, even though climate change as a campaign issue did not became prominent at all (Tripathi 2019). To sum up, the BJP used almost exclusively masculinised demands while the INC pushed also social demands which are in most cases feminised.

After receiving 37% of the vote and thus winning the 2019 election by a large margin (India Votes 2022), the BJP again provided the minister of the MoEFCC. Prakash Javadekar (BJP), who had already led the ministry from 2014 to 2016, was once again appointed environment minister. It is therefore interesting to ask whether Javadekar would continue his earlier hegemonic project of development, or join Modi in—as already described—striving for India's significant role on the global climate stage. Various opportunities arose in 2019 to continue to push the latter goal as an encompassing demand. For example, in September, India hosted the 14th United Nations Convention to Combat Desertification in Delhi. In addition, also in 2019, Modi announced the launch of the Coalition for Disaster Resilient Infrastructure (CDRI) at the UN Climate Action Summit in New York. In his speech, Modi named CDRI in a series with ISA and invited all states to join (Modi 2019). Today the CDRI has 31 members, including larger countries such as the USA, Australia and Brazil, as well as the EU and other European nations, right through to some Small Island States (Coalition for Disaster Resilient Infrastructure 2022). Although the initiative's annual report gives the impression that it is in its infancy and primarily a forum for knowledge exchange, India has once again made its claim to play an important role with regard to climate-induced disaster risk and related research, not least due to the fact that the CDRI's secretariat is Delhi-based (Coalition for Disaster Resilient Infrastructure 2021). Modi's argumentation explaining the need to launch the CDRI is remarkable:

> *"What we need is a global people's movement to bring about behavioural change...therefore India is here today to present a practical approach and roadmap. In order to make our infrastructure resilient in the face of disasters, India is launching a Coalition for Disaster Resilient Infrastructure." (Modi* 2019*)*

Following the first strategeme he once again implicitly equivalised the demands for a *low carbon lifestyle* as well as the *culturally-rooted closeness to nature by India* with a *significant Indian role* directed towards a much-needed global movement and the initiative taken by his government (Modi 2019). Of course, pointing to both the need for a low carbon lifestyle by the masses and the Indian cultural roots caring for the environment also splits the discursive space configuring an 'us' and 'them'. The repeated reference back to a *significant Indian role* suggests that Modi's concern continues to be to establish this demand as an empty signifier and thus to build on his previous term in office. This impression is also supported by the annual MoEFCC report where ISA and CDRI are highlighted as "India's pioneering global initiatives" (Ministry of Environment, Forest and Climate Change 2021, 256). The additional announcement in New York that Modi would inaugurate the Indian-sponsored solar panels on the roof of the United Nations the next day (Modi 2019) suggests that the significant Indian role also includes or at least pushes for a certain pioneering role.

Whether these measures, such as the solar panels or the introduction of the CDRI, are able to fulfil the demand for a significant role in the sense of a well-established discursive hegemony will have to be examined in due course, as these initiatives may also remain unsuccessful attempts if the hegemonic project lacks a critical number of socio-political forces supporting the demand.

A first contradiction to the more significant role of India being an actual leadership role can be found in Javadekar's closing sentence in his speech in Madrid in 2019 at COP25:

> *"This is the time for ownership and this is the time for responsible action. India has and will continue to do its bit—expecting commensurate multilateral action with developed countries taking the lead." (Javadekar* 2019*)*

It becomes clear that Javadekar sees the significant role of India primarily in the fact that India will play an active role within the scope of its possibilities, but not the sole leading role. A significant role of India in climate change as the empty signifier seems to work, as pointing to the responsibility of the 'radical other'—the developed states—opens up a huge empty space 'filling' the expectations of the empty signifier *significant Indian role.* Javadekar uses different strategemes to rearticulate and temporarily fix this encompassing demand as the empty signifier of the Indian discourse of that time. He divides the discursive space using strategeme II. In addition to demanding historical responsibility from the developed states, he also requests them to provide *financial resources* and *technical support* (Javadekar 2019). Furthermore, Javadekar rearticulates the demands to install a vital *loss and damage mechanism* and *climate justice* (Javadekar 2019), thereby again following strategeme II. In addition to the division of the discursive space, Javadekar pushes for further strategemes. He uses strategeme I equivalising various demands to reconstruct the Indian chain of equivalence in his speech in Madrid. Besides the demands already mentioned, the demands for *adaptation in India, renewable energy* and the familiar old demands of the Modi government, namely the *recognition of India as culturally-rooted close to nature,* are also equivalised. Relatively new is the demand for *multilateralism,* which is also equivalised, going hand-in-hand with the central demand for a *significant Indian role.* A closer look at this speech suggests the shape, or rather the void, of the empty signifier. It seems that the Indian government, at least in the person of Javadekar, does not necessarily favour a significant role in the sense of a "leader by example" (Malnes 1995), adopting concrete initiatives and actions. Rather, Javadekar favours a significant role in the context of a functioning multilateralism in which the historically responsible actors of the West in particular take on a leading role in implementing ambitious measures.

Nevertheless, Javadekar insists that India has a significant role, pointing out that "Only 6 countries are on track to meet their NDCs announced in Paris. We are leading the pack" (Javadekar 2019). Thus, the empty signifier *significant Indian role* is constructed to represent leadership within a smaller, defined group, while not claiming an absolute leading position. Interestingly, the sentence quoted is followed by the reference to a "Sustainable lifestyle" which "is a part of the ethos of India" (Javadekar 2019). By linking the *significant Indian role* with the promotion of a *sustainable lifestyle*, which India directly attests to itself, the identity shaped by the government at this point again refers to the Indian *culturally-rooted closeness to nature*. Here Javadekar uses strategeme V showing a variety of readings of the empty signifier. To identify whether the government follows strategeme V on further occasions as well as strategeme VI—(Re-) Articulation of subject positions for politico-societal actors, extending the hegemonic project's reach—further contributions of the climate change discourse will also be deconstructed in the following.

Although some of the Indian non-governmental actors did participate in COP25 in Madrid, whether they supported the hegemonic project of the Indian government or actively worked against it cannot be reconstructed. The All India Women's Conference (AIWC), for example, reports its participation in rearticulating the demands for *climate justice* in the form of *loss and damage,* and also refers to successes, for example with regard to the stronger participation of women embedded in the 5-year Gender Action Plan adopted in Madrid (Nair 2019, 19). Usha Nair also points to the speech of the eight-year-old Indian girl Licypriya Kangujam who insisted on intragenerational fairness:

> *"I'm fighting for my future. Losing my future is not like losing an election or losing a few points on the stock market. I'm here to speak for my generation and all the generations to come" (Kangujam quoted in Nair* 2019, *19).*

As the AIWC explicitly highlights and supports this voice, the organisation puts forward demands other than those of the government, namely *intragenerational equity*, and *greater participation of women.* Still, as the report makes no reference to India either playing a significant role, let alone welcoming it, it cannot be decided whether the AIWC supported the hegemonic project of the Indian government or was given a subject position in the Indian chain of equivalence in the sense of the strategeme VI. Rather, it will be interesting to see later whether the AIWC's demands are taken into account in the Indian chain of equivalence or whether their counter-hegemonic voices are ignored.

Interestingly, Greenpeace India, on the other hand, emphasises in its annual report of 2019 that India "has the potential to play the role of a climate leader not only in Asia but also globally" (Greenpeace India 2019, 9), although this sentence is preceded by pointing out that the Indian government acted against GPI as bank accounts had been frozen again. Here the appearance of the empty signifier can be observed. GPI describes the potential of the participation of the people on the ground. For instance, reference is made to various environmental protection activities in the field of plastic pollution, as well as political successes achieved in the fields of air pollution and the mere integration of environment as a topic into election manifestos (Greenpeace India 2019). Linking the demands for *environmental protection* and *the protection of the people on the ground* with the *leading Indian role* shows that strategeme V works very well here. Admittedly, this argument by Greenpeace sounds grotesque, bearing in mind their persecution by the Indian government. However, this is an ideal demonstration of how empty an identity-representing signifier can be and yet still capture counterparts in the process of producing a common identity. And of course, the other way round is also possible: GPI can appropriate this empty signifier, claiming a role in the Indian climate discourse. Regardless of the underlying utility, it must be noted that GPI, following strategeme V, expands the meaning of the empty signifier.

GPI together with AIWC also stands out from other actors in terms of the gendered dimension of the discourse. They focus less on masculinised demands such as growth and more on feminised demands such as protecting the environment and the weaker members of society. Even though it must be found out in the end whether they are able to make these demands important ones that shape the Indian identity, we can see that feminised aspects are articulated within the Indian climate discourse.

The identity claimed by the Indian government was often linked to the mitigation measures already delivered and the voluntary commitments still to be achieved following Copenhagen. The fact that India is well on its way to reaching its solar energy and energy efficiency NDCs has been emphasised by the government on various occasions (Javadekar 2019; Ministry of Environment, Forest and Climate Change 2021, 256). This progress might have been jeopardised by the effects of the Corona pandemic, which meant long-lasting social and economic lockdowns in spring 2020 and beyond with consequences that are now to be cushioned by growth programmes. Although it is too early to assess these recovery measures, the crisis and its consequences mark an important point in

India's climate change discourse, as they may trigger essential changes. In particular, the question is whether the hegemonic project of a significant Indian role continues to be pursued by the government or whether the Corona crisis has put this aspiration on the back burner.

Slowly recovering from the first Corona wave, the government initiated a far-reaching reform of the agricultural market, which did respond specifically to the Corona pandemic. Demonstrating against these controversial new laws, farmers blocked entire streets in Delhi in late 2020 and 2021 (Krishnamurthy 2021, 1410). The accusation was that the reform would neoliberalise the agricultural system and deprive farmers of the security of a governmentally regulated market. It can be assumed that such a large resistance with all its associated public attention was not expected, as Amita Baviskar and Michael Levien (2021, 1342) indicate, mentioning that even the famous singer Rihanna and the famous climate activist Greta Thunberg both supported the movement. The Indian Fridays for Future (FFF) group also showed solidarity with the farmers' protests. FFF India created a freely accessible, largely anonymous online sheet where people could find out which hashtags to use in social media or read about other forms of supportive protest (Deutsche Welle Asia 2021). The well-known Indian FFF activist Disha Ravi was therefore detained in February 2021 and accused of sedition (Deutsche Welle Asia 2021). Besides trying to keep the farmers' protests under the radar, the Indian government showed once again how it deals with growth-critical actors, as just months before this incident, FFF's website had been briefly blocked due to "objectionable" content, which is why Delhi Police accused FFF of an "unlawful or terrorist act" (Agarwal 2020). Although the ban was quickly lifted after a wave of protest, it shows how much pressure grassroots initiatives are facing from the BJP-led government. Furthermore, this illustrates that the government is trying to keep the circle of publicly-recognised subject positions in the Indian climate discourse small—particularly if the actors trying to shape the discourse are growth critical and instead pushing feminised ideas such as degrowth.

While the farmers' protest went ahead during the Corona crisis, the canon of climate talks was interrupted due to the outbreak of the global pandemic. Of course, the Corona crisis exacerbated other—often pre-existing—or further crises such as malnutrition or living conditions in the slums of the Indian megacities (Ghosh, Nundy, and Mallick 2020, 4–5). Moreover, Covid-19 was the all-dominating topic, especially because the containment measures taken, such as the first very long and sudden lockdown, had severe consequences. In addition, the already-stretched health system had to deal with the countless Corona patients and the first lockdown caused unimaginable chaos and huge problems especially for vulnerable groups such as migrant and casual workers (Ghosh, Nundy, and Mallick 2020, 4). After several dramatic months, the government began to implement measures to revive the economy in addition to supporting social programmes (International Monetary Fund 2021). The focus of the former was the manufacturing sector. In this context the Centre for Science and Environment (CSE) (2021, 13) complained that the MoEFCC lowered emission norms for pandemic reasons, which made it worthwhile to continue operating old coal-fired power plants and thus endangered the climate targets. After ad-hoc crisis management the CSE demanded the government to think economic reconstruction together with climate change and further challenges such as public health (Centre for Science and Environment 2020, 5). Moreover, the connection between environmental degradation as well as climate change and pandemics and other health risks were increasingly discussed, not least because the lockdowns led to a short-term recovery of the environment such as better air quality in Delhi (Ghosh, Nundy, and Mallick 2020, 6).

Regardless of these factors and debates, neither the demands of the individual actors nor the hegemonic project of the government have changed. At the 2020 digital Climate Ambition Summit, the government's familiar lines of argument were repeated, working towards a significant Indian role pointing to "India's pioneering global initiatives of International Solar Alliance (ISA) and the Coalition for Disaster Resilient Infrastructure (CDRI)" (Ministry of Environment, Forest and Climate Change 2021, 256).

This discourse contribution once again shows that in the Indian climate discourse, the masculine-connoted main concerns of economic growth are given priority and feminised demands such as increased environmental protection are in doubt. The debate about the Corona aid further illustrated this basic configuration of the Indian discourse, which didn't change by the stimulus of the pandemic, although it emphasised the importance of care in general and it would not have been surprising to spark a debate on care for the environment in this context.

As the Indian Union held one of the non-permanent seats of the United Nations Security Council in 2021 and 2022, a specific Indian contribution to the climate change discourse in the framework of the most powerful international institution will be analysed. In 2021, non-permanent UNSC member Germany initiated a discussion on the climate crisis as a threat to world peace. One could say that the idea was to officially securitise the climate crisis.[23] No draft resolution was submitted, partly due to rejection by India. The speech given by

---

[23] For an introduction to the poststructuralist concept of securitisation, see Buzan, Waever, and Wilde 1998. In summary, it is assumed that in the course of securitising a discourse, for example the climate discourse, powerful political measures can be taken more easily and justified by framing them as a security threat. For more detailed case studies of the securitisation of climate change policy by deconstructing discourses of the US, Germany and Mexiko see Lucke 2020.

the then environment minister Prakash Javadekar (BJP) shaped the Indian climate discourse, providing insight into the Indian line of argumentation when it comes to connections of climate change and global security. Javadekar (2021) began his speech in February 2021 with a mantra outlining that environment is a part of everyone which is why climate change is a threat affecting everyone. Demanding *global action* as well as *climate justice*, Javadekar split the discursive space, insinuating the Global North's responsibility in both cases and following strategeme II. He used this strategeme again when he pointed to the need for *financial support from the developed states* and also highlighted their *historical responsibility* to fight the climate crisis. Javadekar focused on manifold different aspects concerning global climate policy such as *gender equality* due to women's vulnerability stressing their responsibility caring for daily life (Javadekar 2021). Equivalising these demands with *multilateralism* and the promotion of a *sustainable lifestyle of all*, he followed already established discursive patterns within the Indian government. Since at this time the CDRI was still new, he also pointed to its inauguration by India and further Indian initiatives to underline the significant Indian role (Javadekar 2021). In this speech it becomes clear that the *significant Indian role* serves perfectly as empty signifier, as all other demands can sit within this encompassing demand. Besides the deconstructed strategemes in this speech, a further new strategeme can be reconstructed in the Indian climate change discourse, as Javadekar uses strategeme IV, the super-differential demarcation of a specific discourse's limits. Since Javadekar said at several points in his speech that in his eyes scientists don't already know enough about the relation of climate change and security threats, he makes clear that in his view the global climate change discourse bears no responsibility for caring about or even interacting with the global security discourse. The strategeme used not only sheds light on Javadekar's understanding of the climate discourse's limit but was also a rejection of the German initiative.

The first major international climate change summit to be held in person was COP26, in Glasgow 2021. There, Modi followed strategeme I, equivalising inter alia the demands for *poverty eradication in India, financial support by the developed states,* and *multilateralism,* which he sees as the solution to the manifold climate-change-induced challenges. Furthermore, he demanded *Indian mitigation actions* and equivalised these with *a significant Indian role* stressing that "Today the whole world believes that India is the only big economy which has delivered in letter and spirit on the Paris Commitment. We are making every effort with determination; and we are working hard and showing results" (Modi 2021). As in many other passages in the discourse, a significant Indian role is not demanded literally here. Rather, identifying the demand for a significant Indian role is the interpretation result from the discursive context and the historical contingency of the discourse. The fact that this is a historical contingency is apparent because even during Singh's tenure, India attempted to transform its own role into a proactive, strong role, mostly with reference to mitigation measures already taken or planned (cf. Sects. 3.2.1 and 3.2.2). This reading corresponds to the leadership understanding of 'leading by example', recapitulated earlier in this chapter. Modi also uses this ductus at the beginning of his speech, interweaving the additional demand that India be recognised *as a developing state*, when he points out that he is "happy that a developing country like India (…) has left no stone unturned to show that it has fulfilled its obligation" (Modi 2021). This statement shows that India still claims to be a developing country in the context of climate policy, but at the same time it wants to be perceived as playing a significant role going beyond the reading of a developing country. At first glance, this is an almost impossible balancing act. Yet, it succeeds here and in other contributions since the empty signifier of the *significant role* is open to the demand of recognising India as part of the group of developing states because it is equivalised with pioneering initiatives such as ISA. India's financial support for other developing states, for example, is presented by Indian officials as pioneering work (Mukherjee 2017, 179), since India—especially in contrast to states of the Global North—seems to succeed in these multiple roles, on the one hand as a donor, and on the other hand demanding financial support for the global fight against the consequences of climate change from the Global North. The representation of the Indian climate change identity by the empty signifier *significant role* enables both the demand for *mitigation efforts* and also *recognition as a developing state* with its associated demands such *as poverty eradication* or *support by the industrialised countries.* Thus, a clear definition of what is meant by Modi's *significant role* as a hegemonic project is almost impossible. Nevertheless, strategemes I and III can be demonstrated in Modi's speech in Glasgow. Strategeme II can also be reconstructed, since Modi achieves the division of the discursive space by demanding *climate justice* while distancing India from the industrialised countries. The Prime Minister points to the numbers: India is only responsible for five percent of worldwide emissions while it is the home of over 17 percent of the world's population (Modi 2021).

Once again following strategemes I and II, two more previously-identified demands can be reconstructed. These are, on the one hand, the reference to India's *culturally-rooted closeness to nature* and, on the other hand, the demand for the *universal adoption of a sustainable lifestyle.* The second demand is rearticulated by the announcement of a new Indian initiative. Modi proposes the "One-Word Movement (…) LIFE…L, I, F, E, which means Lifestyle For Environment" insisting that this kind of lifestyle should be adopted by all people (Modi

2021).[24] Since Modi linked the demand for a more *sustainable lifestyle* to *India's culturally-rooted closeness to nature*, we can assume once again that Modi thinks of the sustainable lifestyle demand as a genuine Indian demand reiterating its historical roots. Therefore, the use of strategeme VII—the occasional but deliberate breach of the antagonistic line of division—can here be attested, as Modi tries to implement the Indian demand in the antagonistic chain of equivalence belonging to the industrialised countries. Indeed, adopting a sustainable way of living to protect the climate is not a genuine Indian idea or demand. But referring to the historical roots of the Indian close-to-nature culture Modi plants exactly this understanding of the demand for a sustainable lifestyle for all and is at least in some cases successful, as other actors such as his then environmental minister Bhupender Yadav later followed exactly the same strategeme. Interestingly, Modi refers almost exclusively to aspects of daily living such as dietary choices, packaging or water management forming his vision of a sustainable lifestyle (Modi 2021). This is an indication that the demand is closely associated with the private sphere, where these daily living choices are made. Hence, the demand is connotated with femininity, as the private is based in the same dominated sphere as care and femininity due to century old norms (Aggestam and True 2021, 400). To slightly exaggerate: climate protection is to be privatised and transferred to the feminised sphere, while business as usual can take place in the public sphere, which is connoted with masculinity.

Besides the LiFE initiative, the Indian Prime Minister made a second announcement in his speech: by 2070, India should have reached Net-Zero. Modi called this and four further mitigation goals which are all (renewable) energy-related the "five nectar elements, 'Panchamrit'" (Modi 2021).[25] Panchamrit is a symbol used in a traditional form of Hindu and Jain worship (Down to Earth 2021). It contains milk, honey, curd, ghee and sugar and is offered to the devotees praising the goddess. Analogising the five mitigation announcements with praising one of the most important Hindu gods shows two things: How Modi wants to culturally root and simultaneously exaggerate Indian climate change engagement. In fact, both motives serve the hegemonic project of a *significant Indian role* as they constitute greater leeway for interpreting the symbolic equivalent of the universal, which is presented by the empty signifier (strategeme V). The

[24] The LiFE initiative was launched in June 2022 by Modi with some prestigious guests, such as World Resource Institute President Aniruddha Dasgupta, Bill Gates and UNDP Head Achim Steiner (All India Radio 2022).

[25] Modi used the Panchamrit metaphor as early as 2015 when he described his five pillars of foreign policy: dignity, engagement, shared prosperity, regional and global security, as well as cultural and civilizational linkages (Basrur 2017, 11).

pattern of creating references to India's traditional culture and therefore transferring them into the present has been reconstructed here on several occasions, and it will be discussed again in the following chapter against the backdrop of the offensive-hegemonic strategy pursued by the government.

These five "Panchamrit" pledges can be seen as an antecedent revision of the NDCs (Senthilkumar 2021); they were supposed to be more ambitiously revised and submitted in an updated version in 2020 based on the agreements that emerged from the Talanoa Dialogue.[26] As with most other countries, India didn't pledge new climate goals for 2020 or 2021. However, the Panchamrit announcement led to recognition from different actors, as the net-zero pledge in particular was not expected (Vaughan 2021). Unsurprisingly, Indian parliamentarians fired accusations at the government that these goals were neither widely discussed nor previously announced. Nevertheless, there was some support, both for the Panchamrit pledges and for India's significant role in Glasgow. Of course, it was parliamentarians of the BJP supporting Modi and his environment minister regarding the significant Indian role on the climate change stage as well as the pledged goals (Kotak 2021; Gavit 2021). While most opposition politicians also supported the announced new climate targets or wanted to enhance them, Adhir Ranjan Chowdhury (INC) summed up his doubts about the demand for a strong role by referring to the concrete outcome of the Glasgow negotiations:

> *"... you are saying that India took the lead. But no, it is not right. After objection by India, China, Iran, Venezuela, Cuba and other developing countries, the amended paragraph asked for all countries to escalate a force towards the phase-down instead of earlier phase-out [of coal]" (Chowdhury* 2021*).*

To understand this criticism, a brief look is needed at the text that was finally adopted at the Glasgow Climate Pact, and its genesis. In the Pact, the 'phase-out' of coal is no longer mentioned. Even though it was in the drafted text during the summit (Singh, Sheldrick, and Browning 2021), it was replaced by "phasedown of unabated coal power" (United Nations Framework Convention on Climate Change 2021a, 3) following an Indian "last minute intervention" (Singh Khadka 2021) which was supported by China (Singh, Sheldrick, and Browning 2021). This and a further linguistic attenuation concerning fossil fuel subsidies reproduced Modi's hegemonic project of a significant role in climate change governance—signalling that there is no getting around India. Hence, strategeme V is once again followed, putting further meaning into the empty signifier besides

[26] The Talanoa Dialogue was initiated in Bonn in 2017 by COP host Fiji and represents the first stock taking round agreed in the Paris Agreement, which should initiate the raise of the ambitions of the NDCs as early as 2018 (Zürn 2017).

'leading by example' with regards to mitigation or the culturally-rooted sustainable lifestyle. Bhupender Yadav (BJP), who had been the Indian environmental minister since July 2021, negotiated the meaningful 'phasedown' of coal in the Glasgow Climate Pact. In different blog posts presenting his view of the summit, Yadav reproduced the well-known aspects of the Indian hegemonic project, in particular pointing to India's significant role as it has "pioneered global initiatives" such as ISA as well as CDRI (Yadav 2021c). In the same text, Yadav states:

> *"It is a matter of pride for us Indians to note that we have not just been fulfilling our responsibilities as a global power in the fight against climate change, we are also taking the lead in the fight, emerging as a model worth emulating." (Yadav* 2021c*)*

The words "global power" and "taking the lead" show that in Yadav's eyes India's climate change identity should be seen as more than that of a developing country trying its best to combat climate change. Blogging from COP26, Yadav went even further:

> *"Modi has once again shown that India is a responsible global power house and is at the forefront to fight climate change with climate action in the true spirit of climate justice by proposing a five-point climate agenda." (Yadav* 2021b*)*

By portraying India not only as leading by example, but as a "global power house" fighting against climate change "at the forefront" (Yadav 2021b), Yadav gives the empty signifier an almost dazzling character.

In contrast, Yadav recapped the whole summit some days later without pointing again to an Indian leadership role. Instead, he emphasised the needs of developing countries regarding adaptation, insinuating that India is one of them (Yadav 2021a). This discursive development during the summit should be understood in the context of Yadav's narrative for India's intervention concerning the phase-out of coal. My assumption is confirmed by Yadav's later statements that India had to initiate the weakening of the paragraph on coal particularly because of its development needs and as a part of the group of developing countries for whom he insisted on speaking (Express News Service 2021). Even if it cannot be conclusively clarified whether it was solely an Indian idea to weaken the language on coal and other fossil fuels subsidies, India had to explain it and is therefore closely associated with this political foray. Although the revised text presumably also arose from final negotiations between the summit leader UK along with China and the USA, it was brought into the final summit discussion by Yadav, as requested by the summit leader UK (Nandi 2021). Indeed, it was an interesting start for the new environment minister on the global stage, shaping the narrative of the Indian identity in climate change policy as a significant actor. It is abundantly clear here that masculinised demands and responses to climate change are being maintained because the end of fossil fuel subsidies has been averted so that fossil-based development can continue.

To summarize, the first years of Modi's second term mark a change in the government's hegemonic project: From *India as a developing country* to *India as a significant actor*. Although 'leading by example' is emphasised at various points, which speaks for the understanding of India as a pioneer, it must rather be assumed that the Indian government is pursuing the formation of an identity as a significant actor. The empty signifier of a significant actor certainly leads the way as a pro-active partner, as the initiatives regarding disaster-resilient infrastructure or the rapid expansion of solar energy show. The events at the summit in Glasgow illustrate that a significant role also means that India understands its identity as 'there is no way around the state' when it comes to global climate policy. This Indian identity is, of course, supported by patterns that have already been pursued within the Indian climate change discourse, particularly the economy-related demands which are highly masculinised. Furthermore, the feminised demand to follow the Indian long-established sustainable lifestyle does also help to anchor the identity. The repeated references by the Indian government to the Indian culturally-rooted closeness to nature is another important supportive demand anchoring the empty signifier. In deconstructing the discourse, various strategemes have been demonstrated, including for the first time the clear demarcation of the climate discourse from another discourse, in this case security.

Overall, it is possible to identify an offensive-hegemonic strategy for each Indian government. Starting with the identity of India as *a state of sustainable development* in the tenure of Singh (until 2014), the Modi I government (2014–2019) shaped the empty signifier of *India as a culturally-rooted developing state*. In the time period examined in this chapter, 2019–2021, the identity representing the empty signifier was reshaped by the hegemonic actors following the hegemonic project of *India as a culturally-rooted significant actor*. The following Table 3.1 provides an overview of the hegemonic projects, the resulting hegemonies and the actors (re-) producing them.

Table 3.1 Overview of different hegemonic projects, (re-)producing actors and hegemonies of the Indian discourse

Hegemonic project(s)	**(Re-)producing actors** (selected)	**Hegemony** Intersubjectively produced and temporary stable identity
(1) 2010–2014		
Sustainable Development	• PM Singh • Environmental Minister Ramesh • TERI • Greenpeace India • Federation of Indian Chambers of Commerce and Industry (FICCI) • Ministry of Environment, Forests and Climate Change	Sustainable Development
Climate Justice	• Minister of Environment Natarajan • CSE • CSE-trained journalists	
(2) 2014-max. 2019		
Developing state culturally-rooted close to nature	• PM Modi • BJP • Environmental Minister Javadekar • Environmental Minister Vardhan • Government of India (INDCs) • Ministry of Environment, Forests and Climate Change Commission	Developing state culturally-rooted close to nature
(3) 2019–2021		
Significant role culturally-rooted close to nature	• PM Modi • BJP • Environmental Minister Yadav • Ministry of Environment, Forests and Climate Change, • (GPI; trying to reshape it)	Significant actor culturally-rooted close to nature

The question of how gendered discourse patterns in the different demands led to or helped to shape these identities and to what extent these identities are gendered will be taken up again when the gendered discourse patterns are discussed and analysed with an ecofeminist lens for the entire period under study (cf. Sect. 3.3). First, however, a final foundation is to be laid for this, namely the precise methodological classification of the use of the demand for the recognition of India's *culturally-rooted closeness to nature* and how this is also part of the offensive hegemonic strategy as an additional new strategeme emerging from the results of the study at hand.

3.2.8 Strategeme X: Rethinking the Offensive-Hegemonic Strategy

The reconstruction of the Indian climate change discourse had the particular aim of revealing the discursively-produced identity and the demands used, which will be further analysed for their gendered patterns in the following chapter. For this purpose, the strategemes followed by various actors were also identified. The strategemes anchor numerous articulated demands forming the chain of equivalence and fix—of course, only temporarily—the empty signifier. As Nonhoff (2006, 212, 2010, 312) has pointed out, the set of strategemes used to establish a hegemonic project is not closed and is variable and expandable depending on the subject of the project, not least as further strategemes might be deductively identified structuring the outcomes of the study. This could be the case here: A Strategeme X might have emerged.

In analysing the Indian climate discourse and the offensive-hegemonic strategy employed, most strategemes (i.e. I–VII) elaborated theoretically and empirically by Nonhoff were reconstructed. Furthermore, the strategemes of the second level of the discourse (strategemes VIII and IX), were not searched for due to the fast-moving character of the analysed discourse (cf. Sect. 2.2.4), proved by the fact that the hegemonic projects have changed several times during the 11-year research period.

Besides the yet-to-be discussed gendered patterns, the analysis highlighted two striking aspects regarding the character of the identified strategemes. On the one hand, it is significant that some strategemes were only used in exceptional cases, such as strategeme IV when Javadekar demarcated the climate change discourse's limits from that of security policy (cf. Sect. 3.2.7) or strategeme VII when the government of Singh imported the demand for mitigation measures from the antagonistic chain of equivalence into the Indian chain (cf. Sect. 3.2.1).

On the other hand, most demands were mentioned in the context of various strategemes such as the demands for *climate justice* or *energy security*. Another demand that was repeatedly articulated using various strategemes is the appreciation of *India's culturally-rooted closeness to nature*. In fact, this demand has been omnipresent in its different facets since the BJP-led government assumed office in 2014: in the call to implement an international Yoga Day, in citing ancestral prayers and mantras when giving speeches about climate change, or in dressing up India's climate goals in the garb of the five nectars 'Panchamrit' (cf. Sect. 3.2.7). The demand to appreciate *India's culturally-rooted closeness to nature* is regularly equivalised with other demands and used to divide the discursive space. Moreover, pointing to India's cultural roots and their closeness to nature expands the meaning of the empty signifier. Hence, strategemes I, II and V using this demand have been reconstructed. Interestingly, this particular demand is sometimes explicitly gendered when it is traditionally charged, reinforcing old-fashioned gender stereotypes. This aspect will be discussed further in the next chapter.

First, however, we need to look at how this demand was used beyond the strategemes mentioned above and whether an additional, hitherto unknown, strategeme can be identified. The ecofeminist hegemony analysis has shown that the demand to appreciate *India's culturally-rooted closeness to nature* was strategically articulated by the Modi government. As the demand was not previously part of the Indian chain of equivalence, it seemed to have appeared out of a nowhere. Of course, it is not surprising that the Hindu nationalist BJP constantly tries to integrate the demand into the Indian chain of equivalence, due to the BJP's political attitude emphasising its own roots in Hindu traditions and their importance for both the society as well as the party. However, the decades-long abstinence from this demand as a central component of India's identity shaped by and within its climate discourse suggests that the discourses of climate policy and cultural policy have been clearly separated.

This does not necessarily mean that traditional Hindu environmental conservation practices, for example, were discontinued and thus do not form part of the discourse. Rather, these practices were not anchored or repeated in the framework of the demand to appreciate India's culturally-rooted closeness to nature by hegemonic practices trying to make this aspect part of the Indian identity. Prior to the research period, Indira Gandhi, for instance, put environmental protection on the agenda several times during her term as Prime Minister and initiated various highly visible projects, among them the very first Indian tiger reserve (Rangarajan 2009, 304). But while she referenced Indian culture in the process and invoked the unity of the nation state as a prerequisite for protecting the environment

(Rangarajan 2009, 308), she did not refer back to Indian traditions. Similarly, environmental NGOs made use of explicitly spiritual and religious practices in their efforts to protect the environment (Misra 2007, 142). But although the Chipko movement, for instance, cited Hindu scriptures praising nature (Misra 2007, 142), they were merely used as a tool by Chipko calling for the protection of nature to secure the livelihood needs of the people living in the forests (Bandyopadhyay 1999)—and not invoked as the epitome of any kind of culturally-rooted closeness to nature. In stark contrast, the Indian ecofeminist movement around the scholar activist Vandana Shiva had explicitly drawn on the cultural Indian roots and its traditional closeness to nature (Shiva 1988, 55–58) when calling for environmental protection. But even if the movement succeeded in occasionally making the demand for the appreciation of *India's culturally-rooted closeness to nature* an at times visible part of the discussion on the national level (foremost in the 1970s and 1980s), Indian ecofeminists did not succeed in making it a *hegemonic* demand of the climate change discourse, as the reconstruction of the discourse has shown for the period 2010–2014 (cf. Sects. 3.2.1 and 3.2.2 above). Accordingly, *India's culturally-rooted closeness to nature* can hardly be said to have occupied a relevant place in the Indian chain of equivalence of the climate change discourse—but this assessment only held true until the BJP-led government assumed office and worked towards integrating this demand into the chain.

Of course, this result must be discussed in the context of political developments from 2014 onwards. The discourse of Hindu culture and tradition has gained greater significance since the Modi government came into power, which was to be expected not least because of Modi's prior tenure as Chief Minister of Gujarat where the recourse to Hindu nationalist persuasions was even more aggressively implemented (Sud 2022). After initial restraint, today the BJP acts in the framework of the Hindutva ideology. Hindutva can be described as an ideology serving Hindu nationalism (Shani 2021, 264). It uses different religious ideas, traditions and rituals of Hinduism (Sahgal 2020, 20) as well as cultural aspects (Hansen and Roy 2022b, 1) in striving for a Hindu state ruled by the Hindu majority.[27] Since 2014, different aspects of the rich Hindu traditions have also been invoked, and sometimes exaggerated, in discourses other than the Indian culture discourse and contexts—implicitly justifying controversial new legislative

[27] Researchers agree that the concept or ideology of Hindutva has changed significantly over time (Pai and Kumar 2015, 130). Hindutva is furthermore understood differently by different political forces and groups, and thus pursued with varying degrees of radicalism. For an in-depth discussion of the nexus of nationalism, religion and Hindutva in India see Shani 2021.

acts, for example the highly-criticised Citizenship Amendment Act (CAA).[28] But what does this have to do with the culturally-rooted closeness to nature in the Indian climate change discourse?

The first contributions articulating the demand to appreciate India's culturally-rooted closeness to nature showed that it is generated from the larger context of Hindu traditions. Modi's speech at the UN General Assembly particularly illustrated how the demand in relation to nature and India's responsibility in climate protection was derived from the larger context of Indian culture. In further steps, Modi and the other members of the government integrated the demand to appreciate India's culturally-rooted closeness to nature into the Indian climate discourse citing traditional Hindu prayers in their speeches or referring to Holy Mother India. Since the appearance of this demand, which was for a long time solely part of the Indian culture discourse, the supposedly closed space of the Indian climate change discourse has been opened up to weave in the demand in the chain of equivalence supporting the hegemonic project. Hence, the analysis has identified an additional strategeme, used by the hegemonic actors of the Indian climate change discourse: *Strategeme X: Occasional but deliberate breach of a specific discourse's limits.*

One could argue that strategeme X is the inversion of the basic strategeme IV *Super-differential demarcation of a specific discourse's limits.* Nonhoff (2006, 231) emphasises that reasonably stable limits of the discourse are essential for a hegemony to take effect. Therefore, if necessary, the hegemonic actors clearly close the discourse's limits. This was demonstrated when Javadekar articulated that climate policy is not responsible for security policy (cf. Sect. 3.2.7). The newly identified strategeme X seems to follow the diametrical opposite—and therefore to become a grotesque function. If action is taken contrary to the basic strategeme and the discursive space is opened to import outside demands, a short-term or even potentially permanent latent destabilisation of the outer limits of the discourse must be assumed. Therefore, strategeme X must be critically reflected as it probably signifies the destabilisation of the discourse's limits and therewith the destabilisation of the hegemonic project itself. Of course, this is not in the interest of the hegemonic actors striving for a successful hegemony.

---

[28] The CAA grants "fast-track citizenship to all non-Muslims persecuted in other South Asian countries" (Sinha 2021, 330). Combined with the announced National Population Register, the CAA is perceived as a serious threat to Muslims' electoral as well as residency rights (Sinha 2021, 330), legitimised via Hindu cultural values and the figure of whether you are ready to "assimilate" or remain as an enemy (Shani 2021, 271).

However, a glimpse into the Indian climate change discourse shows why strategeme X can actually *help* the hegemonic project to succeed. Integrating demands of the cultural or Hindu discourse into the chain of equivalence of the Indian climate change discourse, the current Indian government faces support by most of the people, who are in favour of the government's Hindutva politics. Nevertheless, the BJP cannot be entirely sure of its support from society; but concerning its climate change policy it seems that the government faces little opposition concerning the narrative of India's traditional closeness to nature.

Of course, not all actors are satisfied with the Modi government's climate policy, but interestingly the narrative of the culturally-rooted India is not the subject of criticism. Rather, it seems that the voices questioning this narrative find it difficult to argue against the cultural roots, as the government's Hindutva course enjoys broad support by the people.[29] Thus, speaking against the instrumentalisation of Indian cultural roots or rituals would cost sympathy, particularly if the critique is perceived as targeting the Hindu traditions, rituals or values. This is especially true regarding the fact that various initiatives that criticise mining in territories of indigenous groups, for example, are known for their respect for traditional lifestyles and cultures. In any case, it might be difficult for critics to walk the thin line between criticising the instrumentalisation of cultural roots in the context of a Hindu nationalist project and criticising centuries-old traditions that are part of the social practices and even the identity of various groups. However, the instrumentalisation component does have implications for the gendered Indian identity and will therefore be discussed in depth in Sect. 3.3.2.

Returning to the theoretical architecture of strategeme X, these empirical findings mean that new demands of a totally other discourse are integrated into the discourse we are analysing. As the discourse's limits must therefore be opened—at least at specific points in time—the importing of another discourse demand can dilute the success of the hegemonic project or lead to a change of the chain of equivalence. Both consequences are possible—changes in the chain of equivalence which are not in the sense of the hegemonic actors, or simply the failure

[29] Shani (2021, 271) asserts that the middle class in particular appreciates the government's Hindutva course. Hansen and Roy (2022b, 1) furthermore attest that Hindutva currently "enjoys an unprecedented level of mainstream acceptance". Indeed, there is also criticism and opposition. Marginalised groups such as the Muslims or more liberal groups such as students actively oppose the Hindutva course, for example in the context of the protests against the CAA (Shani 2021, 273–74). In the context of the climate discourse, however, the reference to Hindu traditions is not (yet) addressed as threatening or divisive across society.

of the hegemonic project, as the hegemonic actors have to deal with possibly additional demands planted into the discursive space or the chain of equivalence by other actors using the discourse's porous limits. Accordingly, following strategeme X appears to be quite risky as it could destabilise the hegemony or the hegemonic actor's project. To discuss this danger, we will take a brief look at the stability of the offensive-hegemonic strategy.

Using various strategemes, the Modi government was able to help the respective hegemonic project to succeed and to secure it over a certain period of time. Almost at no point did members of the government find themselves in the situation of having to fend off an attack on the discursive limits of climate policy. Hence, we can assume that the limits of the discursive space were quite stable and thus it is possible to import the demand from the other discourse.

Thus, strategeme X is characterised on the one hand by the fact that it can only be used if the discursive space we are studying is relatively closed and the hegemonic actors can assume that they can ward off possible interventions. On the other hand, the use of strategeme X indicates that the discourse into which the demand should be integrated is well-anchored in a larger political context, as otherwise the social acceptance of the new demand would be highly unlikely. In the specific case of integrating India's cultural roots and traditions into the Indian climate change discourse, we can attest that it builds to some extent on the mainstream acceptance of Hindutva in India (Hansen and Roy 2022a, 1) as well as the Hindutva discourse influencing different policy fields as shown above. Accordingly, Hansen and Roy (2022b, 20) describe the risk taken by the Modi government as follows:

> *"More than any other political force, new Hindutva has fashioned Indian politics into an emotional drama running on fear, anger, and resentment. This makes the BJP government vulnerable the day the emotional wave turns against them."*

Relying on the stability of Hindutva is therefore fraught with risk when it is particularly dependent on "emotional waves". At this point, it is absolutely valid to argue that demanding the appreciation of India's culturally-rooted closeness to nature, introduced from the cultural discourse, is not accompanied by the aforementioned emotional dramas. However, the hegemony analysis has shown that this demand is often accompanied by other feelings, such as pride in Indian traditions or even the feeling of superiority. Superiority is generated as Indian government officials never seem to tire of emphasising that these centuries-old traditions and values have to be learnt by other states first, referring to the necessary introduction of sustainable lifestyles by the Global North.

Yet the benefit of integrating the demand to appreciate India's culturally-rooted closeness to nature is obviously considered greater than the risk of destabilising the respective discourse due to the expansion of its limits. Even though this calculation might not always work out, several scholars confirm that the BJP follows the idea of Hindutva to justify its political strategies and decisions in different fields (Sinha 2021, 328; Chandra and Walton 2020, 186). Moreover, various fields, such as social and economic schemes, are clearly influenced by the Hindutva ideology (Chacko 2019, 402). Modi and his government are thus also trying to benefit from the increasing entrenchment of Hindutva in the climate discourse. This also validates Narlikar's (2017, 105) point that the emphasis on the spiritual and cultural roots of Indian society and its nature conservation traditions supports the social acceptance of future Indian mitigation measures. Of course, the success of this tactic has been confirmed by the previous hegemony analysis; both hegemonic projects of the BJP-led government have been achieved, inter alia through the newly introduced demand. The establishment of India's identity as both a *culturally-rooted developing state* as well as a *culturally-rooted significant actor* has been reached through referring to aspects of the Hindutva discourse.

Reflecting on the chapter, I have been able to prove two things: first, strategeme X must be seen as an additional component of the offensive-hegemonic strategy pursued by the BJP-led government. Second, this strategeme cannot be used without risks. After this theoretical/empirical completion of the set of strategemes of the offensive-hegemonic strategy, Table 3.2 shows all the reconstructed strategemes identified in the respective years.

In the next chapter, I will analyse the gendered discourse patterns used in shaping the Indian climate identity. Henceforth, the ecofeminist approaches will be used as a foil discussing the diverse results of the hegemony analysis, to finally identify the significance of gendered discourse patterns for the Indian climate identity which serves to justify the chosen answers fighting climate change.

Table 3.2 Strategemes used by the various actors involved in the respective hegemonic project or hegemony of India

Years 2010–2021		10	11	12	13	14	15	16	17	18	19	20	21
I	Articulation of equivalences between different demands made with regard to the universal	+	+	+	+	+	+	+	+	+	+	+	+
II	Antagonistic division of the discursive space	+	+	+	+	+	+	+	+	+	+	+	+
III	Representation	+	+	+	+	+	+	+	+	+	+	+	+
IV	Super-differential demarcation of a specific discourse's limits	–	–	–	–	–	–	–	–	–	–	–	+
V	Gradual constitution of greater leeway for interpreting the symbolic equivalent of the universal	–	–	+	–	–	+	–	–	–	+	–	+
VI	(Re-)Articulation of subject positions for politico-societal forces	–	+	+	–	–	–	+	–	–	–	–	–
VII	Occasional but deliberate breach of the antagonistic line of division	+	–	+	–	–	+	–	–	–	–	–	+
VIII	Strategeme of true championship	–	–	–	–	–	–	–	–	–	–	–	–

(continued)

Table 3.2 (continued)

Years 2010–2021		10	11	12	13	14	15	16	17	18	19	20	21
IV	Strategeme of true meaning	–	–	–	–	–	–	–	–	–	–	–	–
X	Occasional but deliberate breach of a specific discourse's limits	–	–	–	–	+	+	–	+	–	+	+	+

3.3 The Gendered Indian Climate Policy Identity: Between Sustainable Development and a Significant Indian Role

To reconstruct India's gendered climate identity, the nuanced understanding of the operational mechanisms behind the offensive-hegemonic strategy and the articulated demands elaborated in detail in the previous chapters is imperative. Insights from ecofeminism, notably the deconstruction of the mostly implicitly rearticulated dichotomous world order and its impact on climate policy, serve as the normative foil and filter to discerningly assess the outcomes of the hegemony analysis. Beyond merely identifying gendered patterns, this chapter is dedicated to a meticulous examination and critical reflection upon the findings, elucidating their profound implications for the landscape of Indian climate policy.

Determining to what extent gendered discourse patterns help the hegemonic projects to succeed and asking generally whether they have an impact will finally answer how the Indian climate change identity is gendered in its influence on the Indian climate change policy. Referencing different aspects of ecofeminist theorising introduced at the beginning (cf. Sect. 2.3), I try to find out which gendered demands or greater patterns were used within the framework of the offensive-hegemonic strategy followed in the various periods of Indian climate change discourse already reconstructed and discussed. The demands that have been strategically positioned with the help of the strategemes—especially the core strategemes I-III and strategeme V, which fills the empty signifier with multiple meanings—are therefore examined for their gendered patterns.

In concrete terms, the table elaborated in Sect. 2.3.3, following Plumwood and Gaard, is placed as a foil on the results of the previous reconstruction of the offensive-hegemonic strategy. Comparing these insights with aspects of gendered patterns used by the actors to achieve the respective identity also shows the relevance of these patterns. Not only do the socially-produced dichotomies play a role, but so do the dominance relations that resonate within them, as the theoretically derived Table 2.4 in Sect. 2.3.3 reminds us.

In the following chapters, the focus will now lie on the second dimension of gendered discourses, the justification of responses given regarding how the climate catastrophe should be dealt with, and the gendered share of the respective climate policies. The gendered subject positions of various participants in the discourse, which constitute the first dimension of gendered discourses, will also be discussed here. The discourses will thus be analysed regarding their gendered dimensions along the timeline and therefore divided into sections covering the Singh government (Sect. 3.3.1) and the Modi government (Sect. 3.3.2).

3.3.1 Masculinised Sustainable Development for All

After the summit in Copenhagen, the INC-led government pushed the Indian climate identity as a state committed to *sustainable development (SD)* (cf. Sect. 3.2.1). To overcome the old and yet strongly anchored antithesis of growth and environmental protection, SD was followed by the Singh government as its hegemonic project. The empty signifier SD represented the Indian identity in this specific period of time, not least by relying on gendered discourse patterns. How were (implicitly) gendered discourse elements used to implement this successful hegemonic project? This question will be discussed first regarding Jairam Ramesh's term and second regarding Jayanthi Natarajan's term as environment ministers in the Singh administration.

The newly-pledged *mitigation* efforts, integrated into the Indian chain of equivalence, are an important gendered demand of the empty signifier SD. This becomes particularly clear when we look at the understanding of SD through a gender lens. At the time of Singh, SD was characterised foremost by its first commitment to mitigation targets, which of course are interlinked with a more pro-active Indian role. According to Ramesh and Singh, the mitigation goals were to be achieved primarily through advances in technology (cf. Sect. 3.2.1). Both demands, techno-scientific solutions and mitigation, are masculinised in their connotation (Djoudi et al. 2016, 257; MacGregor 2014, 626). Besides classical economic growth, which was constructed as a co-benefit of technological progress, a new focus was placed on science to improve SD and therewith mitigation and growth efforts. All these aspects are culturally located in the sphere in which masculinity and dominance are located (cf. Sect. 2.3.3). Furthermore, the politicians were always interested with the male-connoted natural sciences, which should generate innovation, and not with the social sciences, which investigate how society deals with the climate crisis. During Ramesh's term in office, the Singh government promoted the triad of economic success, technical solutions and innovation formulating the Indian climate policy in the spirit of ecological modernisation (Bäckstrand and Lövbrand 2006, 52–53) which resonates with the concept of weak sustainability. Following the ecofeminist approach of Hultman and Anshelm (2017, 32), ecomodern masculinity can be attested here, which supports and reproduces technology, innovation and economic solutions as the main answers to combat the climate crisis due to strongly anchored masculinised norms overruling alternatives such as degrowth or sufficiency ideas.

When it comes to development-related growth, it stands to reason that feminised discourse patterns could be used to lend weight to one's development demands, as women and reproductive needs which are linked to them as a homogenous group are often associated with questions of development. Furthermore, in the framework of the Rio process for example, women were seen as the key to development due to their socially constructed, but nevertheless important roles concerning small-scale economies as well as subsistence and daily life routines (cf. Sect. 3.1.1). These fall into the area of reproductive work, which Val Plumwood assigns to the feminised sphere (cf. Sect. 2.3.3). In the case of Ramesh and Singh, however, hardly any such patterns can be reconstructed. Individual exceptions can be found in the Five-Year Plan 2013, whose chapter "Women's Agency and Child Rights" contains a short passage on women in development programmes as well as in adaptation and mitigation. However, as already shown, the Five-Year Plan's priorities in the context of SD were not focused on feminised demands.

Rather, masculinised demands were articulated, anchoring the hegemonic project SD via different strategemes. One example of masculinised discourse patterns is the demanded *stronger role* that draws India as part of the solution through trying to limit climate change instead of being seen as the naysayer. The wish to play a stronger role in global climate change policy was particularly reflected in equivalising it with SD, which shows that the demand for a strong Indian role is an integral part of the—consciously or unconsciously followed—offensive-hegemonic strategy of that time. Striving for a stronger, more active role can be classified as a masculinised demand due to gendered dichotomisations operating in societies to this day (Bruff and Wöhl 2016, 96). These gendered dichotomisations are sketched in the matrix according to Val Plumwood and Greeta Gaard (cf. Sect. 2.2.4) which has already shown that the active doing, the commitment anchored in the public sphere, is assigned to the same sphere as masculinity. Since meaning is produced particularly through differentiation, *the active* as a decisive aspect of a more important Indian role in climate change policy is distinguished from *the passive*, which is associated with the feminised sphere, and therefore dominated by the opposite categories (Plumwood 1993, 50).

Linked to the new strong role and to mitigation—in the sense of technical solutions generating economic growth as a co-benefit—is the demand to strengthen Indian climate change sciences. As already discussed, scholars found that science as reason and rationality or the mind in general are often discursively constructed as diametrically opposed to emotion, animality or nature. The former triad is assigned to masculinity, whereas emotion and animality as well as nature are assigned to the feminised sphere—dominated by the former. Taking a closer look at the government's new interest in sciences, we can see that they focused on technical aspects of climate change, exemplified by the INCCA's (Indian Network for Comprehensive Climate Change Assessment) Assessments (cf. Sect. 3.2.1). Ecofeminist research shows that relying solely on technical solutions is a masculinised discourse pattern (Hultman and Anshelm 2017). Mastering nature using the mind and producing technological solutions are "qualities traditionally appropriated to men" (Plumwood 1993, 44) and therefore assigned to the masculinised sphere. Thus, the demand for *progress of Indian climate change sciences* also represents a masculinised discourse pattern.

The gendered demands under discussion were especially articulated by Environment Minister Ramesh. But even when Natarajan who focused increasingly on climate justice became Environment Minister, we found that the government still stuck to the hegemonic project SD. Did the gendered structure of the discourse nevertheless change? Natarajan notably drew on the old Indian demand for climate justice and reactivated allies such as the CSE (cf. Sect. 3.2.2). How the demand for CJ is gendered is relatively obvious. First, Natarajan linked the much-needed fight against the widespread poverty in India to CJ. Or as Samir Saran and Aled Jones (2017, 106) attest with regard to Natarajan's period in office: "The pendulum had swung back and the poverty narrative made a strong return". Indeed, poverty is often associated with women in India due to their socially constructed roles—which is one aspect of why actors interpellated them as victims of climate change and thereto induced poverty. In demanding CJ by pointing to the need for protection of vulnerable groups such as women, Adivasi groups or poor people in general, actors often refer to the impossibility of the vulnerable to protect themselves, thereby creating a picture of inactive objects. I of course do not want to imply that all actors do not attribute or trust agency to the victims of climate change. Rather, what is perceived in the discourse and how different demands or aspects of the Indian climate change discourse are dichotomised is under discussion here. As already examined, the inactive object is dichotomised to the masculinised subject. Besides that, India as a state is also constructed as an entity in need of protection, which is why Natarajan and CSE argue that binding Indian mitigation efforts are not affordable—especially against the background of the enormous responsibility of the developed states.

Although CJ as Natarajan's particular interest does not become the empty signifier, it is nevertheless clear that she uses gendered discourse patterns to strengthen her demand. She tries to highlight other pressing needs such as development, thereby overshadowing possible mitigation- or other activities to protect the climate. Without saying that India is a weak actor, insisting on its vulnerability she tries to write India and vulnerable groups as the protection-needing, inactive part of the climate change story. This inactive-ness is clearly assigned to the sphere which also is associated with femininity, the queer, and the oppressed. This happens on two levels of gendered discourses: On the first level, the poor, who are very often women, are portrayed as particularly vulnerable and therewith foremost interpellated as victims of climate change. On the second level of gendered discourses, the response to the manifold problem of climate change is pushed as being solely the task of the historically responsible states while India is painted as a rather passive state in need of protection.

However, the analysis showed that SD found much more support, particularly from governmental actors (cf. Sect. 3.2.2). At this point, masculinised aspects have once again asserted themselves hegemonically in the battle for the best, or rather the most recognised, answer to climate change in the face of ideas with feminine connotations. The masculinised idea of growth was clearly able to assert itself against CJ because of the prevailing values in society. Nevertheless, it can be assumed that the share of CJ in SD was still pertinent and strengthened. In order to determine to what extent CJ was a relevant aspect of the Indian identity and to what extent it was gendered during Singh's time in office, SD must be examined more closely in the light of ecofeminist insights.

The analysis revealed that SD serves as an umbrella term that includes several foci as well as different interpretations by different contributors of the discourse. Indeed, the various interpretations depend on the context, time and space of the articulation within the discourse. Therefore, it is not surprising that SD is gendered in multifaceted ways. The distinction between two dimensions of gendered climate discourses is particularly relevant here. We will start with the first dimension of gendered discourses: Which of the subject positions are discursively gendered in the context of the demand for SD during the Singh period? The answer is both complex and simple: All subject positions are gendered.

Plumwood theoretically derived the conclusion that masculinity is set as *the universal* while *the particular* is constructed as the feminine. These dualisms are "variants of a gendered reason/nature contrast" (Plumwood 1993, 45) as outlined in the chapter on poststructuralist strands in ecofeminism (cf. Sect. 2.4.1 and 1.3). Put simply, this would mean that all subject positions which are not interpellated as the particular or a deviation from the universal, are considered the universal ones and are thus produced by means of masculinised discourses. The complex picture emerges from the empirical findings. When it comes to SD during the Singh term, clearly-gendered subject positions can be identified within the Five-Year Plan discussing women's agency and needs in poverty reduction and development programmes (cf. Sect. 3.2.1) as well as in the context of the SAPCCs, when SD meant in a lot of cases that women should be won for work (cf. Sect. 3.2.5)[30]—work that is notably in the fields of adaptation, farming or other facets of caring for the environment in the context of daily life routines and subsistence.

Since women are often drawn as 'natural' allies of nature and environment (cf. Sect. 2.3), in the Five-Year Plan of the Singh government as well as in most of the SAPCCs they are interpellated either as the victims of climate change or as actors caring for the environment in the context of their livelihood roles or responsibilities in forestry, fishery or farming. As both women and nature are assigned to the dominated sphere, they are discursively interpellated into roles which leave only limited leeway and are also clearly assigned to the sphere of reproduction and care work in a wider sense. Of course, other gendered roles constructed in the SAPCCs which are not explicitly assigned to women can also be observed. Even though these roles are hardly explicitly gendered, they are implicitly masculinised, because they are set as the norm or universal.

The second dimension of gendered discourses is aimed at the question of how to deal with the (irreversible) consequences of the climate catastrophe. Some actors tried to formulate answers in the form of various demands such as mitigation or renewable energy production, which are both masculinised (MacGregor 2010, 231; Djoudi et al. 2016, 257). Since the demands of a strong Indian role and the strengthening of Indian climate change research have also been reconstructed as clearly masculinised discourse patterns, as have the financial and technical support demands, 'only' CJ and SD remain as discourse patterns that are not clearly masculinised. This yields two findings: First, the Indian identity of SD during the Singh government cannot be described as discursively produced exclusively using masculinised patterns. Second, the dominance of masculinised discourse

[30] Since the states developed the SAPCCs during both the Singh and Modi governments, the findings from the hegemony analysis are relevant to both terms discussed in this and the following chapter.

patterns is nevertheless enormous and has contributed to the strategic success of the establishment of the empty signifier. Hence, *sustainable development* was successfully anchored as hegemony and thus at least temporarily anchored also as the relatively stable Indian climate policy identity, particularly with the help of masculinised discourse patterns following the idea of ecological modernisation.

What is the significance of these findings beyond the establishment of the hegemony? What do they mean for Indian climate policy? Since hegemony theory assumes that different strategemes are used to help the hegemonic project succeed, we should suppose that—unconsciously or consciously—the gendered discourse patterns are part of the strategy of the hegemonic actors. Furthermore, these patterns also have significance for India's climate policy of the time, an era that gave rise to mitigation goals, demanded climate action from the union states and territories, and sought to overcome the sole insistence on the needs of catch-up development and poverty alleviation. In addition, the Indian government has tried—more or less forcefully—to better protect the rights of tribal groups and their living space, also to the detriment of catch-up development. Out of this confluence came the need for a narrative that had to meet both the overall societal demand for development and growth, as well as greater climate protection—because of India's enormous vulnerability and its aspiration to play a major role in the international arena.

India's identity as a state of sustainable development allowed it to justify all these policies, which the Singh government pushed and implemented to varying degrees. Although giving way internationally—hence the stronger commitment in the field of mitigation which was criticised domestically—the government was able to outshine the criticism with the help of the narrative of a strong, important and also energetic actor striving for economic growth, characterised by sustainable development. At the same time, this narrative of strength also allowed the various protection aspects to be part of the identity—so that the administration could also justify to increasingly consider the interests of those in need of protection. At the beginning, however, when Ramesh was still Environment Minister, it may be that the identity was not yet sufficiently convincing, which is why he also took on a new role in the cabinet; however, with the slightly stronger charging of the SD identity through climate justice—partly with feminised aspects—SD succeeded surprisingly well in justifying the government's climate policy. The new momentum that the Singh government injected to India's climate policy was clearly due to the pro-active, ambitious and yet strong image of an India that keeps the needs of the vulnerable in mind and wants to pursue an innovative mitigation policy—a country that is committed particularly to the masculinised aspects of sustainable development. The predominance of masculinised figures within the hegemonic strategy of the Singh government also had its downsides. As the masculinised answers on climate change were the main solutions peddled, the feminised options only fit into the created image of India to a limited extent, which is why adaptation and degrowth aspects did not take up much space. The climate policy identity SD as defined by the Singh government sees economic growth as the solution and not as part of the problem, which must be criticised from an ecofeminist perspective (Irving and Helin 2018, 269). Overall, the analysis of this period has shown that the Indian government—consciously or unconsciously—used gendered patterns in its strategy to achieve broad support for its policies or at least to explain them adequately. The fact that most actors to a certain extent followed the government's idea of climate change policy, or in short: that the identity and the use of gendered patterns 'caught on', must undoubtedly be attributed in large part to the existing patriarchal society and its prevalent neoliberal idea of economic growth as desirable or indeed the highest goal.

3.3.2 Gendered Cultural Roots Shaping the Indian Climate Change Identity under Modi

India's identity in climate policy changed when Prime Minister Modi came to power. From having identified as a state of sustainable development, it initially reverted to the identity of a development state; however, in contrast to earlier times, this identity was expanded to include and accentuate India's culturally-rooted closeness to nature. The Indian identity then altered slightly towards the second half of Modi's first term from 2017 onwards; from the beginning of Modi's campaign for election in 2019 at the latest, one must speak of a newly adjusted identity: India as *a significant actor still culturally-rooted close to nature*. What role did gendered discourse patterns play in forcing and successfully installing these identities?

It can be assumed that the BJP-led government initially pushed the development demand very strongly for tactical reasons. This attempt killed two birds with one stone: First, as it was the greatest election campaign promise of 2014. Second, it steered the expectations on India's Paris commitments. Generating and anchoring this identity, the hegemonic actors structured the discourse inter alia with the help of gendered discourse patterns. One of these gendered discourse patterns was the shaping of the demand for development, linking it very closely with the masculinised demand for growth. It is masculinised because production and various attributes, such as mind or reason, are all associated with the same sphere as masculinity (Magnusdottir and Kronsell 2015), and these are needed to generate innovation-driven growth as demanded by the BJP. Hence, the Modi government has also followed the idea of ecological modernisation, even though with a higher emphasis on the economic success aspect.

The analysis shows that growth was one of the central demands presented as a response to other demands, such as poverty reduction and energy security. Like the Congress Party, the BJP was committed to a neoliberal understanding of economic growth in its election manifesto. Unlike the Congress Party, the BJP paid little attention to social aspects of the climate problem, as it equivalised the masculinised demands growth, energy security and other economy-related themes first and foremost with each other. Concerning the rights of marginalised groups or caring for the environment, which are both feminised discourse patterns, the BJP hardly formulated any demands in its election programme.

When the BJP got elected, the subsequent actions of the government showed that the hegemonic actors almost exclusively followed masculinised discourse patterns of the second dimension, aggressively demanding development and growth as answers to the challenges of climate change. Clearance procedures were accelerated and the protection aspect of developmental and economic programmes—which is associated with femininity—almost completely bypassed. Hence, the government rearticulated the narrative of the dominated nature which can clearly be assigned to the sphere producing the master identity (cf. Sect. 2.3.3) referring to masculinity. Accordingly, India's climate policy identity was underpinned by masculinised discourse patterns such as 'productivity' and the 'mind' in the quest for technical innovation to master nature. Growth became the all-dominant and simultaneously masculinised task. This most important aspect of the BJP's understanding of the empty signifier of development was clearly anchored in the public sphere; it was supported by different subject positions such as economic actors but also huge parts of the society, manifested in the election outcome.

While generating public support for his growth promise, Modi tried to shift the responsibility for dealing with climate change and its consequences to the private sphere, individualising therewith the need to protect climate and environment. A good example was his statement at the Indian Teacher's Day 2014 when the Prime Minister almost denied climate change and insinuated that only people's lifestyles had changed (cf. Sect. 3.2.3). In doing so, he privatised not only the causes of climate change but also its consequences. As the private sphere is discursively associated with femininity (Prügl 2021, 302; MacGregor 2010; Merchant 1989, ch. 7), environmental protection is feminised and endowed with less attention and fewer financial resources, not least shown by the cut to the MoE-FCC's budget. Pushing climate protection out of the public and into the private sphere, it is furthermore depoliticised (Fraser 2016) and at the same time becomes additional work on the shoulders of the people responsible in the household—who are mostly women. Another very important aspect of the Indian identity at the time was the supposed closeness to nature of Indian culture, which was meant to negate any mitigation demands from outside, given that Indian society was pictured as inherently environment-friendly. The gendered aspects used to implement this figure will be addressed again later.

Primarily committed to growth, Javadekar (2014a, 2014b) stated on several occasions that Indian growth could not be compromised by anyone, insisting on the state's sovereignty. In doing so, he used an additional masculinised discourse pattern because sovereignty and freedom are associated with the same sphere as masculinity (Plumwood 1993, 43). The hegemonic actors thereby tried to push other—supposedly feminised—demands such as adaptation (Djoudi et al. 2016, 257) or environmental protection (Weller 2019, 877) out of the public eye whenever these would compromise the growth imperative. Hence, the outlined feminised demands were woven into the chain of equivalence by the hegemonic actors, but only in homeopathic doses whenever it served the idea of economic growth. With this mix of gendered discourse patterns, the BJP government succeeded in asserting its hegemonic project of an Indian identity as a *developing state recognised for its culturally-rooted closeness to nature* and at the same time universalising this narrative as a service to society. Accordingly, this period of Indian climate discourse was characterised by ecomodern masculinity, although the emphasis on growth was even more aggressive than in the previous Singh administration.

This aggression becomes clear in another gendered pattern of discourse used by the Modi government in dealing with NGOs. The suppression of growth-critical NGOs (cf. Sect. 3.2.4) such as Greenpeace India, which for example allied with marginalised groups, clearly followed the masculinised logic of dominating nature and its protectors. Thus, the 'impoverished' groups or the 'portrayed-as-primitive' nature is mastered and made completely accessible for economic growth following masculinised ideas and norms. It was the anchoring of this utterly one-dimensional development story which prepared the ground for the broad social acceptance of the promotion of unconditional, neoliberal economic growth. This component of the development promise can lead to a violent atmosphere towards environmental activists, discussed in Sect. 3.2.6.

A further impressive example of the strategic use of gendered discourse patterns by the hegemonic actors is how they demanded the recognition of India as culturally-rooted close to nature, pointing to different aspects of Hindu traditions such as ancient prayers or the concept of yoga. The cultural roots, which were supposed to emphasise India's age-old connection with nature, were repeatedly dragged into the public discourse as a central aspect of the Modi government's hegemonic strategy. Indeed, as shown above, even the high-risk step of opening the discursive space of climate change was taken, integrating demands of the separated cultural discourse into the Indian chain of equivalence (strategeme X; cf. Sect. 3.2.8).

Notwithstanding, the demand to recognise India as culturally-rooted close to nature is a challenge for the ecofeminist lens employed in this study. At this point the critical ecofeminist approach of decoding the socially-produced gendered dualisms that structure societies has reached its limits in terms of clearly classifying these dichotomies into masculinised on the one hand and feminised or queered on the other. The fact that the dichotomies, such as reason/nature, public/private or culture/nature are derived from Western societies and the time of enlightenment (Hekman 1990, 5; Tickner 1993, 60–63; Prügl 2021, 302) can henceforth be seen as a restriction of the theoretical foundation of the study—at least if they are blindly adopted without regard to the circumstances of the case. Such a special circumstance exists here and requires a prudent and limited application of the dichotomies where they are not applicable. While the period of colonisation ensured that most gendered stereotypes and the associated power structures dominating nature were consolidated in colonised societies as well (Tickner 1993), 'culture' is gendered differently in the Indian discourse than the ecofeminist approach suggests by principally assigning culture to the masculinised sphere. In contrast, the culture-related demand in India's climate discourse thus incorporates both feminised and masculinised discourse patterns due to the diverse components of Hindu culture. Given that traditional Indian culture is not clearly constructed as a dichotomy to nature, this makes the deconstruction of the gendered aspects of the demand to recognise India as culturally-rooted close to nature therefore more complex and at the same time even more interesting.

However, even in this case it is still valuable to apply the ecofeminist approach deconstructing and criticising dichotomous orders. Because the approach can deconstruct the power structures created or strengthened by specific aspects of the demand very clearly, it can continue to serve as a guard rail even here. In fact, regarding the demand to recognise India as culturally-rooted close to nature, both masculinised and feminised sub-patterns of discourse—partly already deconstructed—can be discerned. India's centuries-old tradition of living with the beat of nature is called a vital part of the Indian culture—which the Indian government seems always to equate with Hinduism.[31] When actors refer to this cultural proximity to nature, a range of traditions are usually associated with it. Alongside subsistence farming in harmony with nature or ancient Hindi prayers these are mainly (1) the veneration of natural heritage such as 'Mother' Ganga and (2) yoga.

(1) Praising natural monuments as part of spirituality is a gendered discourse of the first dimension as it often creates feminised subjects such as 'Mother' Earth or 'Mother' Ganga. Of course, in the sense of the first dimension of gendered discourses, the interpellation of the subjects *Mother Earth* or *Mother Ganga* can be just figuratively attested here, as *Mother Earth* does not represent a subject participating in the discourse in the classical way. But we see in this narrative that the earth is feminised as in the Indian INDCs (Government of India 2015, 1) and thus constructed as a passive entity that must be cared for by the active actors (cf. Sect. 3.2.4). Indeed, besides masculinised safeguarding the feminised earth, this narrative can also produce the idea of legitimately dominating the earth. If both safeguarding and dominating the earth are constructed as the answers to the climate catastrophe, we can speak here about masculinised discourse patterns in the sense of the second dimension of gendered discourses.
(2) When Indian yoga practice is named by the hegemonic actors as an important part of the environmentally protective Indian culture, gendered discourse patterns are used once more. On the occasion of the United Nations General Assembly, Modi demanded the adoption of an International Yoga Day and emphasised: "[Yoga] is not about exercise but to discover the sense of oneness with yourself, the world, and the nature" (Modi 2014, 56). As nature is located in the feminised sphere, the use of feminised discourse patterns seems reasonable. Constructing yoga as an individual contribution to climate protection, it is once again assigned to the private sphere (Miller 2020, 100), which is located in the same sphere as femininity. Interestingly, the hegemonic actors demanding the recognition of India as culturally-rooted close

[31] Moreover, the BJP as a Hindu nationalist party has explicitly pursued the idea of Hindutva since the beginning of the 2019 election campaign (cf. Sect. 3.2.7). As discussed in Sect. 3.2.3, references to Hindutva were not apparent in the first term, although today the overlap of culture-related demands in the climate change discourse and Hindutva ideology can be discerned. Therefore, literature regarding the use of Indian cultural values for the purpose installing a Hindutva state as well as its gendered dimensions will be factored into the following.

to nature via the practice of yoga follow masculinised discourse patterns as well. As Christopher Patrick Miller (2020, 100) has pointed out, yoga is also used by the BJP government to secure its neoliberal project of growth in the climate change discourse. The government uses masculinised discourse patterns to construct subject positions which are interpellated as strong, proactive contributors to economic growth enabled by yoga in terms of individual strength.[32] Hence, we can speak about a double-gendered discourse pattern of individualisation. In addition to the already familiar patterns of locating climate protection in people's private sphere via feminised attributes, masculinised discourse patterns are used to strengthen the labour power of individuals and thus their contribution to as masculine connotated growth which is part of the public sphere. The latter is new, but it fits with current developments in India. As already shown, especially since the second term of Modi's government started, cultural demands are often used to advance the BJP's Hindutva project. According to Catarina Kinnvall (2019, 283), this is also to be achieved through the "masculinisation of Indian politics", including the cultural demands discussed here. Kinnvall furthermore states that "under Modi, modernization, development and neoliberal policies have been combined in an appeal to populist cultural politics." (Kinnvall 2019, 293).

Thus, the demand to recognise India as culturally-rooted close to nature is in multiple ways gendered. It is used as needed in different contexts such as on the global or national level to achieve the goal of anchoring India's identity as a *culturally-rooted developing state*. Indeed, to install this identity the government needs both motivated people striving for growth and the privatisation—or more precisely: the denationalisation—of climate and environmental protection and its accompanying disregard by state actors. At the same time, the hegemonic actors still maintain the image of an environmentally connected society through cultural and sometimes spiritual references—which can hardly be criticised without being branded as anti-national or anti-development (cf. Sect. 3.2.4). On the international stage, Miller (2020, 103) summarises the strategic added value of instrumentalising yoga as follows:

> *"[T]ransnational yoga's self-responsibilizing ethos is weaponized to counter the Global North's climate privilege though nevertheless signals Modi and the BJP's commitment to neoliberal economic ideology".*

[32] Miller (2020) draws on the discourse of biopolitics in India contributing to the reinforcement of the neoliberal project of economic growth, among other things with recourse to physical health through yoga.

This is also echoed by the demand that the developed states in particular should establish low-carbon lifestyles, suggesting that Indian society has been pursuing this for a long time due to their ancient traditions. Hence, the aspect of a culturally-rooted closeness to nature, as part of the Indian identity as a developing state, served different functions to justify India's climate policy, both domestically and externally,[33] and to generate a majority in society for it. This is even more true for India's climate identity in the second Modi term, *India as a culturally-rooted significant actor.* Nevertheless, the demand to recognise India as culturally-rooted close to nature was constantly reproduced and stabilised during *both* BJP terms.

Most of the gendered discourses that have been deconstructed so far belong to the second dimension, which means that they contribute to the discursive establishment of ideas on how to deal with the consequences of the climate catastrophe or to justify the formulated answers to combat climate change. In this context, the first Modi government—with the help of various actors—primarily relied on development and, as already delineated, economic growth. Interpellating specific subject positions helped the BJP to implement its growth interest—relativising the calls for environmental and climate protection. In concrete terms, the first dimension of gendered climate discourses was used for example when Environment Minister Javadekar interpellated women solely as carers for the environment (cf. Sect. 3.2.3) securing the privatisation of climate protection. In the framework of most NAPCCs, the actors tried to win women as workers in their respective fields adapting to climate change or development. Even though women were interpellated in the larger context of development, the discourse analysis showed that the government and its allies did this primarily concerning growth. This was contradicted by NGOs such as the AIWC, which for decades has strived to help women achieve self-determined subject positions in the world of work—but in the sense of sustainable development and growth empowering women. Although the feminised discourse patterns such as developing sustainable work used by the AIWC were regularly demanded, they are marginalised compared to the masculinised patterns focusing on economic growth and pushed by the hegemonic actors. The same happened to other actors interested in more diverse subject positions for women. It can therefore be stated that in the Indian climate discourse, masculinised images are often used against feminised images in order to

[33] Arijit Mazumdar (2018) notes that India's general foreign policy under Modi uses yoga, but also other cultural aspects such as Buddhism, as credible and vital soft power instruments. More specifically tailored to Indian climate policy, Miller (2020) also pursues the soft power argument.

discursively consolidate the identity as a significant and yet culturally anchored actor.

Since the different demands were used in both Modi terms—albeit with different emphases—the analysis here was not completely divided into temporal sections. Nevertheless, the particularities of the second term in relation to gendered discourse patterns should be briefly highlighted. As for electoral and other reasons the empty signifier has changed over time from a *developing country which is culturally-rooted close to nature* to a *significant actor which is still culturally-rooted close to nature*, only the discourse patterns that contributed to this transformation are considered. The focus is therefore on the reference to cultural values, which the BJP government has increasingly reinforced. This culminated in the demand that the other states also cultivate a low-carbon lifestyle, a demand the government was able to make with a certain ease and moral strength by referring to its own culturally-rooted closeness to nature. Since the low-carbon lifestyle restrictions in fact mean restrictions in consumption, this demand is to be assessed as feminised. This moral strength was also secured in particular through masculinised discourse patterns for, as already described and discussed, the demands pursued since the Corona pandemic and then also in Glasgow 2021 are very much tailored to India's significance and strength. Leading in renewable energies, embodied by the ISA, but also taking a leading role in disaster-resilient infrastructure (cf. Sect. 3.2.7) shows an immense significance that can be defined as a masculinised attribute. Even if it seems grotesque, the weakening of the text passage on coal phase-out in Glasgow 2021 illustrates the Indian identity in climate policy that has emerged and is reproduced again and again: *India as a significant actor culturally-rooted close to nature.* While we reconstructed serious criticism of the climate policy actions of the BJP government, the administration has succeeded in deploying and using the various discourse patterns in such a way that, on the one hand, this identity can always be reproduced and thus anchored, while on the other hand the measures taken or not taken can be justified and defended. Whether this policy is consistent with the identity created is, of course, a completely different matter.

In sum, both the Modi and Singh governments have followed the idea of ecological modernisation via ecomodern masculinity. Most of the actors who pushed for the reconstructed identities used particularly masculinised discourse patterns shaping a masculinised Indian identity in climate policy and at the same time using it as a resonance space for the acceptance of their respective climate policies. In most cases, feminised discourse patterns were also used and present, but the analysis showed that these were mostly overshadowed or dominated by masculinised measures and priorities, or that they were used to support the masculinised demands. For example, growth is made possible by being able to pursue it without compromise by discursively relegating climate protection to the private sphere. The image of a significant India, which relies on growth in various facets and can achieve it not least through its own economic or scientific strength was played out—in varying degrees depending on the ruling government and of course its allies.

Indeed, these findings are not fixed. Whether the Modi government's strategy will be questioned more strongly in the future, for example in the form of a stronger pushback against Hindutva ideology, remains to be seen. For climate policy, this would probably mean that the gendered justifications are no longer sufficient. Whether and how the EU uses gendered discourse patterns which shape the EU's identity to justify its chosen responses to the climate crisis will be explored in the following chapter.

References

Aamodt, Solveig, and Iselin Stensdal. 2017. "Seizing policy windows: Policy Influence of climate advocacy coalitions in Brazil, China, and India, 2000–2015." *Global Environmental Change* 46: 114–25. https://doi.org/10.1016/j.gloenvcha.2017.08.006.

Agarwal, Anil, and Sunita Narain. 1991. *Global Warming in an Unequal World: A case of environmental Colonialism.* New Delhi: Centre for Science and Environment. http://www.indiaenvironmentportal.org.in/files/GlobalWarming%20Book.pdf.

Agarwal, Kabir. 2020. "Citing Anti-Terror Law, Delhi Police Block Global Youth Climate Activism Website." *The Wire*, July 23. Accessed May 27, 2022. https://thewire.in/environment/fridays-for-future-website-block-eia-prakash-javadekar-uapa.

Aggestam, Karin, and Jacqui True. 2021. "Political leadership and gendered multilevel games in foreign policy." *International Affairs* 97 (2): 385–404. https://doi.org/10.1093/ia/iiaa222.

All India Radio. 2022. "PM Modi launches Global initiative 'Lifestyle for Environment (LiFE)' Movement; Calls for human centric and collective efforts that further sustainable development." Accessed June 28, 2022. https://newsonair.gov.in/Main-News-Details.aspx?id=442124.

All India Women's Conference. 2013. "Decision 23/CP.18 Submission from All India Women's Conference." Accessed May 22, 2022. https://unfccc.int/resource/docs/2013/smsn/ngo/374.pdf.

Andharia, Janki, and Chandan Sengupta. 1998. "The Environmetnal Movement: Global Issues and the Indian Reality." *The India Journal of Social Work* 59 (1): 422–49.

Atteridge, Aaron. 2013. "The Evolution of Climate Policy in India: Poverty and Global Ambition in Tension." In *Climate Governance in the Developing World*, edited by David Held, Charles Roger, and Eva-Maria Nag, 53–71. Cambridge: Polity Press.

Atteridge, Aaron, Manish K. Shrivastava, Neha Pahuja, and Himani Upadhyay. 2012. "Climate Policy in India: What Shapes International, National and State Policy?" *Ambio* 41 (1): 68–77. https://doi.org/10.1007/s13280-011-0242-5.

Bäckstrand, Karin, and Eva Lövbrand. 2006. "Planting Trees to Mitigate Climate Change: Contested Discourses of Ecological Modernization, Green Governmentality and Civic Environmentalism." *Global Environmental Politics* 6 (1): 50–75. https://doi.org/10.1162/glep.2006.6.1.50.

Bahuguna, Sunderlal. 1980. "Chipko.".

Bandyopadhyay, Jayanta (1999): Chipko movement: Of floated myths and flouted realities. In *Economic and Political Weekly* 34 (15): 880–882.

Banerjee, Subhabrata B. 2012. "A Climate for Change? Critical Reflections on the Durban United Nations Climate Change Conference." *Organization Studies* 33 (12): 1761–86. https://doi.org/10.1177/0170840612464609.

Basrur, Rajesh. 2017. "Modi's foreign policy fundamentals: a trajectory unchanged." *International Affairs* 93 (1): 7–26. https://doi.org/10.1093/ia/iiw006.

Baviskar, Amita, and Michael Levien. 2021. "Farmers' protests in India: introduction to the JPS Forum." *Journal of Peasant Studies* 48 (7): 1341–55. https://doi.org/10.1080/03066150.2021.1998002.

Betz, Joachim. 2012. "India's Turn in Climate Policy: Assessing the Interplay of Domestic and International Policy Change." GIGA Working Papers 190.

Bharatiya Janata Party. 2014. "Election Manifesto 2014." Accessed January 28, 2022. http://library.bjp.org/jspui/bitstream/123456789/252/1/bjp_lection_manifesto_english_2014.pdf.

Bharatiya Janata Party. 2019. "Sankalp Patra." Accessed June 22, 2022. http://library.bjp.org/jspui/bitstream/123456789/2988/1/BJP-Election-english-2019.pdf.

Boykoff, Max. 2010. "Indian media representations of climate change in a threatened journalistic ecosystem." *Climatic Change* 99 (1-2): 17–25. https://doi.org/10.1007/s10584-010-9807-8.

Bretherton, Charlotte. 1998. "Global environmental politics: putting gender on the agenda?" *Review of International Studies* 24 (1): 85–100.

Bruff, Ian, and Stefanie Wöhl. 2016. "Constitutionalizing Austerity, Disciplining the Household: Masculine Norms of Competitiveness and the Crisis of Social Reproduction in the Eurozone." In *Scandalous Economics: Gender and the Politics of Financial Crises*, edited by Aida A. Hozic and Jacqui True, 92–108. Oxford studies in gender and international relations. New York, NY: Oxford University Press.

Burk, Martha, and Heidi Hartmann. 2017. "Corporate Boards and International Policies: Gender Parity on Corporate Boards: A Path to Women's Equality?" In *Gender in the twenty-first century: the stalled revolution and the road to equality*, edited by Davis Shannon, Sarah Winslow, and David J. Maume, 127–146. Oakland: University of California Press.

Buzan, Barry, Ole Waever, and Jaap de Wilde. 1998. *Security: A New Framework for Analysis.* Boulder, Colo. Lynne Rienner.

Carbon Disclosure Project. 2011. "CDP India 200 Report 2011. Accelerating Low Carbon Growth." https://wwfin.awsassets.panda.org/downloads/cdp_india_200_report_2011.pdf.

Centre for Science and Environment. 2009. "Richest Indians emit less than poorest Americans." CSE Factsheet. Accessed January 19, 2021. http://www.indiaenvironmentportal.org.in/files/richest_poorest_emissions.pdf.

Centre for Science and Environment. 2010. "Annual Report 2009–2010." Accessed March 13, 2022. https://cdn.cseindia.org/userfiles/Annual_Report2009-10.pdf.

Centre for Science and Environment. 2011a. "Annual Report 2010–2011." Accessed March 07, 2022. https://cdn.cseindia.org/userfiles/Annual_Report2010-11.pdf.

Centre for Science and Environment. 2011b. "Event: Indian Ministry of Environment and Forests and Centre for Science and Environment side event at COP 17." https://www.cseindia.org/event-indian-ministry-of-environment-and-forests-and-centre-for-science-and-environment-side-event-at-cop-17-3461.

Centre for Science and Environment. 2012a. "Annual Report 2011–2012." Accessed April 02, 2022. https://cdn.cseindia.org/userfiles/Annual%20Report%202011-12.pdf.

Centre for Science and Environment. 2012b. "CSE-MoEF Side Event at CoP 18 Doha." Accessed March 22, 2022.

Centre for Science and Environment. 2014. "Annual Report 2013–2014." Accessed April 05, 2022. https://cdn.cseindia.org/userfiles/annual_report_2013-14.pdf.

Centre for Science and Environment. 2015a. "Annual Report 2014–2015." Accessed April 18, 2022. https://cdn.cseindia.org/userfiles/annual-report-2014-15.pdf.

Centre for Science and Environment. 2015b. "India's INDC is fair, and its renewable energy and forestry targets are ambitious, says CSE." Accessed April 29, 2022. https://www.cseindia.org/indias-indc-is-fair-and-its-renewable-energy-and-forestry-targets-are-ambitious-says-cse-6053.

Centre for Science and Environment. 2016. "Annual Report 2015–16." Accessed April 29, 2022. https://cdn.cseindia.org/userfiles/annual-report2015-16.pdf.

Centre for Science and Environment. 2020. "2019–20 Annual Report: Knowledge-Based Activism." Accessed June 27, 2022. https://www.cseindia.org/static/page/Annual-report2020.pdf.

Centre for Science and Environment. 2021. "2020–21 Annual Report: Knowledge-Based Activism." Accessed June 27, 2022. https://www.cseindia.org/static/page/CSE-annual-report-2021.pdf.

Centre for Science and Environment, Friends of the Earth International, and What next Forum. 2015. "Programme for Global Renewable Energy and Energy Access Transformation (GREEAT)." Accessed May 26, 2022. https://www.cseindia.org/content/downloadreports/6165.

Chacko, Priya. 2019. "Marketizing Hindutva : The state, society, and markets in Hindu nationalism." *Modern Asian Studies* 53 (2): 377–410. https://doi.org/10.1017/S0026749X17000051.

Chakravarty, Shoibal, and M. V. Ramana. 2012. "The Hiding Behind the Poor debate. A syntetic overview." In Dubash 2012, 218–29.

Chandra, Rohit, and Michael Walton. 2020. "Big potential, big risks? Indian capitalism, economic reform and populism in the BJP era." *India Review* 19 (2): 176–205. https://doi.org/10.1080/14736489.2020.1744997.

Chandran, Manoj. 2012. "Study on the Impact of ban on green felling of Chir Pine (Pinus roxburghii Sarg.)." Accessed January 25, 2022. https://www.researchgate.net/public

ation/329970050_Study_on_the_Impact_of_ban_on_green_felling_of_Chir_Pine_Pinus_roxburghii_Sarg.
Chaturvedi, Sanjay, and Timothy Doyle. 2015. *Climate Terror. A Critical Geopolitics of Climate Change.* London: Palgrave Macmillan.
Chowdhury, Adhir R. 2021. "Speech." In Lok Sabha 2021.
Ciecierska-Holmes, Natalia, Kirsten Jörgensen, Lana L. Ollier, and D. Raghundandan, eds. 2020. *Environmental Policy in India.* Abingdon: Routledge.
Ciecierska-Holmes, Natalia, Kirsten Jörgensen, Lana L. Ollier, and D. Raghundandan. 2020. "Introduction: environmental policy in India." In Ciecierska-Holmes, Jörgensen, Ollier, and Raghundandan 2020, 3–16.
Climate Action Tracker. 2015a. "Country Assessment India." Accessed April 25, 2022. https://climateactiontracker.org/media/documents/2018/4/CAT_2015-12-02_CountryAssessment_India.pdf.
Climate Action Tracker. 2015b. "INDCs lower projected warming to 2.7°C: significant progress but still above 2oC." Accessed May 01, 2022. https://climateactiontracker.org/documents/39/CAT_2015-10-01_INDCsLowerProjectedWarming_CATUpdate.pdf.
Climate Action Tracker. 2022. "Fair Share." Accessed May 08, 2022. https://climateactiontracker.org/methodology/cat-rating-methodology/fair-share/#:~:text=The%20Climate%20Action%20Tracker's%20%E2%80%9Cfair,analysis%20to%20close%20data%20gaps.
Coalition for Disaster Resilient Infrastructure. 2021. "Annual Report 2020–2021." Accessed June 24, 2022. https://cdri.world/sites/default/files/publication/CDRI_Annual_Report_2020_21.pdf.
Coalition for Disaster Resilient Infrastructure. 2022. "Members of the Coalition for Disaster Resilient Infrastructure (CDRI)." Accessed June 24, 2022. https://cdri.world/members.
Cohen, Stephen P. 2001. *India. Emerging Power.* Washington: Brookings Institutions Press.
Deutsche Welle Asia. 2021. "Who is Indian activist Disha Ravi?" Accessed May 27, 2022. https://learngerman.dw.com/en/who-is-arrested-india-toolkit-activist-disha-ravi/a-56600025.
Djoudi, Houria, Bruno Locatelli, Chloe Vaast, Kiran Asher, Maria Brockhaus, and Bimbika Basnett Sijapati. 2016. "Beyond Dichotomies: Gender and Intersecting Inequalities in Climate Change Studies." *Ambio* 45 (Suppl 3): 248–62. https://doi.org/10.1007/s13280-016-0825-2.
Down to Earth. 2018. "Is PM Modi really a 'Champion of the Earth'?" *Down to Earth,* September 27. Accessed June 07, 2022. https://www.downtoearth.org.in/news/forests/is-pm-modi-really-a-champion-of-the-earth--61739.
Down to Earth. 2021. "CoP26: Modi offers 'Panchamrita' concoction for climate conundrum at Glasgow." *Down to Earth,* November 2. https://www.downtoearth.org.in/news/climate-change/cop26-modi-offers-panchamrita-concoction-for-climate-conundrum-at-glasgow-80001.
Dröge, Susanne, and Christian Wagner. 2015. "India's Position in International Climate Negotiations: No Shift under Modi." SWP Comments 14. Accessed April 10, 2021. https://www.swp-berlin.org/publications/products/comments/2015C14_dge_wgn.pdf.
Dubash, Navroz K., ed. 2012. *Handbook of Climate Change and India. Development, Politics and Governance.* Abingdon: Routledge.

Dubash, Navroz K. 2013. "The politics of climate change in India: narratives of equity and cobenefits." *WIREs Clim Change* 4 (3): 191–201. https://doi.org/10.1002/wcc.210.

Dubash, Navroz K., ed. 2019. *India in a Warming World: Integrating Climate Change and Development.* Oxford: Oxford University Press.

Dubash, Navroz K. 2021. "Varieties of climate governance: the emergence and functioning of climate institutions." *Environmental Politics* 30 (1): 1–25. https://doi.org/10.1080/09644016.2021.1979775.

Dubash, Navroz K., and Anu Jogesh. 2014. "From Margins to Mainstream? State Climate Change Planning in India as a 'Door Opener' to a Sustainable Future." Accessed May 27, 2022. https://papers.ssrn.com/sol3/Delivery.cfm/SSRN_ID2474518_code2275529.pdf?abstractid=2474518&mirid=1.

Dubash, Navroz K., and Neha B. Joseph. 2016. "Evolution of Institutions for Climate Policy in India." 3. Accessed January 20, 2022. https://cprindia.org/sites/default/files/articles/Dubash_Joseph_Evolution_of_Institutions_for_Climate_Policy_in_India.pdf.

Dubash, Navroz K., Radhika Khosla, Ulka Kelkar, and Sharachchandra Lele. 2018. "India and Climate Change: Evolving Ideas and Increasing Policy Engagement." *Annual Review of Environment and Resources* 43 (1): 395–424. https://doi.org/10.1146/annurev-environ-102017-025809.

Ergas, Christina, and Richard York. 2012. "Women's Status and Carbon Dioxide Emissions: a Quantitative Cross-National Analysis." *Social science research* 41 (4): 965–76. https://doi.org/10.1016/j.ssresearch.2012.03.008.

Express News Service. 2021. "Phase-down vs phase-out at COP: Bhupender Yadav stresses 'national circumstances'." *The Indian Express*, November 24. Accessed July 03, 2021. https://indianexpress.com/article/india/phase-down-vs-phase-out-at-cop-bhupender-yadav-stresses-national-circumstances-7638091/.

Falkner, Robert, Hannes Stephan, and John Vogler. 2010. "International Climate Policy after Copenhagen: Towards a 'Building Blocks' Approach." *Global Policy* 1 (3): 252–62. https://doi.org/10.1111/j.1758-5899.2010.00045.x.

Federation of Indian Chambers of Commerce and Industry. 2012. "Impacts, Governance and Future of CDM. Indian Industry Perspective." Accessed March 16, 2022. https://ficci.in/spdocument/20158/CDM-FICCI%20-Report.pdf.

Fernandes, Denise, Kirsten Jörgensen, and N. C. Narayanan. 2020. "Factors shaping the climate policy process in India." In Ciecierska-Holmes, Jörgensen, Ollier, and Raghundandan 2020, 158–73.

Foreign Contribution (Regulation) Act. Ministry of Law and Justice. 1976. Accessed May 09, 2022. https://legislative.gov.in/sites/default/files/A1976-49.pdf.

Foster, Emma A. 2017. "Gender, Environmental Governmentality, and the Discourses of Sustainable Development." In *Routledge international handbook of gender and environment*, edited by Sherilyn MacGregor, 216–28. Routledge international handbooks. London: Routledge.

Fraser, Nancy. 2016. "Contradictions of capital and care." *New Left Review* (100): 99–117.

Fröhlich, Franziska. 2022. Interview by A. Zürn. 2022. Würzburg.

Fujiwara, Noriko. 2010. "The Political Economy of India's Climate Agenda." CEPS Working Document 325. Accessed January 08, 2021. https://www.ceps.eu/ceps-publications/political-economy-indias-climate-agenda/.

Ganguly, Sunayana. 2020. "Civil society and state interaction in environment policy in India." In Ciecierska-Holmes, Jörgensen, Ollier, and Raghundandan 2020, 60–85.
Gavit, Heena V. 2021. "Speech." In Lok Sabha 2021.
Ghosh, Aritra, Srijita Nundy, and Tapas K. Mallick. 2020. "How India Is Dealing with COVID-19 Pandemic." *Sensors international* 1. https://doi.org/10.1016/j.sintl.2020.100021.
Ghosh, Padmaparna. 2009. "I want to position India as a proactive player: Jairam Ramesh." Accessed January 23, 2022. https://www.livemint.com/Politics/h97ogi3qEaToXuP9T8YTuI/I-want-to-position-India-as-a-proactive-player-Jairam-Rames.html.
Gieg, Philipp, Timo Lowinger, Manuel Pietzko, Anja Zürn, Ummu S. Bava, and Gisela Müller-Brandeck-Bocquet, eds. 2021. *EU-India Relations*. Cham: Springer International Publishing.
Giese, Lucas J. 2017. "The Role of NGOs in International Climate Governance: A Case Study of Indian NGOs." All College Thesis Program, 2016–2019 36. https://digitalcommons.csbsju.edu/cgi/viewcontent.cgi?article=1037&context=honors_thesis.
Global Witness. 2012. "A hidden crisis? Increase in killings as tensions rise over land and forests." Accessed June 08, 2022. https://www.globalwitness.org/en/campaigns/environmental-activists/hidden-crisis/.
Global Witness. 2013. "Deadly Environment: The dramatic rise in killings of environmental and land defenders." Accessed June 08, 2022. https://cdn2.globalwitness.org/archive/files/library/deadly%20environment.pdf.
Global Witness. 2015. "How Many More? 2014's deadly environment: the killing and intimidation of environmental and land activists, with a spotlight on Honduras." Accessed June 08, 2022. https://www.globalwitness.org/documents/17882/how_many_more_pages.pdf.
Global Witness. 2016. "On Dangerous Ground: 2015's deadly environment: the killing and criminalization of land and environmental defenders worldwide." Accessed June 08, 2022. https://www.globalwitness.org/en/campaigns/environmental-activists/dangerous-ground/.
Global Witness. 2017. "Defenders of the Earth: Global killings of land and environmental defenders in 2016." Accessed June 08, 2022. https://www.globalwitness.org/documents/19122/Defenders_of_the_earth_report.pdf.pdf.
Global Witness. 2018. "At what Cost? Irresponsible business and the murder of land and environmental defenders in 2017." Accessed June 08, 2022. https://www.globalwitness.org/documents/19595/Defenders_report_layout_AW2_lowres_5jP01UZ.pdf.
Global Witness. 2019. "Enemies of the State?" Accessed May 02, 2022. https://www.globalwitness.org/documents/19766/Enemies_of_the_State.pdf.
Global Witness. 2020. "Defending tomorrow: The climate crisis and threats against land and environmental defenders." Accessed June 08, 2022. https://www.globalwitness.org/documents/19938/Defending_Tomorrow_EN_high_res_-_July_2020.pdf.
Global Witness. 2021. "Last line of defence: The industries causing the climate crisis and attacks against land and environmental defenders." Accessed June 08, 2022. https://www.globalwitness.org/documents/20191/Last_line_of_defence_-_high_res_-_September_2021.pdf.
Gogoi, Elizabeth. 2019. "State Climate Change Planning: Has It Reached the Mainstream?" In Dubash 2019, 370–80.
Gogoi, Gaurav. 2014a. "Parliament Speech." In Lok Sabha 2014.

Gogoi, Gaurav. 2014b. "Parliament Speech." In *Lok Sabha Debates: 03.12.2014*, edited by Lok Sabha. New Delhi. https://eparlib.nic.in/bitstream/123456789/758798/1/0312_16_III.pdf. Accessed April 21, 2022.

Government of Arunachal Pradesh. 2011. "Arunachal Pradesh State Action Plan on Climate Change." Accessed May 22, 2022. http://www.nicra-icar.in/nicrarevised/images/State%20Action%20Plan/Arunachal-Pradesh-SAPCC.pdf.

Government of Bihar. 2015. "Bihar State Action Plan on Climate Change: 'Building Resilience through Development'." Accessed May 29, 2022. https://moef.gov.in/wp-content/uploads/2017/08/Bihar-State-Action-Plan-on-Climate-Change-2.pdf.

Government of Chhattisgarh. 2014. "State Action Plan on Climate Change: Inclusive Growth for Improved Resilience." Accessed May 22, 2022. https://moef.gov.in/wp-content/uploads/2017/08/Chhattisgarh.pdf.

Government of Gujarat. 2014. "State Action Plan on Climate Change." Accessed May 22, 2022. http://gujenvis.nic.in/PDF/Gujarat-SAPCC.pdf.

Government of India. 1972. "Wild Life (Protection) Act." https://legislative.gov.in/sites/default/files/A1972-53_0.pdf.

Government of India. 2008. "National Action Plan on Climate Change." https://archivepmo.nic.in/drmanmohansingh/climate_change_english.pdf.

Government of India. 2014. "Government reconstitutes the Prime Minister's Council on Climate Change." Accessed April 17, 2022. https://pib.gov.in/newsite/printrelease.aspx?relid=111090.

Government of India. 2015. "India's Intended Nationally Determined Contribution: Working Towards Climate Justice." Accessed April 24, 2022. https://www4.unfccc.int/sites/ndcstaging/PublishedDocuments/India%20First/INDIA%20INDC%20TO%20UNFCCC.pdf.

Government of India. 2016. "Views from the Government of India on possible elements and guiding principles for continuing and enhancing the two-year Lima Work Programme on gender." In *Possible elements and guiding principles for continuing and enhancing the Lima work programme on gender*, edited by United Nations Framework Convention on Climate Change (UNFCCC), 17–18. https://unfccc.int/resource/docs/2016/sbi/eng/misc02.pdf. Accessed May 22, 2022.

Government of India. 2018. "Submission by India on SBI Agenda item 20: Gender and Climate Change." Accessed May 22, 2022. https://www4.unfccc.int/sites/SubmissionsStaging/Documents/201804051550---India%20Gender%20submission.pdf.

Government of Kerala. 2014. "Kerala State Action Plan on Climate Change." Accessed May 22, 2022. https://envt.kerala.gov.in/wp-content/uploads/2019/10/Kerala-State-Action-Plan-on-Climate-Change-KSAPCC-2014-August.pdf.

Government of Madhya Pradesh. 2014. "Madhya Pradesh State Action Plan on Climate Change: Integrating Concerns—Converging Possibilities." Accessed May 22, 2022. http://www.climatechange.mp.gov.in/sites/default/files/ongoing-projects/Madhya%20Pradesh%20State%20Action%20Plan%20on%20Climate%20Change.pdf.

Government of Nagaland. 2012. "Nagaland State Action Plan on Climate Change: Achieving a Low Carbon Development Trajectory." Accessed May 22, 2022. https://moef.gov.in/wp-content/uploads/2017/09/Nagaland.pdf.

Government of Odisha. 2018. "Odisha Climate Change Action Plan (For the Period 2018–23)." Accessed May 22, 2022. http://climatechangecellodisha.org/pdf/State%20Action%20Plan%20on%20Climate%20Change%202018-23.pdf.

Government of Punjab. 2014. "Punjab State Action Plan on Climate Change." Accessed May 22, 2022. https://moef.gov.in/wp-content/uploads/2017/09/Punjab.pdf.

Government of Tamil Nadu. 2014. "Tamil Nadu State Action Plan for Climate Change." Accessed May 22, 2022. https://moef.gov.in/wp-content/uploads/2017/09/Tamilnadu-Final-report.pdf.

Government of Uttarakhand. 2014. "Uttarakhand Action Plan on Climate Change." Accessed May 22, 2022. https://moef.gov.in/wp-content/uploads/2017/08/Uttarakhand-SAPCC.pdf.

Greenpeace India. 2007. "Hiding behind the poor: A report by Greenpeace on Climate injustice." Accessed January 19, 2021. https://www.environxchange.com/images/article/166/hiding-behind-the-poor.pdf.

Greenpeace India. 2012. "Annual Report 2011." Accessed March 16, 2022. https://www.greenpeace.org/static/planet4-india-stateless/2018/06/Annual-Report-2011.pdf.

Greenpeace India. 2014. "Annual Report 2013." Accessed March 16, 2022. https://www.greenpeace.org/static/planet4-india-stateless/2018/06/Annual-Report-2013.pdf.

Greenpeace India. 2016. "Annual Report 2014–2015: The Epic Fight Back." Accessed May 08, 2022. https://www.greenpeace.org/india/en/publication/1056/annual-report-2014-2015/.

Greenpeace India. 2019. "You can't sink a rainbow. You can't seize the sunrise: Annual Report 2018–2019." Accessed June 26, 2022. https://www.greenpeace.org/static/planet4-india-stateless/2020/07/aea1005b-gp-annualreport_web.pdf.

Gupta, Himangana. 2015. "Women and Climate Change: Linking Ground Perspectives to the Global Scenario." *Indian Journal of Gender Studies* 22 (3): 408–20. https://doi.org/10.1177/0971521515594278.

Hallding, Karl, Marie Jürisoo, Marcus Carson, and Aaron Atteridge. 2013. "Rising powers: the evolving role of BASIC countries." *Climate Policy* 13 (5): 608–31. https://doi.org/10.1080/14693062.2013.822654.

Hansen, Thomas B., and Srirupa Roy, eds. 2022a. *Saffron Republic: Hindu Nationalism and State Power in India.* Metamorphoses of the political. Cambridge: Cambridge University Press.

Hansen, Thomas B., and Srirupa Roy. 2022b. "What Is New About 'New Hindutva'?" In *Saffron Republic: Hindu Nationalism and State Power in India*, edited by Thomas B. Hansen and Srirupa Roy, 1–24. Metamorphoses of the political. Cambridge: Cambridge University Press.

Hekman, Susan J. 1990. *Gender and Knowledge: Elements of a Postmodern Feminism.* Boston: Northeastern Univ. Press.

Huang, Jennifer. 2017. "Climate Justice & The Paris Agreement." *Journal of Animal and Environmental Law* 9 (1): 23–59.

Hultman, Martin, and Jonas Anshelm. 2017. "Masculinites of Global Climate Change: Exploring Ecomodern, Industrial and Ecological Masculinity." In *Climate Change and Gender in Rich Countries: Work, Public Policy and Action*, edited by Marjorie G. Cohen, 19–34. Routledge studies in climate, work and society. London: Routledge.

Indian Network for Climate Change Assessment (INCCA). 2010. "Climate Change and India: A 4x4 Assessment. A Sectoral and Regional Analysis for 2030s." Accessed

April 02, 2022. https://web.archive.org/web/20111018233211/http://moef.nic.in/downloads/public-information/fin-rpt-incca.pdf.

India. 2012. "Increasing Ambition Level under Durban Platform for Enhanced Actions." https://digitallibrary.un.org/record/724258.

India Votes. 2022. "2019 Lok Sabha / Parliamentary Election Results." Accessed June 24, 2022. https://www.indiavotes.com/lok-sabha/2019/0/17/0.

Indian National Congress. 2014. "Your Voice. Our Pledge." Accessed April 14, 2022. https://cdn.inc.in/manifestos/pdf_documents/000/000/001/original/Manifesto-2014.pdf?1506426027.

Indian National Congress. 2019. "Congress Will Deliver." Accessed April 14, 2022. https://manifesto.inc.in/pdf/english.pdf.

Intelligence Bureau. 2014. "Concerted efforts by select foreign funded NGOs to 'take down' Indian development projects." Accessed May 08, 2022. https://archive.nyu.edu/bitstream/2451/40704/2/NGO%20report.pdf.

Intergovernmental Panel on Climate Change. 2007. "Climate Change 2007: Impacts, Adaptation and Vulnerability." Accessed March 11, 2022. https://www.ipcc.ch/site/assets/uploads/2018/03/ar4_wg2_full_report.pdf.

International Monetary Fund. 2021. "Policy Responses to Covid-19." Accessed June 27, 2022. https://www.imf.org/en/Topics/imf-and-covid19/Policy-Responses-to-COVID-19#I.

International Solar Alliance. 2022a. "About ISA." Accessed May 01, 2022. https://isolaralliance.org/about/background.

International Solar Alliance. 2022b. "Director General." Accessed May 01, 2022. https://isolaralliance.org/about/director-general.

Irving, Simon, and Jenny Helin. 2018. "A World for Sale? An Ecofeminist Reading of Sustainable Development Discourse." *Gender, Work & Organization* 25 (3): 264–78. https://doi.org/10.1111/gwao.12196.

Isaksen, Kari-Anne, and Kristian Stokke. 2014. "Changing climate discourse and politics in India. Climate change as challenge and opportunity for diplomacy and development." *Geoforum* 57: 110–19. https://doi.org/10.1016/j.geoforum.2014.08.019.

Jaffrelot, Christophe. 2021. *Modi's India: Hindu Nationalism and the Rise of Ethnic Democracy.* Princeton: Princeton University Press.

Jaffrelot, Christophe, and Gilles Verniers. 2020. "The BJP's 2019 election campaign: not business as usual." *Contemporary South Asia* 28 (2): 155–77. https://doi.org/10.1080/09584935.2020.1765985.

Jaremko Bromwich, Rebecca. 2020. "Mother Earth in Environmental Activism: Indigeneity, Maternal Thinking, and Animism in the Keystone Pipeline Debate." In *Environmental activism and the maternal: Mothers and Mother Earth in activism and discourse*, edited by Rebecca Bromwich, Noemie Richard, Olivia Ungar, Melanie Younger, and Maryellen Symons, 53–71. Bradford: Demeter.

Javadekar, Prakash. 2014a. "Speech in Parliament." In *Lok Sabha Debates: 22.12.2014*, edited by Lok Sabha. New Delhi. https://eparlib.nic.in/bitstream/123456789/758797/1/2212_16_III.pdf. Accessed April 21, 2022.

Javadekar, Prakash. 2014b. "Statement." Accessed April 17, 2022. https://unfccc.int/files/meetings/lima_dec_2014/statements/application/pdf/cop_20_hls_india.pdf.

Javadekar, Prakash. 2015. "SUO Motu Statement by Environment Minister on 21st Conference of Parties at Paris." Accessed April 25, 2022. https://pib.gov.in/newsite/PrintRelease.aspx?relid=133555.
Javadekar, Prakash. 2019. "Statement of Union Environment Minister at UNFCCC COP25." Accessed May 22, 2022. https://pib.gov.in/Pressreleaseshare.aspx?PRID=1595787.
Javadekar, Prakash. 2021. "Statement by Hon'ble Minister for Environment, Forests and Climate Change at UN Security Council Open VTC Debate "Maintenance of international peace and security: Addressing climate-related risks to international peace and security." Accessed June 29, 2022. https://mea.gov.in/Speeches-Statements.htm?dtl/33562/Statement_by_Honble_Minister_for_Environment_Forests_and_Climate_Change_at_UN_Security_Council_Open_VTC_Debate_Maintenance_of_international_peace_and_.
Jayaram, Dhanasree. 2018. "From 'Spoiler' to 'Bridging Nation': The Reshaping of India's Climate Diplomacy." *Revue internationale et stratégique* 109 (1): 181. https://doi.org/10.3917/ris.109.0181.
Jayaram, Dhanasree. 2021a. *Climate Diplomacy and Emerging Economies: India as a Case Study*. Milton: Routledge.
Jayaram, Dhanasree. 2021b. "India's Climate Diplomacy Towards the EU: From Copenhagen to Paris and Beyond." In Gieg, Lowinger, Pietzko, Zürn, Bava, and Müller-Brandeck-Bocquet 2021, 201–26.
Jogesh, Anu. 2012. "A change in climate? Trends in climate change reportage in the Indian print media." In Dubash 2012, 266–86.
Jogesh, Anu. 2019. "Looking Out, Looking In: the Shifting discourse on Climate Change in the Indian Print Media." In Dubash 2019, 301–25.
Jörgensen, Kirsten, Arabinda Mishra, and Gopal K. Sarangi. 2015. "Multi-level climate governance in India: the role of the states in climate action planning and renewable energies." *Journal of Integrative Environmental Sciences* 12 (4): 267–83. https://doi.org/10.1080/1943815X.2015.1093507.
King, Ed. 2014. "Whither India: Modi faces climate choice as 2015 deal nears." *Climate Home News*, August 11. Accessed April 17, 2022. https://www.climatechangenews.com/2014/08/11/whither-india-modi-faces-climate-choice-as-2015-deal-nears/.
Kinnvall, Catarina. 2019. "Populism, ontological insecurity and Hindutva: Modi and the masculinization of Indian politics." *Cambridge Review of International Affairs* 32 (3): 283–302. https://doi.org/10.1080/09557571.2019.1588851.
Kotak, Manoj. 2021. "Speech." In Lok Sabha 2021.
Krishnamurthy, Mekhala. 2021. "Agricultural market law, regulation and resistance: a reflection on India's new 'farm laws' and farmers' protests." *Journal of Peasant Studies* 48 (7): 1409–18. https://doi.org/10.1080/03066150.2021.1998001.
Kumar, Vineet. 2018. "Coping with Climate Change: An Analysis of India's State Action Plans on Climate Change." Accessed May 27, 2022. http://cdn.cseindia.org/attachments/0.40897700_1519110602_coping-climate-change-volII.pdf.
Latvia and the European Commission. 2015. "Intended Nationally Determined Contribution of the EU and its Member States." Accessed April 29, 2022. https://www4.unfccc.int/sites/ndcstaging/PublishedDocuments/European%20Union%20First/LV-03-06-EU%20INDC(Archived).pdf.
Lele, Sharachchandra. 2012. "Climate change and the Indian environmental movement." In Dubash 2012, 208–17.

Lok Sabha, ed. 2014. *Lok Sabha Debates: 06.08.2014*. New Delhi. Accessed April 21, 2022. https://eparlib.nic.in/bitstream/123456789/758770/1/0608-1_16_II.pdf.

Lok Sabha, ed. 2021. *Further discussion on climate change raised by Smt. Kanimozhi Karunanidhi on 8th December 2021*. Accessed July 02, 2022. http://loksabhaph.nic.in/Debates/Result17.aspx?dbsl=7736.

Lowinger, Timo. 2019. "The India that can't say yes? Weltinnenpolitik und die theorieimmanente Exklusion des Globalen Südens." *S+F* 37 (4): 203–8. https://doi.org/10.5771/0175-274X-2019-4-203.

Lowinger, Timo, Anja Zürn, Philipp Gieg, and Manuel Pietzko. 2021. "Norm Contestation in EU Strategic Partnerships: The Cases of Civil Society Involvement and Climate Justice in EU–India Relations." In Gieg, Lowinger, Pietzko, Zürn, Bava, and Müller-Brandeck-Bocquet 2021, 31–66.

Lucke, Franziskus von. 2020. *The Securitisation of Climate Change and the Governmentalisation of Security*. Cham: Springer International Publishing AG.

MacGregor, Sherilyn. 2010. "'Gender and climate change': from impacts to discourses." *Journal of the Indian Ocean Region* 6 (2): 223–38.

MacGregor, Sherilyn. 2014. "Only Resist: Feminist Ecological Citizenship and the Post-politics of Climate Change." *Hypatia* 29 (3): 617–33. https://doi.org/10.1111/hypa.12065.

Magnusdottir, Gunnhildur L., and Annica Kronsell. 2015. "The (In)Visibility of Gender in Scandinavian Climate Policy-Making." *International Feminist Journal of Politics* 17 (2): 308–26. https://doi.org/10.1080/14616742.2014.896661.

Malnes, Raino. 1995. "'Leader' and 'Entrepreneur' in International Negotiations." 1 (1): 87–112. https://doi.org/10.1177/1354066195001001005.

Matejova, Miriam, Stefan Parker, and Peter Dauvergne. 2018. "The politics of repressing environmentalists as agents of foreign influence." *Australian Journal of International Affairs* 72 (2): 145–62. https://doi.org/10.1080/10357718.2017.1421141.

Mazumdar, Arijit. 2018. "India's Soft Power Diplomacy under the Modi Administration: Buddhism, Diaspora and Yoga." *Asian Affairs* 49 (3): 468–91. https://doi.org/10.1080/03068374.2018.1487696.

Mehra, Malini. 2014. "The miseducation of Narendra Modi on climate change." *Climate Home News*, September 8. Accessed April 17, 2022. https://www.climatechangenews.com/2014/09/08/the-miseducation-of-narendra-modi-on-climate-change/.

Mehrishi, Mira. 2012. "Statement 18th COP." Accessed April 04, 2022. https://unfccc.int/resource/docs/cop18_cmp8_hl_statements/Statement%20by%20India%20%28COP%2018%29.pdf.

Member of an Indian Development and Climate NGO (Interviewee 2). 2019. Interview by A. Zürn. September 26. New Delhi.

Merchant, Carolyn. 1989. *Ecological Revolutions: Nature, Gender, and Science in New England*. Chapel Hill: University of North Carolina Press.

Michael, Arndt. 2013. *India's foreign policy and regional multilateralism*. Critical Studies of the Asia Pacific Series. Houndmills, Basingstoke, Hampshire: Palgrave Macmillan.

Michaelowa, Katharina, and Axel Michaelowa. 2011. "India in the international climate negotiations: from traditional nay-sayer to dynamic broker." 70. Accessed January 12, 2022. https://www.files.ethz.ch/isn/133820/WP70.pdf.

Michaelowa, Katharina, and Axel Michaelowa. 2012. "India as an emerging power in international climate negotiations." *Climate Policy* 12 (5): 575–90. https://doi.org/10.1080/14693062.2012.691226.

Mies, Maria, and Vandana Shiva. 1993. *Ecofeminism*. London: Zed Press.

Miller, Christopher P. 2020. "Soft Power and Biopower: Narendra Modi's 'Double Discourse' Concerning Yoga for Climate Change and Self-Care." *DHARM* 3 (1): 93–106. https://doi.org/10.1007/s42240-020-00068-w.

Ministry of Environment and Forests. 2011. "Annual Report 2010–2011." Accessed March 11, 2022. https://moef.gov.in/wp-content/uploads/2018/04/AR-EngVol2-1.pdf.

Ministry of Environment and Forests. 2012. "Annual Report 2011–12." Accessed March 24, 2022. https://moef.gov.in/wp-content/uploads/2018/04/AR-11-12-En.pdf.

Ministry of Environment, Forest and Climate Change. 2015. "Parampara: India's Culture of Climate Friendly Sustainable Practices." Accessed May 10, 2022. https://moef.gov.in/wp-content/themes/moef-green/ebook/parampara-ebook/Parampara_2016_Ebook.html#page/1.

Ministry of Environment, Forest and Climate Change. 2017. "Annual Report 2016–2017." Accessed May 22, 2022. http://ismenvis.nic.in/Database/MoEFCC_Annual_Report_2016-2017_17509.aspx.

Ministry of Environment, Forest and Climate Change. 2019. "Annual Report 2018–19." Accessed May 22, 2022. https://moef.gov.in/wp-content/uploads/2019/08/Annual-Report-2018-19-English.pdf.

Ministry of Environment, Forest and Climate Change. 2021. "Annual Report 2020–21." Accessed May 22, 2022. https://moef.gov.in/wp-content/uploads/2021/03/Environment-AR-English-2020-21.pdf.

Ministry of Environment, Forests and Climate Change. 2015. "Annual Report 2014–2015." Accessed April 18, 2022. https://moef.gov.in/wp-content/uploads/2018/04/EnvironmentAnnualReportEng.pdf.

Misra, Shalini. 2007. "Spirituality, Culture and the Politics of Environmentalism in India." *The Journal of Entrepreneurship* 16 (2): 131–45. https://doi.org/10.1177/097135570701600201.

Mitra, Amit. 1993. "Chipko: an unfinished mission." Accessed January 18, 2021. https://www.downtoearth.org.in/coverage/chipko-an-unfinished-mission-30883.

Modi, Narendra. 2011. "Convenient Action: Gujarat's Response to Challenges of Climate Change." Accessed April 10, 2022. https://www.narendramodi.in/ebooks/convenient-action-2.

Modi, Narendra. 2014. "Speech at United Nations General Assembly." http://mea.gov.in/PM_SpeechBook_2014-25/?page=46.

Modi, Narendra. 2015a. *Convenient action: Continuity for change*. Revised and expanded edition. Haryana, India: Lexis Nexis.

Modi, Narendra. 2015b. "PM's statement at the United Nations Summit for the adoption of Post-2015 Development Agenda." Accessed April 25, 2022. https://www.narendramodi.in/text-of-pm-s-statement-at-the-united-nations-summit-for-the-adoption-of-post-2015-development-agenda-332923.

Modi, Narendra. 2015c. "The rich world must take greater responsibility for climate change." *Financial Times*, November 29. Accessed May 22, 2022. https://www.ft.com/content/03a251c6-95f7-11e5-9228-87e603d47bdc.

Modi, Narendra. 2015d. “Speech India Pavilion at COP 21.” http://mea.gov.in/PM_SpeechBook_2014-25/?page=260.

Modi, Narendra. 2015e. “Statement by Prime Minister at COP21 Plenary.” Accessed April 29, 2022. https://unfccc.int/files/meetings/paris_nov_2015/application/pdf/cop21cmp11_leaders_event_india.pdf.

Modi, Narendra. 2017. “Culture and music can play a vital role in connecting the country: PM Modi.” Accessed May 22, 2022. https://www.narendramodi.in/pm-modi-to-inaugurate-spic-macay-conference-at-iit-delhi-via-vc-535764.

Modi, Narendra. 2019. “PM’s remarks at Climate Action Summit 2019 during 74th session of UNGA.” Accessed June 24, 2022. https://www.pmindia.gov.in/en/news_updates/pms-remarks-at-climate-action-summit-2019-during-74th-session-of-unga/.

Modi, Narendra. 2021. “National Statement by Prime Minister Shri Narendra Modi at COP26 Summit in Glasgow.” Accessed May 22, 2022. https://www.mea.gov.in/Speeches-Statements.htm?dtl/34466/National+Statement+by+Prime+Minister+Shri+Narendra+Modi+at+COP26+Summit+in+Glasgow.

Mohan, Aniruddh. 2017. “From Rio to Paris: India in Goobal Climate Politics.” ORF Occastional Paper 130. Accessed April 12, 2022. https://www.orfonline.org/wp-content/uploads/2017/12/ORF_Occasional_Paper_130_Climate_Mohan.pdf.

Mohan, Aniruddh, and Timon Wehnert. 2019. “Is India pulling its weight? India’s nationally determined contribution and future energy plans in global climate policy.” *Climate Policy* 19 (3): 275–82. https://doi.org/10.1080/14693062.2018.1503154.

Mukherjee, Rohan. 2017. “India’s Inernational Development Program.” In *The Oxford handbook of Indian foreign policy*, 173–87. Oxford: Oxford University Press.

Müller-Brandeck-Bocquet, Gisela, Manuel Pietzko, and Anja Zürn. 2018. “Die Zukunft der Europäischen Union: Potenziale besser nutzen.” Würzburger Jean-Monnet-Papers 1.

Nachiappan, Karthik. 2019. “Agenda-setting from behind: India and the framework convention on climate change.” *India Review* 18 (5): 552–67. https://doi.org/10.1080/14736489.2019.1703365.

Nag, Ratna de. 2014. “Speech.” In Lok Sabha 2014, 205–7.

Nagel, Joane. 2016. *Gender and Climate Change: Impacts, Science, Policy.* New York: Routledge Taylor & Francis Group.

Nair, Usha. 2019. “COP 25 Meeting, Madrid, Spain: A Resume.” *Roshni.* http://www.aiwc.org.in/roshni/ROSHNI-oct-dec-2019.pdf.

Nandi, Jayashree. 2021. “Didn’t introduce ‘phase down’ in COP26 statement, clarifies India.” *Hindustan Times*, November 18. Accessed July 03, 2022. https://www.hindustantimes.com/india-news/didnt-introduce-phase-down-in-cop-26-statement-clarifies-india-101637196023549.html.

Narain, Sunita. 2010a. “All is not well: climate negotiations in a new avatar.” Accessed February 25, 2022. https://www.downtoearth.org.in/blog/all-is-not-well-climate-negotiations-in-a-new-avatar-139.

Narain, Sunita. 2010b. “It’s a deal by and for the rich polluters.” *The Times of India*, December 17. https://timesofindia.indiatimes.com/home/science/its-a-deal-by-and-for-the-rich-polluters/articleshow/7115146.cms.

Narain, Sunita, and Chandra Bhushan. 2015. “Capitan America: US climate goals: a reckoning.” Accessed April 29, 2022. https://cdn.cseindia.org/userfiles/Capitan-America-Report.pdf.

Narasimham, Thota. 2014. "Speech." In Lok Sabha 2014, 152–62.

Narlikar, Amrita. 2017. "India's role in global governance: a Modification?" *International Affairs* 93 (1): 93–111. https://doi.org/10.1093/ia/iiw005.

Natarajan, Jayanthi. 2011. "Remarks by Smt. Jayanthi Natarajan, Hon'ble Minister for Environment & Forests." Accessed March 18, 2022. https://www.cseindia.org/-indian-environment-minister-jayanthi-natarajan-gives-hard-hitting-speech-receives-standing-ovation-3556.

Natarajan, Jayanthi. 2013. "Opening Statement by Smt. Jayanthi Natarajan. Minister Environment and Forests India." https://unfccc.int/files/meetings/warsaw_nov_2013/statements/application/pdf/cop19_hls_india.pdf.

Never, Babette. 2012. "Who Drives Change? Comparing the Evolution of Domestic Climate Governance in India and South Africa." *The Journal of Environment & Development* 21 (3): 362–87. https://doi.org/10.1177/1070496512449821.

Never, Babette. 2015. *Knowledge Systems and Change in Climate Governance*. London: Routledge.

NITI Aayog. 2017. "INDIA. Three Year Action Agenda 2017–18 to 2019–20." Accessed January 30, 2022. https://www.niti.gov.in/sites/default/files/2018-12/India_ActionAgenda.pdf.

Nonhoff, Martin. 2006. *Politischer Diskurs und Hegemonie: Das Projekt »Soziale Marktwirtschaft«*. Bielefeld: transcript Verlag.

Nonhoff, Martin. 2010. "Hegemonieanalyse: Theorie, Methode und Forschungspraxis." In *Handbuch sozialwissenschaftliche Diskursanalyse*, edited by Reiner Keller. 4. Aufl., 299–331. Wiesbaden: VS Verlag für Sozialwissenschaften.

Nonhoff, Martin. 2019. "Hegemony Analysis: Theory, Methodology and Research Practice." In *Discourse, Culture and Organization: Inquiries into Relational Structures of Power*, edited by Tomas Marttila, 63–104. Cham: Springer International Publishing.

Office of the Press Secretary. 2014. "U.S.-China Joint Announcement on Climate Change." News release. November 12. Accessed April 24, 2022. https://obamawhitehouse.archives.gov/the-press-office/2014/11/11/us-china-joint-announcement-climate-change.

Pachauri, Rajendra K. *Himalayan Glaciers Bucking the Trend*. With the assistance of P. Bagla, 2009. NDTV. Accessed March 11, 2022. https://www.youtube.com/watch?v=bnYmQjFoNCs.

Pachauri, Rajendra K. 2010. "Speech Delhi Sustainable Development Summit 2010." https://wsds.teriin.org/2010/RKP_Speech.pdf.

Pai, Sudha, and Avinash Kumar. 2015. "Understanding the BJP's Vicotry in Uttar Pradesh." In *India's 2014 elections: A modi-led BJP sweep*, edited by Paul Wallace, 119–38. New Delhi: SAGE.

Pandey, Kundan. 2019. "Finally, both BJP and Congress manifestos talk climate change." https://www.downtoearth.org.in/news/general-elections-2019/finally-both-bjp-and-congress-manifestos-talk-climate-change-63913.

Pathak, Subhash. 2021. "Bihar govt to work on action plan to deal with climate crisis." *The Hindustan Times*, August 17. Accessed May 22, 2022. https://www.hindustantimes.com/cities/patna-news/bihar-govt-to-work-on-action-plan-to-deal-with-climate-crisis-101629187292657.html.

Pathak, Swapna, and Christie Parris. 2021. "India's diplomatic discourse and development dilemma in the international climate change regime." *India Review* 20 (1): 1–28. https://doi.org/10.1080/14736489.2021.1875699.

Pillai, Aditya V., and Navroz K. Dubash. 2021. "The limits of opportunism: the uneven emergence of climate institutions in India." *Environmental Politics* 30 (sup1): 93–117. https://doi.org/10.1080/09644016.2021.1933800.

Plagemann, Johannes, and Miriam Prys-Hansen. 2020. "'Responsibility', change, and rising powers' role conceptions: comparing Indian foreign policy roles in global climate change negotiations and maritime security." *International Relations of the Asia-Pacific* 20 (2): 275–305. https://doi.org/10.1093/irap/lcy028.

Planning Commission Government of India. 2008. "11th Five Year Plan 2007–2012. Inclusive Growth." Accessed February 08, 2022. https://niti.gov.in/planningcommission.gov.in/docs/plans/planrel/fiveyr/11th/11_v1/11th_vol1.pdf.

Planning Commission Government of India. 2013a. "12th Five Year Plan (2012–2017). Social Sectors." Accessed January 30, 2022. https://niti.gov.in/planningcommission.gov.in/docs/plans/planrel/fiveyr/12th/pdf/12fyp_vol3.pdf.

Planning Commission Government of India. 2013b. "Twelth Five Year Plan (2012–2017). Faster, More Inclusive and Sustainable Growth." Accessed January 30, 2022. https://niti.gov.in/planningcommission.gov.in/docs/plans/planrel/fiveyr/12th/pdf/12fyp_vol1.pdf.

Plumwood, Val. 1993. *Feminism and the Mastery of Nature.* Reprinted 1997. Transferred to digital print 2003. Feminism for today. London: Routledge.

Press Information Bureau. 2015. "India's Intended Nationally Determined Contribution is Balanced and Comprehensive: Environment Minister." News release. October 2. Accessed April 24, 2022. https://pib.gov.in/newsite/printrelease.aspx?relid=128403.

Press Information Bureau. 2018. "PM inaugurates first assembly of the International Solar Alliance." Accessed May 22, 2022. https://pib.gov.in/Pressreleaseshare.aspx?PRID=1548295.

Priyadarshini, Priya, and Purushothaman C. Abhilash. 2019. "Promoting tribal communities and indigenous knowledge as potential solutions for the sustainable development of India." *Environmental Development* 32. https://doi.org/10.1016/j.envdev.2019.100459.

Prügl, Elisabeth. 2021. "Untenable dichotomies: de-gendering political economy." *Review of International Political Economy* 28 (2): 295–306. https://doi.org/10.1080/09692290.2020.1830834.

Pulver, Simone. 2012. "Corporate responses to climate change in India." In Dubash 2012, 254–65.

Qi, Xinran. 2011. "The rise of BASIC in UN climate change negotiations." *South African Journal of International Affairs* 18 (3): 295–318. https://doi.org/10.1080/10220461.2011.622945.

Raghunandan, D. 2020. "India's climate policy. Paris Agreement, NDC and after." In Ciecierska-Holmes, Jörgensen, Ollier, and Raghundandan 2020, 203–24.

Raina, Gautam, and Sunanda Sinha. 2019. "Outlook on the Indian scenario of solar energy strategies: Policies and challenges." *Energy Strategy Reviews* 24: 331–41. https://doi.org/10.1016/j.esr.2019.04.005.

Raina, V. K. 2009. "Himalayan Glaciers A State-of-Art Review of Glacial Studies, Glacial Retreat and Climate Change." Accessed March 11, 2022. http://www.indiaenvironmentportal.org.in/files/MoEDiscussionPaper.pdf.

Raj, Kalyani. 2015. "Low Carbon Development Strategies." Accessed May 26, 2022. https://www.inforse.org/europe/pdfs/Conf_15_COP21_11_LCD_Women_Kalyani_Raj_AIWC_India.pdf.

Rajamani, Lavanya. 2009. "India and Climate Change: What India Wants, Needs, and Needs to Do." *India Review* 8 (3): 340–74. https://doi.org/10.1080/14736480903116842.

Ramachandran, R. 2012. "India in the Intergovernmental Panel on Climate Change." In Dubash 2012, 63–77.

Ramesh, Jairam. 2010. "Speech by Mr. Jairam Ramesh, Minister of Environment & Forests, India and Leader of Indian Delegation." Accessed March 10, 2022. https://unfccc.int/files/meetings/cop_16/statements/application/pdf/101209_cop16_hls_india.pdf.

Ramesh, Jairam. 2015. *Green Signals: Ecology, Growth, and Democracy in India.* New Delhi: Oxford Univ. Press.

Rangarajan, Mahesh. 2009. "Striving for a Balance: Nature, Power, Science and India's Indira Gandhi, 1917-1984." *Conservation & Society* 4 (4): 299–312.

Rastogi, Namrata P. 2011. "Winds of Change: India's Emerging Climate Strategy." *The International Spectator* 46 (2): 127–41. https://doi.org/10.1080/03932729.2011.576179.

Rathee, Dhruv. 2019. "Environment is the Modi government's most under-reported disaster." Accessed May 09, 2022. https://ecologise.in/2019/04/14/dhruv-rathee-environment-is-the-modi-governments-most-under-reported-disaster/.

Rodrigues, Valerian. 2014. "Elections and Civil Society in India." Würzburger Arbeitspapiere zur Politikwissenschaft und Sozialforschung 4. https://opus.bibliothek.uni-wuerzburg.de/opus4-wuerzburg/frontdoor/deliver/index/docId/10500/file/WAPS4_Elections_Civil_Society_India_pdfa.pdf.

Sahgal, Gita. 2020. "Hindutva Past and Present." *Feminist Dissent*, no. 5: 19–49. https://doi.org/10.31273/fd.n5.2020.757.

Saldanha, Leo. 2018. "Ajit Naik's murder: Urgent need to investigate forces behind cycle of violence in Karnataka." Accessed June 08, 2022. https://counterview.org/2018/08/09/ajit-naiks-murder-urgent-need-to-investigate-forces-behind-cycle-of-violence-in-karnataka/.

Sandilands, Catriona. 1999. *The Good-Natured Feminist: Ecofeminism and the Quest for Democracy.* Minneapolis: University of Minnesota Press.

Sangomla, Akshit. 2021. "PM's climate change council has not met in almost 7 years. Who is, then, advising the PM for his climate goal announcements?" Accessed January 19, 2022. https://www.downtoearth.org.in/news/climate-change/pm-s-climate-change-council-has-not-met-in-almost-7-years-80369.

Saran, Samir, and Aled Jones. 2017. *India's Climate Change Identity.* Cham: Springer International Publishing.

Sarkar, Urvashi. 2018. "UN's Green Award for Modi Comes Despite Criticism of Environmental Record." *The Wire*, October 3. Accessed June 07, 2022. https://thewire.in/environment/un-green-award-narendra-modi-controversial-environmental-record.

Saryal, Rajnish. 2018. "Climate Change Policy of India." *South Asia Research* 38 (1): 1–19. https://doi.org/10.1177/0262728017745385.

Sauer, Natalie. 2019. "India elects: Congress takes on Modi with green promises." Accessed June 24, 2022. https://www.climatechangenews.com/2019/04/10/india-elects-congress-takes-modi-green-promises/.

Saxena, K. B. 2019. "Compensatory Afforestation Fund Act and Rules: Deforestation, Tribal Displacement and an Alibi for Legalised Land Grabbing." *Social Change* 49 (1): 23–40. https://doi.org/10.1177/0049085718821766.
Schreurs, Miranda, and Elim Papadakis. 2009. *The a to Z of the Green Movement*. With the assistance of E. Papadakis. Lanham: Scarecrow Press.
Schroeder, Heike, and Heather Lovell. 2012. "The role of non-nation-state actors and side events in the international climate negotiations." *Climate Policy* 12 (1): 23–37. https://doi.org/10.1080/14693062.2011.579328.
Sengupta, Sandeep. 2012. "International climate negotiations and India's role." In Dubash 2012, 101–17.
Sengupta, Sandeep. 2019. "India's Engagement in Global Climate Negotiations from Rio to Paris." In Dubash 2019, 114–41.
Sengupta, Sucharita. 2018. "The climate change discourse in India: Is a social movement possible?" In *Intractable Conflicts in Contemporary India: Narratives and Social Movements*, edited by Savyasaachi, 219–33. London: Routledge India.
Senthilkumar, S. 2021. "Speech." In Lok Sabha 2021.
Sethi, Nitin. 2009. "Jairam for Major Shift at Climate Talks." Accessed January 24, 2022. https://timesofindia.indiatimes.com/india/jairam-for-major-shift-at-climate-talks/articleshow/5136979.cms.
Sethi, Nitin. 2010. "India takes step to dilute climate change stand." *The Economic Times*, December 11. Accessed March 15, 2022. https://economictimes.indiatimes.com/news/politics-and-nation/india-takes-step-to-dilute-climate-change-stand/articleshow/7080770.cms?from=mdr.
Sethi, Nitin. 2013a. "A fading shade of green." *The Economic Times*, April 25. Accessed April 06, 2022. https://economictimes.indiatimes.com/news/environment/developmental-issues/a-fading-shade-of-green/articleshow/20236552.cms.
Sethi, Nitin. 2013b. "Emission impossible: Weather's turning on climate change." *The Economic Times*, June 5. Accessed April 06, 2022. https://economictimes.indiatimes.com/news/environment/global-warming/emission-impossible-weathers-turning-on-climate-change/articleshow/20436278.cms.
Shani, Giorgio. 2021. "Towards a Hindu Rashtra: Hindutva, religion, and nationalism in India." *Religion, State and Society* 49 (3): 264–80. https://doi.org/10.1080/09637494.2021.1947731.
Shiva, Vandana. 1988. *Staying Alive: Women, Ecology and Development*. London: Zed.
Shiva, Vandana, and Jayanto Bandyopadyay. 1986. *Chipko: India's civilisational response to the forest crisis*. Delhi: INTACH.
Shrivastava, Sambhav K. 2014. "Too little in environment ministry's kitty." Accessed May 09, 2022. https://www.downtoearth.org.in/news/too-little-in-environment-ministrys-kitty-45226.
Singh, Chandni, Divya Solomon, and Nitya Rao. 2021. "How does climate change adaptation policy in India consider gender? An analysis of 28 state action plans." *Climate Policy* 21 (7): 958–75. https://doi.org/10.1080/14693062.2021.1953434.
Singh, Manmohan. 2009. "PM's statement in Lok Sabha on the debate on the PM's recent visits abroad on July 29, 2009." Accessed January 08, 2021. https://archivepmo.nic.in/drmanmohansingh/content_print.php?nodeid=777&nodetype=2.

Singh, Manmohan. 2010. "PM's Speech at the Inauguration of the 10th Delhi Sustainable Development Summit." https://wsds.teriin.org/2010/PM_Speech.pdf.

Singh, Manmohan. 2011. "PM's address at the 11th Delhi Sustainable Development Summit." Accessed March 25, 2022. https://archivepmo.nic.in/drmanmohansingh/speech-details.php?nodeid=979.

Singh, Manmohan. 2012. "PM's address at the12th Delhi Sustainable Development Summit." https://archivepmo.nic.in/drmanmohansingh/content_print.php?nodeid=1131&nodetype=2.

Singh, Manmohan. 2013. "PM's Address at the Inaugural Session of Delhi Sustainable Development Summit." Accessed April 06, 2022. https://wsds.teriin.org/2013/pdf/pm_speech2013.pdf.

Singh, Shivani, Aaron Sheldrick, and Noah Browning. 2021. "'Down' and 'out'? COP26 wording clouds way ahead on climate." *Reuters*, November 15. Accessed June 28, 2022. https://www.reuters.com/business/cop/business-usual-global-fossil-fuel-firms-now-after-un-climate-deal-2021-11-15/.

Singh Khadka, Navin. 2021. "COP26: Did India betray vulnerable nations?" *BBC*, November 16. Accessed July 02, 2022. https://www.bbc.com/news/world-asia-india-59286790.

Singh Panwar, Tikender, and Aravind Unni. 2019. "'Urban' Vision in 2019 Manifestoes of Parties Leaves a Lot to be Desired." Accessed June 24, 2022. https://www.newsclick.in/urban-vision-2019-manifestoes-parties-leaves-lot-desired.

Sinha, Subir. 2021. "'Strong leaders', authoritarian populism and Indian developmentalism: The Modi moment in historical context." *Geoforum* 124: 320–33. https://doi.org/10.1016/j.geoforum.2021.02.019.

Statista. 2022. "Solar energy capacity in India from 2009 to 2021." Accessed September 05, 2022. https://www.statista.com/statistics/865760/india-solar-energy-capacity/.

Subramanian, Samanth. 2015. "India's war on Greenpeace." *The Guardian*, August 11. Accessed May 08, 2022. https://www.theguardian.com/world/2015/aug/11/indias-war-on-greenpeace.

Sud, Nikita. 2022. "The Actual Gujarat Model: Authoritarianism, Capitalism, Hindu Nationalism and Populism in the Time of Modi." *Journal of Contemporary Asia* 52 (1): 102–26. https://doi.org/10.1080/00472336.2020.1846205.

Sudhi, K. S. 2021. "Action Plan on climate change not yet ready." *The Hindu*, October 21. Accessed May 27, 2022. https://www.thehindu.com/news/national/kerala/action-plan-on-climate-change-not yet-ready/article37112274.ece.

Sule, Supriya. 2014. "Speech." In Lok Sabha 2014, 180–83.

Swarnakar, Pradip. 2019. "Climate Change, Civil Society, and Social Movement in India." In Dubash 2019, 253–72.

Swarnakar, Pradip, Rajshri Shukla, and Jeffrey Broadbent. 2022. "Beliefs and Networks: Mapping the Indian Climate Policy Discourse Surrounding the Paris Climate Change Conference in 2015." *Environmental Communication* 16 (2): 145–62. https://doi.org/10.1080/17524032.2021.1973528.

Swenden, Wilfried, and Rekha Saxena. 2020. "Environmental competencies in India's federal system." In Ciecierska-Holmes, Jörgensen, Ollier, and Raghundandan 2020, 17–38.

Thaker, Jagadish, and Anthony Leiserowitz. 2014. "Shifting discourses of climate change in India." *Climatic Change* 123 (2): 107–19. https://doi.org/10.1007/s10584-014-1059-6.

The Economic Times. 2014. "India not supporting ex-ante review of INDCs 'regressive': Centre for Science and Environment." *The Economic Times*, December 10. Accessed April 18, 2022. https://economictimes.indiatimes.com/news/politics-and-nation/india-not-supporting-ex-ante-review-of-indcs-regressive-centre-for-science-and-environment/articleshow/45459484.cms.

The Economic Times. 2015. "India must show how it has preserved environment over ages: PM Narendra Modi." *The Economic Times*, April 24. Accessed May 22, 2022. https://economictimes.indiatimes.com/news/politics-and-nation/india-must-show-how-it-has-preserved-environment-over-ages-pm-narendra-modi/articleshow/47043842.cms.

The Energy and Resources Institute. 2022. "About WSDS." https://wsds.teriin.org/about-wsds.php.

The Guardian. 2010. "Rajendra Pachauri says glaciers mistake in IPCC report was 'human failure'." May 14. https://www.theguardian.com/environment/2010/may/14/rajendra-pachauri-defends-ipcc-report.

The Times of India. 2015. "PM leaves for Paris to attend global climate change meet." *The Times of India*, November 29. Accessed May 26, 2022. https://timesofindia.indiatimes.com/india/pm-leaves-for-paris-to-attend-global-climate-change-meet/articleshow/49970181.cms.

The Wire. 2017. "What Trump's Decision to Pull the US Out of the Paris Climate Deal Means." *The Wire*, June 2. Accessed May 22, 2022. https://thewire.in/environment/trump-paris-agreement-climate-change-figueres.

Tickner, J. A. 1993. "States and Markets: An Ecofeminist Perspective on International Political Economy." *International Political Science Review* 14 (1): 59–69. https://doi.org/10.1177/019251219301400104.

Times of India. 2015. "'Balanced' Paris climate deal an 'important achievement' for India: Prakash Javadekar." *Times of India*, December 12. Accessed May 26, 2022. http://timesofindia.indiatimes.com/articleshow/50153857.cms?utm_source=contentofinterest&utm_medium=text&utm_campaign=cppst.

Tripathi, Sibu. 2019. "Lok Sabha elections 2019: This poll season the environment focus of parties is foggy at best." Accessed June 24, 2022. https://indianexpress.com/elections/lok-sabha-elections-narendra-modi-rahul-gandhi-bjp-congress-manifesto-pollution-environment-5651539/.

UNFCCC. 2012. "Promoting gender balance and improving the participation of women in UNFCCC negotiations and in the representation of Parties in bodies established pursuant to the Convention or the Kyoto Protocol." https://unfccc.int/sites/default/files/cop18_gender_balance.pdf.

UNFCCC. 2014. "Report of the Conference of the Parties on its nineteenth session, held in Warsaw from 11 to 23 November 2013." https://unfccc.int/resource/docs/2013/cop19/eng/10a01.pdf.

United Nations. 1972. "Report of the United Nations Conference on the Human Environment." http://www.un-documents.net/aconf48-14r1.pdf.

United Nations Conference on Environment & Development. 1992. "Agenda 21." Accessed January 11, 2021. https://sustainabledevelopment.un.org/content/documents/Agenda21.pdf.

United Nations Environment Programme. 2022. "Narendra Modi—Policy Leadership: Prime Minister of India." Accessed May 22, 2022. https://www.unep.org/championsofearth/node/50.
United Nations Framework Convention on Climate Change. 2000. "List of Participants: Conference of the Parties 6." Accessed May 10, 2022. https://unfccc.int/sites/default/files/resource/docs/cop6/inf02.pdf.
United Nations Framework Convention on Climate Change. 2001. "List of Participants: Conference of the Parties 7." Accessed May 10, 2022. https://unfccc.int/sites/default/files/resource/docs/cop7/inf04.pdf.
United Nations Framework Convention on Climate Change. 2002. "List of Participants: Conference of the Parties 8." Accessed May 10, 2022. https://unfccc.int/sites/default/files/resource/docs/cop8/inf02.pdf.
United Nations Framework Convention on Climate Change. 2003. "List of Participants: Conference of the Parties 9." Accessed May 10, 2022. https://unfccc.int/sites/default/files/resource/docs/cop9/inf01.pdf.
United Nations Framework Convention on Climate Change. 2004. "List of Participants: Conference of the Parties 10." https://unfccc.int/sites/default/files/resource/docs/cop10/inf03.pdf.
United Nations Framework Convention on Climate Change. 2005. "List of Participants: Conference of the Paries 11." Accessed May 10, 2022. https://unfccc.int/sites/default/files/resource/docs/2005/cop11/eng/inf02p01.pdf.
United Nations Framework Convention on Climate Change. 2006. "List of Participants: Conference of the Parties 12." Accessed May 10, 2022. https://unfccc.int/sites/default/files/resource/docs/2006/cop12/eng/inf01.pdf.
United Nations Framework Convention on Climate Change. 2007. "List of Participants: Conference of the Parties 13." Accessed May 10, 2022. https://unfccc.int/sites/default/files/resource/docs/2007/cop13/eng/inf01p01.pdf.
United Nations Framework Convention on Climate Change. 2008. "List of Participants: Conference of the Parties 14." Accessed May 10, 2022. https://unfccc.int/sites/default/files/resource/docs/2008/cop14/eng/inf01p01.pdf.
United Nations Framework Convention on Climate Change. 2009. "List of Participants: Conference of the Parties 15." Accessed May 10, 2022. https://unfccc.int/sites/default/files/resource/docs/2009/cop15/eng/inf01p01.pdf.
United Nations Framework Convention on Climate Change. 2010. "List of Participants: Conference of the Parties 16." Accessed May 10, 2022. https://unfccc.int/sites/default/files/resource/docs/2010/cop16/eng/inf01p01.pdf.
United Nations Framework Convention on Climate Change. 2011. "List of Participants: Conference of the Parties 17." Accessed May 10, 2022. https://unfccc.int/sites/default/files/resource/docs/2011/cop17/eng/inf03p01.pdf.
United Nations Framework Convention on Climate Change. 2012. "List of Participants: Conference of the Parties 18." Accessed May 10, 2022. https://unfccc.int/sites/default/files/resource/docs/2012/cop18/eng/inf02.pdf.
United Nations Framework Convention on Climate Change. 2013. "List of Paricipants: Conference of the Parties 19." Accessed May 10, 2022. https://digitallibrary.un.org/record/768925?ln=en.

United Nations Framework Convention on Climate Change. 2014. "List of Participants: Conference of the Parties 20." Accessed May 10, 2022. https://unfccc.int/sites/default/files/resource/docs/2014/cop20/eng/misc01p01.pdf.

United Nations Framework Convention on Climate Change. 2015. "List of Participants: Conference of the Parties 21." Accessed May 10, 2022. https://unfccc.int/sites/default/files/resource/docs/2015/cop21/eng/inf03p01.pdf.

United Nations Framework Convention on Climate Change. 2016. "List of Participants: Conference of the Parties 22." Accessed May 10, 2022. https://unfccc.int/resource/docs/2016/cop22/eng/inf03p01.pdf.

United Nations Framework Convention on Climate Change. 2017. "List of Participants: Conference of the Parties 23." Accessed May 10, 2022. https://unfccc.int/sites/default/files/resource/inf04_0.pdf.

United Nations Framework Convention on Climate Change. 2018. "List of Participants: Conference of the Parties 24." Accessed May 10, 2022. https://unfccc.int/sites/default/files/resource/i03.pdf.

United Nations Framework Convention on Climate Change. 2019. "List of Participants: Conference of the Parties 25." Accessed May 10, 2022. https://unfccc.int/sites/default/files/resource/cp_inf4.pdf.

United Nations Framework Convention on Climate Change. 2021a. "Glasgow Climate Pact." Accessed June 28, 2022. https://unfccc.int/sites/default/files/resource/cop26_auv_2f_cover_decision.pdf.

United Nations Framework Convention on Climate Change. 2021b. "List of Participants: Conference of the Parties 26." Accessed May 10, 2022. https://unfccc.int/sites/default/files/resource/cp2021_inf03p01.pdf.

United Nations General Assembly. 2015. "Transforming our world: the 2030 Agenda for Sustainable Development." Accessed March 17, 2022. https://www.un.org/en/development/desa/population/migration/generalassembly/docs/globalcompact/A_RES_70_1_E.pdf.

United States of America. 2016. "Intended Nationally Determined Contribution." Accessed April 29, 2022. https://www4.unfccc.int/sites/ndcstaging/PublishedDocuments/United%20States%20of%20America%20First/U.S.A.%20First%20NDC%20Submission.pdf.

Vanhala, Lisa, and Cecilie Hestbaek. 2016. "Framing Climate Change Loss and Damage in UNFCCC Negotiations." *Global Environmental Politics* 16 (4): 111–29. https://doi.org/10.1162/GLEP_a_00379.

Vardhan, Harsh. 2017. "Climate Change is The Result of Unsustainable Lifestyles and Consumption Patterns; India's Glorious Traditions have Stood for Sustainable Lifestyles." Accessed May 22, 2022. https://pib.gov.in/PressReleasePage.aspx?PRID=1510045.

Vardhan, Harsh. 2018. "India Country Statement For COP-24." Accessed June 06, 2022. https://unfccc.int/sites/default/files/resource/INDIA_cop24cmp14cma1-3.pdf.

Vaughan, Adam. 2021. "COP26: Why India's 2070 net-zero pledge is better news than it sounds." *New Scientist*, November 2. Accessed July 02, 2022. https://www.newscientist.com/article/2295762-cop26-why-indias-2070-net-zero-pledge-is-better-news-than-it-sounds/.

Vihma, Antto. 2011. "India and the Global Climate Governance: Between Principles and Pragmatism." *The Journal of Environment & Development* 20 (1): 69–94. https://doi.org/10.1177/1070496510394325.

Vishnoi, Anubhuti, and Dipanjan Roy Chaudhury. 2017. "Dismissing Donald Trump's snarl, India vows to stay on the green path." *The Economic Times*, June 3. Accessed May 22, 2022. https://economictimes.indiatimes.com/news/politics-and-nation/dismissing-donald-trumps-snarl-india-vows-to-stay-on-the-green-path/articleshow/58969162.cms?from=mdr.

Wagner, Christian. 2019. "Der Anschlag in Kaschmir und seine Folgen." Accessed April 10, 2022. https://www.swp-berlin.org/publikation/der-anschlag-in-kaschmir-und-seine-folgen.

Warren, Karen J. 1988. "Toward an Ecofeminist Ethic." *Studies in the Humanities* 15 (2): 140–56.

Weller, Ines. 2019. "Gender & Environment: Geschlterforschung im Kontext Ökologie, Umwelt, Nachhaltigkeit." In *Handbuch Interdisziplinäre Geschlechterforschung*, edited by Beate Kortendiek, Birgit Riegraf, and Katja Sabisch, 873–80. Geschlecht und Gesellschaft. Wiesbaden: Springer Fachmedien Wiesbaden.

World Commission on Environment and Development. 1987. "Our Common Future." Accessed January 11, 2021. file:///C:/Users/anz22ss/Downloads/A_42_427-EN.pdf.

Yadav, Bhupender. 2021a. "COP Diary | Time to unite against climate change." Accessed July 02, 2022. https://bhupenderyadav.in/blog/time-to-unite-against-climate-change.

Yadav, Bhupender. 2021b. "COP26 Diary 01–11–2021." Accessed July 03, 2021. https://bhupenderyadav.in/blog/cop26-diary-01-11-2021.

Yadav, Bhupender. 2021c. "COP26 Diary 02–11–2021 | How India can meet its challenges." Accessed July 02, 2021. https://bhupenderyadav.in/blog/cop26-diary-02-11-2021.

Ylä-Anttila, Tuomas, and Pradip Swarnakar. 2017. "Crowding-in: How Indian Civil Society Organizations Began Mobilizing Around Climate Change." *The British journal of sociology* 68 (2): 273–92. https://doi.org/10.1111/1468-4446.12251.

Zürn, Anja. 2017. "COP 23: Fortschritte zur Umsetzung des Pariser Klimaabkommens." Accessed March 12, 2021. https://maineuropa.de/cop-23-bonn-fortschritte-zur-umsetzung-des-pariser-klimaabkommens-talanoa.

Dismantling the EU's Gendered Climate Policy Identity

4

The EU climate change policy was gender-blind for a long time (Allwood 2014, 3, 2022, 8) and to a large extent it still is (Heffernan et al. 2021, 5). However, as we already reconstructed in this project, this does not automatically mean that an actor's discursively produced identity and justification of its climate change policy is therefore not gendered. Rather, Sohvi Kangasluoma (2020) has shown, gender structures and gendered value systems can indeed shape climate discourse and are used—consciously or unconsciously—when it comes to justifying climate policy. What exactly that means for the EU's climate change policy and which role gendered discourse patterns play in constructing the EU's climate identity, will be explored in this chapter.

The EU's identity is constructed and represented by the society that negotiates it. Although the EU cannot be equated with a nation state such as India, countless studies show that it can still develop an identity (Krzyżanowski 2015; Risse 2010; Diez 2005; Wæver 1996). As already shown, the unfulfilled society that represents the identity of a state should not be regarded as homogeneous (cf. Sect. 2.2.3)—which particularly applies to the EU and its *sui generis* character. The sum of the hegemonic articulations of the various actors, which will be derived later, should instead be understood as such a society.

The study of the Indian climate discourse has shown that it is worthwhile to deconstruct the discourse by means of a hegemonic analysis and to discuss its significance through the ecofeminist lens when trying to reconstruct an actor's climate policy identity. The methodology of reconstructing the strategemes used by hegemonic actors when articulating their demands in order to address both

---

Supplementary Information The online version contains supplementary material available at https://doi.org/10.1007/978-3-658-46774-6_4.

A. Zürn, *Of Climate and Gender*, https://doi.org/10.1007/978-3-658-46774-6_4

the content and the strategic dimension of the EU climate discourse will therefore be applied again (cf. Sects. 4.2.1–4.2.9). Afterwards, Sect. 4.3 will discuss the question of how the EU climate identity is gendered and to what extent these structures are simultaneously used to justify the specific EU climate policy when formulating answers to combat climate change—while other ideas are not deliberated on. This chapter also contains a table that gives an overview of the results of the analysis, that is: of different hegemonic projects, (re-)producing actors and hegemonies (cf. Table 4.3). The applied critical ecofeminist approach will not only open the view to the mostly implicit gender structures but also provide starting points for a more inclusive climate policy.

First, however, the position of the EU, its priorities and the political environment—within the EU and externally—needs to be considered in order to identify the key demands and important actors of the EU climate discourse up to the beginning of the research period in 2010. This results in the overview of the actors to be analysed (for a list of the actors analysed, see the appendix in the electronic supplementary material), although the circle may well expand in the course of the research period, as actors may become increasingly important or new actors may emerge, such as the grassroots movement Fridays for Future.

The genesis of the EU climate change policy was determined by several important factors: The climate change discussions and negotiations on the global level, non-governmental actors, and of course the member states as well as EU institutions (Skjærseth 2017, 47). It is the responsibility of these EU institutions to implement the EU's climate policy in the framework of its multi-level-governance system characterised by both its supranational and intergovernmental elements. Of course, the governance structure of EU climate policy is much more differentiated today than when questions of climate change first began to be dealt with. Politically, the climate sector has been separated from the environmental sector, which was initially also the umbrella for aspects of climate change policy. Therefore, examining the multi-level-governance system of the Union is important in deducing the central actors of the discourse under study.

Reconstructing the EU's climate discourse in Sects. 4.2.1 to 4.2.9 with the help of hegemony analysis makes a reconstruction of the historical context and the main actors of the discursive contributions necessary (cf. step 4 described in Sect. 2.3). Hence, the EU climate policy prior to the research period and the current governance structures will be reconstructed in the following chapters.

4.1 The EU's Climate Governance and Policy until 2010

The reconstruction of the EU's climate policy and the governance structures shaping it requires several steps. Global climate policy and the climate targets set within this framework often result in knock-on effects for EU climate policy and its actors who discuss and attempt to achieve the goals. To analyse the current EU climate discourse, these aspects are examined with a focus on major past events which are still relevant for the period under study. Hence, an overview will first be generated covering the most important milestones in global climate policy paving the way for the understanding of the EU's role and its key decisions up to Copenhagen (Sect. 4.1.1). The EU has always followed the logic of accepting or pledging targets at the global level and then translating these internally into concrete measures and differentiated targets with the aim of burden-sharing. The resulting initiatives and measures are reconstructed in Sect. 4.1.2 because they form an important part of the context of the EU discourse. To understand the actions taken and compromises made to fight climate change, the EU's multi-level-governance will be reconstructed in Sect. 4.1.3. This chapter allows the creation of an overview of the manifold actors trying to shape the EU climate change discourse. Their contributions will then be scrutinised in the subsequent hegemony analysis.

For experts of EU climate policy, most aspects of the following three chapters will not be new, but they are necessary in order to understand the discussion of the findings of the ecofeminist hegemony analysis applied in Sects. 4.2.1 to 4.2.9. However, not every detail of the EU's demands, ambitions and developments in global and internal climate change policy is essential to dismantle its use of gendered discourse patterns. Therefore, I will focus on the central aspects; first, the historical context of the international climate change policy and the EU's role will be outlined.

4.1.1 The EU at the Global High Table of Climate Governance

Like India, member states of today's EU[1] have been part of the international talks and negotiations on global environmental protection from the very beginning. Since international climate policy has been decoupled from international environmental policy, new global structures have also been formed. The Intergovernmental Panel on Climate Change (IPCC) was founded in 1988 with the USA playing a major role, while scientists in particular—for instance of the Stockholm Environmental Institute—and EU member states such as Western Germany were also pushing for its introduction (Agrawala 1998, 608–11). To this day, the EU draws on the IPCC's expertise, deriving proposals for combating climate change from its renowned Assessment Reports. Even better known and central to the governance dimension of climate change was the creation of the UN Convention, setting the framework for future international climate policy.[2] For the Rio Conference in 1992, where the UN Framework Convention on Climate Change (UNFCCC) was finalised, the EU developed a common position (Tröltzsch 2017, 314) represented by the EU delegation at the summit. When the UNFCCC was adopted in Rio, the EU called successfully for an anchoring of sustainable development, the precautionary principle as well as the recognition of scientific findings (von Lucke et al. 2021, 36). Furthermore, the EU favoured embedding industrialised countries' binding mitigation targets as a central asset of the UNFCCC (Tröltzsch 2017, 311), an idea the US—among others—opposed, and which was not followed. Also refused by the US was the principle of common but differentiated responsibilities (CBDR), which was eventually—due not least "to the EU's influence" (von Lucke et al. 2021, 37), but also due to lobbying by India in the G77 framework—anchored in the UNFCCC. In summary, the EU was "instrumental in setting up the framework in 1992 and managed to place some of its core preferences on the agenda" as von Lucke et al. put it (2021, 36).

---

[1] In 1993, the European Community became the European Union. Although international climate governance began to emerge at the end of the 1980s, with a huge involvement by the then European Community, I will call the European integration organisation hereafter EU for consistency reasons.

[2] Various European states, such as France, Netherlands, and Norway, attempted to forge early alliances to create an international instrument to politically address climate change. Although UNEP and World Meteorological Organisation (WMO) took up this cause under the auspices of the UN General Assembly, the initiatives of these countries were nevertheless relevant in supporting the successful development of the framework convention (Pallemaerts and Williams 2007, 23, 24).

Concerning the question of how the EU as a regional integration organisation managed to have a say, both the Convention of 1992 and the later Kyoto Protocol (KP) "contain[s] the now classic provision allowing 'regional economic integration organisations' to become contracting Parties jointly with their Member States, on condition that they are already Parties to the Convention" (Pallemaerts and Williams 2007, 39). From then on, the EU has also been a formal member of these global climate governance structures, although nation states such as the US didn't always recognise the EU as a negotiating partner on the same footing (Lütz et al. 2021, 182).

But this did not mean that the EU was not successful. Negotiating the KP in the framework of the Berlin mandate, which lasted from 1995 until 1997, to operationalise the UNFCCC, the EU was able to assert itself with its demand for binding mitigation targets of Annex1-states. Nevertheless, it was primarily the USA that determined the architecture of the KP, pushing for the policy instruments Joint Implementation (JI) and Clean Development Mechanism (CDM) (Tröltzsch 2017, 317). Interestingly, the EU was initially highly sceptical about these instruments, as was India (cf. Sect. 3.1.1). Yet, the mechanisms helped the EU member states to achieve their climate goals (Schreurs and Tiberghien 2010, 55, 57), which also induced their positive perception following initial scepticism.

While the EU was "at the forefront of international efforts to address climate change" (Schreurs and Tiberghien 2010, 23) when it announced its first voluntary mitigation goals in 1990, it became in parallel an ambitious international driver (Grubb and Gupta 2000) trying to get as many other states' hands on deck as possible. Having in mind that the USA never ratified the Kyoto Protocol, they withdrew as an international climate protection advocate. Hence, it was up to the other—widely accepted as credible—climate actor, the EU, to convince the necessary UNFCCC member states to reach the needed quorum of "not less than 55 Parties to the Convention" accounting "in total for at least 55 per cent of the total carbon dioxide emissions for 1990" (United Nations Framework Convention on Climate Change 1998, 18). The inception of the first international climate regime in 2005 with Russia joining the KP was a major success for the European Union after long years of convoluted diplomatic efforts. These efforts included cooperating with Russia in the context of a package of different fields, among them global climate change policy (Douma 2007). Not only due to its diplomatic achievement of winning Russia to join, the EU once again assumed a pioneering and credible role in global climate change policy as a leader in the late 1990s and early 2000s. The important role of the EU in these times is best described as leading by example (Schreurs 2016, 220; Oberthür and Roche Kelly 2008, 47) due to its early targets and the measures taken (cf. Sect. 4.1.2). This was not always the case. During the infancy of international environmental and climate negotiations, the EU acted merely as a symbolic leader (Dupont and Oberthür 2017, 77)—a role that was slowly growing in a process backed by highly engaged (future) EU member states (Delreux and Happaerts 2016, 46–49).[3]

The several Conferences of the Parties (COPs) between 1998 and 2006, which were mostly dedicated to the implementation of the KP, will not be reconstructed here. Nevertheless, some COPs during this period revealed the difficulties of the EU as an integration organisation with both intergovernmental and supranational elements shaping international climate negotiations. For instance, at COP6 in The Hague 2001 the EU member states had conflicting opinions which led to the mandatory rejection of the Dutch COP-Presidency proposal which tried to fix a dispute with the US (Lütz et al. 2021, 184). The different delegations of the EU member states sometimes had incompatible preferences and interests, which at some points paralysed the EU delegation. However, the other side of the same coin bears an enormous advantage: No entity has such a high level of expertise, negotiation resources and bargaining power as the EU, since each individual member state and the EU itself all send their own delegations. Although the national delegations might have different stands from time to time, they are an important resource for the whole group sent from the EU, as they are in dialogue and able to inform and support each other. Until 2004, the 20-person EU delegation was appointed by the respective Council Presidency and therefore changed every six months. To this day, the Council Presidency still plays an important role in preparing the EU for the negotiations of the COPs, "although the Commission has taken on an enhanced role over the years" (Dupont and Oberthür 2017, 68). Since 2004, to ensure continuity and a high level of institutionalised expertise in the EU delegation, its members have been appointed by the member states and the Commission for a longer term (Lütz et al. 2021, 179).

---

[3] In the extensive literature on leadership in international environmental and climate policy (for an overview see Liefferink and Wurzel 2017), scholars often analysed the EU's ability and form to lead in these fields (Jordan et al. 2011; Wurzel and Connelly 2011; Bäckstrand and Elgström 2013; Oberthür and Dupont 2021). Following Wurzel, Liefferink and Connelly (2017, 12) we distinguish between structural, entrepreneurial, cognitive and exemplary leadership. Structural leadership builds on classical hard power capacities in the fields of the military or economy. With its strong Single European Market, the EU has the opportunity to lead through economic capacities which might serve its climate policy (Wurzel, Liefferink, and Connelly 2017, 10). An entrepreneurial leader uses diplomacy when negotiating its goals which is therefore, in respect to time, limited to negotiation processes (Oberthür and Groen 2018). Cognitive leadership means that an actor spreads or reformulates ideas. Exemplary leadership is defined as leading by setting examples—intentionally or unintentionally (Wurzel, Liefferink, and Connelly 2017, 12).

Although the global climate question lost some momentum over time—not only due to the though ratification process of the KP but also because of various international crises such as global terrorism, the financial crises and the later Euro crises—two certainties brought renewed energy to the international process of climate regime development: The soon-to-expire first commitment period of the KP in 2012, and foremost, the fourth Assessment Report of the IPCC (AR4) in 2007, which painted an increasingly stark and drastic picture of how the climate will change if greenhouse gas emissions continue to rise. AR4 outlines different scenarios such as the two degrees model (Integovernmental Panel on Climate Change 2007) which became an orientation point in the international climate change policy. Impressed by the findings and models of AR4, it was the EU who tried to set the two-degree target as the minimum horizon or norm of climate efforts, both internally and notably also externally (Van Schaik and Schunz 2012, 175). As the global awareness of the climate catastrophe grew again due to AR4, it was clear that a relaunch of the Kyoto Protocol alone would not suffice. Hence, the COP negotiations were also used to discuss a successor agreement. The EU's main goal was to fully involve not only the largest emitters at the time, the US, but also emerging powers like China (which became the largest emitter in 2007) with binding mitigation targets. It was argued that all countries are needed in attendance to have a chance of reaching the two degrees target. Furthermore, vulnerable countries, for example those organised in the Alliance of Small Island States (AOSIS), demanded greater effort not only from industrialised powers but also from the states becoming greater emitters, such as India (cf. Sect. 3.1.1). Additionally, a window of opportunity opened with the new Obama administration in 2009.

The preparatory work for a Kyoto successor was carried out particularly with the Bali Roadmap at COP13 in 2007, where the US was involved as an official participant of the negotiations again (Lütz et al. 2021, 184). Following reformist objectives, the EU pushed for three central demands: A universal and legally binding treaty, "mitigation commitments to ensure global average temperatures rise no more than 2°C above pre-industrial levels" and climate finance linked to a global distribution key (Oberthür and Groen 2015, 1329). Once again, the EU tried to lead by example, adopting as early as 2007 its targets for the Copenhagen Summit, referred to as the 20-20-20 package (European Commission 2008, 2) which was "[i]nspired by the IPCC's 2 degrees emission reduction scenario" (Van Schaik and Schunz 2012, 175). The EU pledged to reduce at least 20% of its GHGs by 2020, to achieve a 20% share of renewables by 2020 and to increase its energy efficiency by 20%. Interestingly, the EU committed to raising its ambitions to 30% GHG reduction "if there is an international agreement committing other developed countries to 'comparable emission reductions and economically more advanced developing countries to contributing adequately according to their responsibilities and respective capabilities'" (European Commission 2008, 2). Although other states were invited to follow the EU's example in making binding commitments, history proved the EU wrong, as the package didn't in the end serve as "an incentive to get others to accept binding targets" (Parker, Karlsson, and Hjerpe 2017, 247), particularly regarding the main emitters.

Nevertheless, the international community had set itself the goal of adopting a successor agreement in Copenhagen in 2009. Although it was already foreseen that the negotiations would become very intricate, Copenhagen indeed turned out to be the rock bottom of EU climate diplomacy. As already discussed in Sect. 3.1.1, the diminishing importance of the EU was clearly evident when it was not involved in the final behind-closed-doors negotiations. The top-down mitigation targets, which were favoured by the EU, did not correspond with the ideas of the emerging powers or the USA (Müller-Brandeck-Bocquet, Pietzko, and Zürn 2018, 12). BASIC and the USA therefore turned their backs on the EU's preferences and a bottom-up mechanism was anchored in the non-binding Copenhagen Accord (cf. Sect. 3.1.1), which actually meant that the "USA, together with China, India and Brazil, refused to make any binding commitments" (Müller-Brandeck-Bocquet 2022, 121) at that time.

These developments as well as the partly redefined climate policy architecture introduced with the Lisbon Treaty set the tone of EU climate policy from 2010 onwards, which will be the starting point of the ecofeminist hegemony analysis of the EU climate change discourse. For this purpose, Sect. 4.1.2 reconstructs the most important climate policy initiatives that the EU has taken with the aim of reaching the EU's internationally adopted goals. Afterwards, I will develop an overview of the central actors of EU climate change policy and therefore of the important participants of the discourse. Some of them will already be touched on in reconstructing the milestones of EU climate policy until 2010.

4.1.2 Milestones of EU Climate Policy until 2010

The institutional foundation for EU environmental and thus climate policy was laid in the Single European Act, which came into force in 1987. Moreover, the EU set itself the emblematic goal as early as 1990 that by the year 2000 GHG emissions should be stabilized at the 1990 level (Tröltzsch 2017, 311). Vis-à-vis the obligations assumed by the Kyoto Protocol, EU-wide debates followed on how to achieve the targets. The approach of burden-sharing, also called bubbling, to find the needed solutions has shaped the EU climate policy to this day. The EU has been advocating for this solution at the international level from the very beginning. Under the first commitment period of the KP, the ambitious EU target of a 15% reduction was initially considered. After several more years of international discussions, the EU commitment of 8% GHG mitigation by 2012—compared to the reference year 1990—was adopted. Internally, the targets were distributed in the framework of the 1998 Environment Council (ENV) Conclusion (C/98/205) which stipulated that industrialised states such as Luxembourg (−28%) or Germany (−21%) bear the main burden of reducing their emissions, while other member states such as Portugal (+27%) were allowed to increase emissions (ENV 1998, 10; Pallemaerts and Williams 2007, 40).[4]

How did the EU want to achieve the shared mitigation obligations? The measures taken, briefly summarised in Table 4.1, clearly show that the EU placed the focus on market-based initiatives and the promotion of renewable energy. From 2000 onwards, it was up to the Commission of the then 15-member Union to negotiate and introduce one of the core instruments to achieve the KP commitment: the EU emissions trading system (ETS) (Rayner and Jordan 2016, 6). The ETS was the world's first mechanism in the form of a trading platform that enables operating with CO_2 certificates (Schreurs and Tiberghien 2010, 24); various CO_2-emitting companies from different industries such as energy production, the cement industry or steel processing had to hold emission allowances and either buy them if their emissions exceeded the basic allocation or alternatively sell their remaining certificates to other participants in the trading scheme.[5] In this way, the EU wanted to create incentives to save CO_2, for example through technical innovation. Although the ETS has been criticised from different sides as insufficient—for example because of "the 'windfall profits' from free allocation and the alleged competitive distortions resulting from different member-state rules for allocation" (Ellerman, Marcantonini, and Zaklan 2016, 92), and also because of the dramatic low CO_2 prices in its first phase—it can still be seen as the central element of EU mitigation measures. Since 2013, further GHGs and aviation have also been included. Today, the EU is discussing further improvements regarding the current phase IV (2021–2030) in the framework of the Fit for 55 package (cf. Sect. 4.2.9).

[4] For a more detailed overview of the commitments made in the framework of the EU bubble initiated in the wake of Kyoto as well as the reasons for the highly diverging commitments, see Lütz et al. 2021, 189–191.

[5] The ETS system was initially primarily focused on the allocation of allowances and then increasingly developed further in the direction of auctions of allowances. For the precise genesis of the instrument during its different phases see Ellerman, Marcantonini, and Zaklan 2016.

As a pioneering mechanism, the ETS has already served as a blueprint for new trading schemes in other regions of the world (Delreux and Happaerts 2016, 213). It was created as one piece of the mosaic to reach the EU's Kyoto goals as the ETS-linked Kyoto mechanisms CDM and JI (Schreurs and Tiberghien 2010, 24) are additional elements. Furthermore, the first implementation steps towards a noticeable increase of energy efficiency, the significant promotion of renewable energies and (partly voluntary) threshold values were also part of the mitigation endeavours of the EU in the early 2000s (Tröltzsch 2017, 310).

Unlocking the potential of renewable energy was seen by most of the actors on the EU level and by most of the member states as an industrial opportunity to export European technology (von Lucke et al. 2021, 35; Rayner and Jordan 2016, 5). Consequently, "the German Government advocated [for] the priority treatment of measures to promote energy efficiency and renewable energies" in JI and CDM projects (Lütz et al. 2021, 190). This idea corresponds to the findings of the 1995 Eurobarometer survey showing that the society of the European Union largely supported the statement that "economic development should be ensured at the same time as protecting the environment" with 72% (European Commission 1995, 13). Hence, one can speak of a rather positive atmosphere towards renewable energies, not least as "unlike many other EU policy areas, [environmental concerns] generates relatively strong public support" (Rayner and Jordan 2016, 4).

Of course, there was also resentment by some member states, for example because of the higher mitigation ambitions. Furthermore, the national implementation of the ETS was very complex due to differing preconditions and due to the efforts of most member states to secure their domestic industries (Campins Eritja 2007, 70) which eventually frustrated the European Commission. However, one advantage of taking decisions at the EU level—in contrast to solely at national level—soon became apparent (Rayner and Jordan 2016, 5): If, for example, stricter thresholds are adopted in the greater framework of the EU, a multi-level blame game can be used to dispel resentments 'back home', an angle which might also become relevant in the following hegemonic analysis.

Although there were of course many different actors to mobilise within the EU in order to achieve the KP goals, all with specific interests and opportunities, the targets were even overachieved, with the EU reducing its GHGs by more than 11% (European Commission 2022). Indeed, the EU bubble allowed states like Denmark, Germany, Sweden or UK to bear the brunt of the burden, while other EU member states were permitted to increase their emissions or failed to meet their individual goals (European Environment Agency 2014, 10). However, given plummeting GHG figures in 2008 (European Environment Agency 2014,

7), it can be assumed that besides the success of the EU efforts, the economic crisis of this and the following years also played a role in achieving the goals.[6]

In the following, a brief overview of the EU's GHG-reducing measures and policies until 2009, compiled by Rayner and Jordan, will be presented. They distinguish between three categories: GHG emissions, renewable energy and energy efficiency. In all three categories, the EU has taken action under the European Climate Change Programme (ECCP) and related activities to fulfil the KP commitments (1998–2006). Table 4.1 follows the same structure, presenting the measures of the Climate and Energy package (2007–2010) manifesting as the famous 20/20/20 package until 2020.

In addition to the overview of measures taken, a trend towards more stringent or binding policies and projects can be observed over the decades. In the case of the agreement with the car industry to save GHGs, the EU changed its strategy from voluntary agreements (1996) to vehicle emissions regulation (2009). The following table lists the key laws. It does not go into detail, but rather provides an overview of the industries and branches which are interlinked with the cross-cutting climate change policy and are necessary for its success.

Of course, EU climate policy also affects many other areas, as the Green Deal initiated in 2019 and analysed in Sects. 4.2.8 and 4.2.9 demonstrates. This includes, for example, the housing sector—which has a high share of GHG emissions—or the economic policy of the EU (European Commission 2019b). However, the aim of the ecofeminist hegemony analysis is to reconstruct the climate discourse in the narrow sense and not to examine the individual policy areas affected by the EU's climate change policy in depth. Therefore, the selection of the discourse contributions to be examined is rather oriented towards their significance for the entire climate discourse instead of the claim to cover all involved policy fields, since climate policy is by nature a cross-cutting theme.

With this point in mind, I reconstructed the main policies, which are discussed, adopted and introduced by manifold institutions and actors—whose exact function and influence on the EU climate change policy can be derived inter alia from their competencies in the complex interplay of the multi-level governance system. This will be reconstructed in the following chapter to deduce who are the main actors to focus on in the later hegemony analysis. Besides the policymakers, the institutions and actors trying to influence the policymaking process and the EU's stand on global climate change policy will be outlined to get an impression of who shapes the EU climate change discourse.

[6] For an overview of member states' individual performance regarding their burden-sharing goals and whether they reached them see Tröltzsch 2017, 311.

Table 4.1 EU climate policy instrument mix. (as compiled by Rayner and Jordan 2016, 6–7)

ECCP and related actions (1998–2006)	**Climate and energy package and additional action (2007–2010)**
Principal policy focus: **GHG emissions**	
EU ETS Directive (2003)	EU ETS review (2008/9)
Fluorinated gases Regulation (2006)	Fluorinated gases Regulation review
Voluntary agreement with car manufacturers (1998)	Vehicle emissions Regulation (2009)
	Effort Sharing Decision (national non-ETS targets) (2009)
	Carbon Capture and Storage Directive (2009)
Principal policy focus: **Renewable energy**	
Renewable Energy Directive (2001)	Renewable Energy Directive (2009)
Directive on the promotion of biofuels and other renewable fuels for transport (2003)	Fuel quality Directive (2009)
Principal policy focus: **Energy efficiency**	
Energy Services Directive (2006)	Energy Efficiency Directive (2012)
Combined Heat and Power Directive (2004)	Further implementation/legislative review of existing legislation
Ecodesign of Energy-Using Products Directive (2005)	
Energy Labeling Directive (1992)	
Energy Performance of Buildings Directive (2002)	

4.1.3 The Multi-Level Governance System of EU Climate Policy

Although the EU developed manifold measures to reduce its GHG emissions, and tried to act with united forces at the global climate high table, it was—and often still is—internally divided on the question of how to deal with climate change (Van Schaik 2013, 358). This is hardly surprising given the EU's multi-level governance system with its different entities and personalities, let alone the currently 27 member states and their diversity in terms of energy production, economic structures (Geden and Fischer 2014) as well as percipience of the urgency of climate change.[7] The discrepancies and therefore sometimes incongruent EU climate change policies can be explained inter alia by the different preferences of the national governments and can thus be an expression of the multi-level governance system in which the EU's climate policy is negotiated and implemented. This system *sui generis* contains both intergovernmental as well as supranational elements and thus generates many compromises. Hence, to deconstruct the EU climate change discourse, its governance system must be reconstructed. Since this project examines the climate discourse from 2010 onwards, this chapter focuses on the governance system resulting from the Lisbon Treaty (European Union 2007).[8] As already described above, the EU uses the mechanism of burden-sharing. The common and internationally-adopted EU targets, which are discussed and approved at the level of the heads of state and government, and which must additionally be consented ex post by the EP if they are part of legally binding international treaties (Art. 218, par. 6 of the Treaty on the Functioning of the European Union, TFEU, cf. European Union 2012), are implemented EU-internally in the form of specific measures and legislation. Accordingly, the legislative process and its actors, or more precisely their discursive contributions, are important aspects shaping the EU climate discourse.

Within the framework of the ordinary legislative procedure—in which climate policy must be implemented—the institutional triangle consisting of the Council, in the composition of the Environment Council (ENV), the European Parliament and the European Commission is central. It is up to the Commission to take

[7] The 1995 special Eurobarometer attests that "82% of Europeans—which is 3 points less than in 1992 but 8 and 10 points more than in 1988 and 1986—consider that protecting the environment is an 'immediate and urgent problem'" (European Commission 1995, a). Interestingly, the results vary between the different EU member states with support ranging from 97% in Greece to 63% in Belgium. However, the polls are not as divergent as today, which is probably due to the fact that the EU's enlargement provided a greater span of the EU member states compared to 1995. A special Eurobarometer conducted in 2019 found that 76% of EU citizens consider climate change to be a "very serious problem" (European Commission 2020c, 13). This finding varied enormously in the 28 member states surveyed at the time. While in Estonia only 46% or in the Czech Republic 54% of the respondents are likely to think that climate change is a very serious problem, in Spain (90%) and in France (82%) more than eight out of ten respondents agreed. Similar to the overall average, urgency was rated at 73% in Poland or 69% in Germany (European Commission 2020c, 13). Indeed, the reasons explaining these findings might be multi-layered. Nevertheless, as one result, varying degrees of pressure for action can be derived for the respective government, although the European citizens generally welcome EU climate policy (Rayner and Jordan 2016, 4) and recognise climate change as a serious problem.

[8] For a discussion of EU climate policy structures, their advantages and also shortcomings of the former European Community see Chapter 5 of Caspari 1995.

the initiative with a proposal which is then discussed by the Council and the European Parliament. In case of an agreement or renegotiations combined with improvements to achieve a joint text, the laws and commitments must be adopted by both the Council and the Parliament (Art. 294 TFEU, cf. European Union 2012)—an additional competence of the EP, which can certainly develop power (Müller-Brandeck-Bocquet and Rüger 2015, 166). Of course, these actors therefore occupy subject positions within the EU climate discourse. Today, as in the past, the Parliament stands out for its progressiveness and leading ambitions in the field of climate change, even though it had limited legislative power. However, when the co-decision procedure was introduced in 1993, the EP generated more influence besides networking with the Commission's respective responsible Directorate-General (DG) for Environment and further actors in trying to influence the climate agenda (Burns 2017, 53). Indeed, the EP's initiatives showed varying degrees of success (Burns 2017, 54–56), as recent contributions to the EU's climate change discourse, which will be analysed later, also show. Within the EP's Committee on the Environment, Public Health and Food Safety (ENVI), environmental and climate questions are discussed and prepared for decisions in the plenary. Furthermore, the MEPs responsible for climate policy in the respective political groups regularly contribute to the EU climate change discourse. In the context of specific climate-related global events, such as the discussion of COP results, other MEPs also articulate their demands on the EU's climate policy, for example in plenary debates or interviews. Beyond these actions in the framework of the legislative process, the EP plays an increasingly interesting role at the COPs (Biedenkopf 2015). The Parliament sends its own approximately 20-person delegation, traditionally to the second week of the negotiations. The delegation at the COPs typically holds a high number of consultations with international organisations and delegations from other parliaments, while its activities meeting with non-EU member countries' governmental delegations are growing (Delreux and Burns 2019, 346). But more importantly, the EP delegation receives regular briefings by the Commission and the Presidency, and is also involved in some informal discussions (Delreux and Burns 2019, 345). Thus, because the EPs need to consent for the EU's ratification of most international legally binding treaties—a requirement introduced in Lisbon—the role of the EP is strengthened at the global climate policy high table.

The European Commission stands out in the climate field through its right to make proposals which are worked out and implemented by the Directorate-General Climate Action (DG Climate). The Commission's institutional strength in this area has developed over time. Jon Birger Skjærseth (2017, 47) stressed that "With the assistance of the Kyoto Protocol, the Commission took the initiative, built up independent knowledge and crafted support among stakeholders"—which is why Skjærseth characterised the Commission as an "'engine' of climate policy development". After the Copenhagen disaster, with the second term of Commission President Barroso having started in February 2010, the DG Climate and the position of Commissioner for Climate Action were established (Delreux and Happaerts 2016, 209) separating the competencies for climate action from the Commissioner of Environment. With the first term of Jean-Claude Juncker as Commission President in 2014, he restructured the Commission linking the policy fields of climate and energy with Miguel Arias Cañete as new Commissioner responsible for both DG Climate and DG Energy. Cañete was criticised for his former activities in the oil industry and Juncker's further restructuring of the Commission raised questions particularly from external actors. They wondered who the main Commission contact was concerning global climate policy, as Juncker additionally introduced the Commissioner for Sustainability and also a Vice-President (VP) coordinating climate actions (Schoenfeld, Moore, and Gravey 2014). Of course, Juncker's reorganisation, introducing multiple Commissioners led by a VP, could also have been a sign of increased attention paid to climate change and its challenges. In 2019, the Climate Commissioner was once again upgraded: the post was renamed as Executive Vice-President for the European Green Deal. As the VP under Juncker, the first VP for the European Green Deal, Frans Timmermans, led the team of Commissioners working on climate change, coordinating the EU's overall action in this particular field until summer 2023. This illustrates that the main planned focuses of the new Commission President Ursula von der Leyen was the fight against climate change and a sustainable transition (Fischer 2020; cf. Sect. 4.2.7).

Besides the Commission and the responsible Commissioners, the European Council (EUCO) also plays a significant role in shaping the EU's climate change policies. The overarching goals pledged on the global stage are negotiated and shaped decisively by the EUCO. There, the heads of state or government of the 27 EU member states, the permanent president of the European Council and the President of the European Commission meet, setting the EU's overall agenda and priorities by adopting conclusions—which are however not of a binding character, but nevertheless largely implemented. Part of these conclusions is the general agenda of the EU's common climate policy such as its mitigation goals. The EUCO summits are often accompanied by general debates in the run-up, for example at member state level or by specific groups of states formulating their ideas and demands. It can therefore be worthwhile to pick up on these strands of debate in the later analysis.

Being part of the ordinary legislative procedure, co-deciding the EU's climate laws, the intergovernmental Council of the EU also holds decisive power in its hands. ENV, the group of the EU member states' ministers in charge of the environment and climate, discusses the Commission's proposals and votes on them, as does the EP in parallel. Although ENV takes decisions generally in the framework of qualified majority voting, it is still common practice to try and generate consensus. This widely accepted modus vivendi can of course lead to problems and slow down EU climate policy initiatives. But at the same time, it can also ensure that everyone is on board and the EU is therefore able to speak with one voice in international climate negotiations. These ambiguities will be picked up again in the later hegemony analysis. Nevertheless, Rayner and Jordan (2016, 4) remark that the enlargement of the European Union with countries from "formerly communist central and eastern Europe has introduced a new and increasingly problematic dimension for European climate policy". For instance, the enlargement of the EU 2004 included countries known as the Visegrád group. This consists inter alia of Poland, which not least due to the coal-dependency in its energy mix often opposes progressive climate policy initiatives.[9] Of course, enlargement can also produce opportunities for enhancing the situation in these member states adopting EU environmental legislation, or other harmonising effects that improve the new member states' climate conditions.

As already outlined, the Council Presidency has an important role in coordinating the common stance regarding international climate negotiations, because it "acts as the main contact point and spokesperson for the EU" (Dupont and Oberthür 2017, 68). Despite being the principal contact, there have also been cases—such as Copenhagen 2009—where the Swedish Presidency was sidelined by some of the European Heads of State and Government present in parts of the final round of negotiations (Van Schaik 2013, 361). In contrast, UK in 2005 and Germany in 2007, as well as the Dutch in 1992 and 1997 are all regarded as drivers of the climate agenda while they held the Presidency (Schreurs and Tiberghien 2010, 29). Thus, one can conclude that there were both weak and strong council presidencies regarding climate action.

---

[9] For a detailed discussion of how EU climate policy has changed while integrating Central and Eastern European countries, see Bocquillon and Maltby (2017). Marcinkiewicz and Tosun (2015) analysed the significance of the Poland's EU accession for EU climate policy.

All these outlined EU entities can draw on the expertise of the European Environment Agency (EEA). The overall complexity of the EU's multi-level governance system, on the one hand, encompasses a multitude of actors and, on the other hand, requires sometimes lengthy coordination processes. Of course, the negotiation procedure of concrete laws in the ordinary legislative procedure as well as the discussion of general guidelines in the European Council or on the occasion of a newly elected Commission are accompanied by various groups and institutions trying to influence EU climate policy. These include groups from lobby associations, NGOs, academia and think tanks as well as a critical press. Since there is no monolithic EU press, I will later focus on some of the "'European wide' or transnational media" identified by Cristian Nitoiu (2015, 538) which due to "their discourse usually transcends the boundaries of national states and is almost never solely concerned with covering domestic politics within a single member state". He classified the Financial Times, the European Voice and EUObserver.com as European-wide transnational media. Since the European Voice was closed down in 2015 and immediately replaced by POLITICO, both are taken into account as well as the online medium EUObserver. However, this study does not aim to examine the identity represented in the media, but only their contribution to the overall EU discourse. Therefore, voices from the above-mentioned newspapers are only taken up sporadically, not least because only a few opinion contributions in the climate field can be identified that go beyond classic reporting. Yet, the public opinion is not only reflected in media. Surveys, measuring an aggregation of individual opinions, generate an insight into the sentiment of the respondents and therefore represent the public too (Rüger 2013, 52–53). Accordingly, in the discourse analysis, survey results are also used, particularly of course the special Eurobarometer surveys regarding climate change and the environment. Doing so, the awareness of the climate change problem can be measured and based on these polls, the support of planned EU actions can be derived at least to some extent. Furthermore, societal demands which might not already have been taken into account by the politicians can also be detected.

To picture the general context of the EU's climate discourse informing the analysis, it is important to note that the perceived seriousness of climate change by EU citizens has been consistently high from 2008 until 2021, at around 90%, as the following table shows. As already indicated in this chapter, the societies of the Western European countries are more aware of the problem of the climate crisis, although the figures demonstrate that the average awareness of the populations of all EU countries can be considered high.

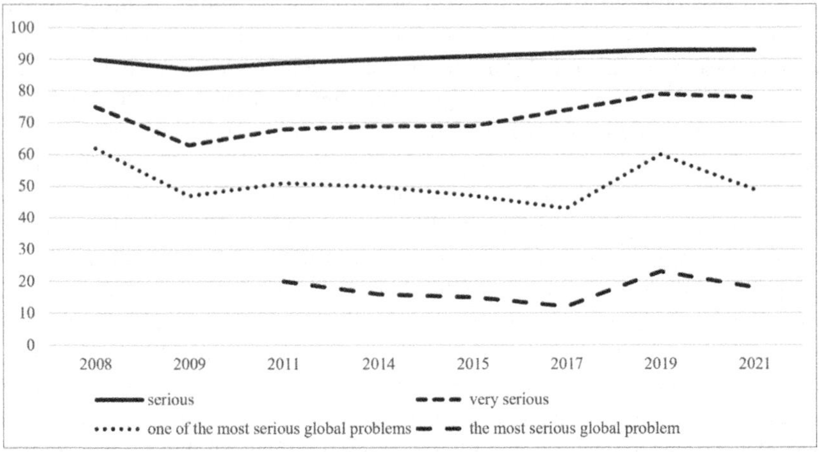

Fig. 4.1 Perceived seriousness of climate change, 2008 to 2021. (Source: European Parliament and European Commission 2008; European Commission 2009, 2011b, 2014a, 2015a, 2017b, 2019a; European Commission 2021c, own figure)

A comprehensive study of a broad database on the environmental conscience of Europeans furthermore found that, as of around 2014, an increasing climate change awareness can be attributed to EU Europeans (Tendero 2022, 178). Another meta-study by Donatella Baiardi and Claudio Morana (2021) also concludes that EU citizens perceive climate change as a problem that needs urgent attention. Although the average awareness is high, the values vary between the people living in the different member states, as diverse Eurobarometer surveys also found. In the later analysis, these figures are used again on an ad-hoc basis, for example when it comes to the increasing societal attention of the climate crisis in the 2019 election year (cf. Sect. 4.2.7). Still, the entire following analysis has to be seen in the context that people living in the EU perceive climate change as a serious problem.

Besides these discursive contributions, further actors articulating demands will also be scrutinised. Regarding NGOs, at least the central actors can be named and later analysed, although the field is of course much larger.[10] From 2000 onwards, the establishment of specialised climate change NGOs in Brussels can

[10] Studying the framing of climate change in the EU, Frank Wendler even stressed that the Environmental Non-Governmental Organisation's (ENGO's) communication would must be "delegated to a separate study" (Wendler 2022b, 90).

be attested (Wurzel, Connelly, and Monaghan 2017, 221). The "four key players" of the Brussels NGOs particularly addressing EU climate change policy are the Climate Action Network Europe (CAN-Europe), Friends of the Earth Europe (FotEE), Greenpeace Europe and World Wide Fund for Nature (WWF) (Wurzel, Connelly, and Monaghan 2017, 222). Interestingly, a few of them receive funds from the European Commission or member states' ministries for specific thematic projects. If this could influence the analysis, it will be addressed accordingly. Other discourse-shaping NGOs or grassroots movements have emerged, for example with Fridays for Future and their famous founder Greta Thunberg; these actors will also be considered in the later analysis.

Further stakeholders trying to shape the EU climate change discourse are business actors. These are of course highly diverse and can therefore hardly be imagined to agree on common climate targets. This fact depends inter alia on their differing branches—which could for instance either be renewable energy or fossil fuels—and diverges regarding further characteristics (Grant 2017, 207). Hence, it is a vast field of different business-oriented stakeholders trying to influence legislation and thus also the climate discourse in Brussels. The Economist (2021) even speaks of an "explosion of activity" in the EU lobbying world. Nevertheless, business actors will also be taken into account in the subsequent analysis, for instance when important associations comment on the EU climate change policy or formulate their demands. Following Hans-Wolfgang Platzer (2017, 602–3), two of the more central actors, BusinessEurope (pan-European employers' association) and Eurochambres (Association of European Chambers of Commerce and Industry.), are considered frequently. Of course, the player in the natural veto role when it comes to business actors is also taken into account, namely the association of the various trade unions at European level. Their and many more actors' contributions will be analysed in the following ecofeminist hegemony analysis, which will reconstruct whether and how the EU's climate identity is gendered and what these gendered discourse patterns can mean for the climate change policy of the EU.

4.2 The EU's Gendered Climate Change Discourse: Leadership Revisited

After the central actors shaping the EU climate discourse, important EU projects of the past and developments on the global stage informing the EU climate governance have been elaborated, the EU climate discourse from 2010 to 2021 will be reconstructed and critically analysed with the help of ecofeminist hegemony analysis. Addressing the research questions, I will follow the remaining stages of the six-step approach outlined in Sect. 2.4.1.

Step 1a, theoretical and methodological preparatory work, has already been done. Step 1b, the preparatory study gaining an overview of central documents and their content—regarding the most current discourse contributions of the central EU institutions—provided first important insights. The following analysis of the EU climate discourse was conducted with the help of the qualitative analysis tool MAXQDA, which was also used in applying step 2, the construction of the discourse corpus, based on the considerations made in Sect. 4.1.3 concerning important actors, central projects, and topics of the EU climate discourse. In parallel with step 2, step 3, the structuring of the corpus, was also carried out in collating which documents need to be discussed in detail at the level of a single text, and which should be reflected in the broader context of the overall discourse—for instance when I search for actors supporting a specific demand shaping the hegemonic project or rearticulating gendered discourse patterns.

Accordingly, the following Sects. 4.2.1 to 4.2.9 are the presentation and discussion of the results of steps 4 to 6—which are the core of the project. As already practised in the case study of India, the individual steps are not delineated from each other, but rather presented in the continuous text. The description of the context (step 4 a) and content of the discursive contributions (step 4 c), the actors articulating them (step 4 b), the reconstruction of the formulated demands and the identification of the strategemes used to help the hegemonic project to succeed (steps 4 d, e and 5) are thus shown and reviewed in parallel in the following analysis. Nevertheless, the focus of the reconstruction of the discourse remains the strategemes used to construct an identity. The content of the discourse has to be understood, but the main task and interest is the offensive hegemonic strategy and thus the gendered identity of the EU. The different demands and strategemes used were coded in MAXQDA and can thus be fanned out and traced in a concise manner. In addition, there is ongoing critical reflection on whether and how the expressed demands that form the chain of equivalence are gendered (step 4 f), hence if they contribute to the dichotomous structure of the world theoretically derived in Sect. 2.3.3. The partially changing discourse coalitions (step 6 a) are also presented in the text, as well as the possible identification of new, or modified, strategemes (step 6 c)—since the set proposed by Nonhoff can be changed and expanded depending on the research subject (Nonhoff 2006, 212, 2010, 312), as the case study of India already showed with the newly introduced strategeme X (cf. Sect. 3.2.8). With the help of the outlined steps, I reconstruct whether and how a hegemony, appearing as a temporarily fixed identity of the EU in climate policy, is discursively produced, whether this is created with the help of gendered discourse patterns and how this identity is used to justify the respective climate policy.

As "In the affluent North, the gender assumptions are more subtle" (MacGregor 2010, 233), the analytical chapters will show that the gendered structure of the EU's climate identity is in most cases only implicitly expressed and needs to be sensitively detected—particularly when it comes to gendered discourse patterns of the second dimension. Moreover, the EU's climate discourse tends to be very technical in nature. Hence, only in a few rare cases can explicit gendered patterns of the first dimension, which interpellate subject positions, be identified. The ecofeminist critique of the EU climate discourse will be examined in detail in Sect. 4.3 with a deep dive into the findings of the analysis dismantling the EU's climate identity using the ecofeminist foil as a normative template.

But first, some specific characteristics of the analysis and its corpus will be pointed out. The variance of the actors whose contributions are examined is dynamic, as different alliances emerge depending on the issues debated, and power constellations—or new actors—enter the stage (cf. Sect. 4.1.3). Most of the discourse contributions examined stem from EU institutions. On the one hand, this is because they are responsible for legislation to implement EU climate policy and because their initiatives are often the stimulus for broader debates. On the other hand, this is due to research pragmatism as their discourse contributions are very well documented and accessible. The discourse is generally more shaped by institutions and groups than by individuals, as it is the case with the Indian discourse. Besides the EU institutions, representative ENGOs and other non-governmental entities such as trade unions and employers' associations will be constantly examined, while further actors, such as the press or scientists, can only be reconstructed in a few cases. It should also be mentioned that the study is confronted with the limitation of not examining the individual member states. Their voices are only taken into account if the respective member state (MS) holds the Council Presidency, or if several MS join together to form a specific group.

Moreover, the results of the above-mentioned Eurobarometer surveys are also part of the corpus. In addition to selecting the important documents of the central actors, other studies of the EU climate discourse were also used in determining the most important debates and documents (see for instance Wendler 2022b). A total of 293 documents are the subject of the study at hand (for an overview of all 293 documents analysed for this case study see the appendix in the electronic supplementary material). Structuring the corpus (step 3), the EP debates were machine-translated into English with the help of the programme DeepL if they were held in one of the 24 official EU languages other than English or German.

Doing so, I follow Jan Krotký (2022, 11), who also used a translation tool in his study analysing EP debates, emphasising today's enormous efficiency of these programmes which is scientifically proven. Indeed, linguistic peculiarities cannot be ascertained, but this is not the interest of this study.

The following analysis is presented chronologically. Although the years are often structured similarly and events such as the Conferences of the Parties, EUCO summits or the EP elections occur at regular intervals, the events during the twelve-year research period are viewed differently depending on the greater context. Accordingly, the EU's performance at the annual COPs, for instance, is reconstructed very accurately in some years and only marginally in others. The results of the analysis presented and discussed are selected against the background of their contribution to the formation of a gendered identity, rather than following a completely balanced template. The subsequent study begins with the year 2010 after the historical disappointment of Copenhagen. In the following we will explore the question what the EU struggle for a common identity looks like and how it is gendered in the period from 2010 until 2021.

4.2.1 (Re-)Orientation after Copenhagen

The international climate, and particularly that within the EU, was chilly after Copenhagen. Indeed, a disappointed and wounded European Union had not only to reflect on its weaknesses and question its role but also reorient itself for the future fight against the climate catastrophe.

In the EP debriefing debate of the Copenhagen results in early 2010, Jo Leinen (Progressive Alliance of Socialists and Democrats; S&D), then Chairman of the EP Committee ENVI, gave a nuanced review of the EU's role at that time and the tasks which were to follow. Out of his frustration regarding former Council debates that showed its disunity, he formulated three future priorities:

> *"The first is that we must regain the leading role in global climate protection. Secondly, we need to show credibility in the commitments that we have made and, thirdly, we urgently need to gain new partners before we get to Mexico." (Leinen* 2010*)*

Leinen's contribution shows that from his point of view the EU clearly lost the leading role in Copenhagen.

Against this background, it is interesting to reconstruct the resolution adopted by the EP on the outcome of COP15 and the hegemonic strategy used therein to establish the EP's hegemonic project. Following strategeme I, the EP created a chain of equivalence demanding inter alia *higher mitigation goals by the EU*, equivalising this demand with *energy security* and the call for a future *green economy* and pointing to earlier successes (European Parliament 2010c, 2). Furthermore, the EP called on the actors involved "to *speak with one voice* in international climate negotiations" and went on to say that this is how the EU's "leading role" can be maintained (European Parliament 2010c, 2; emphasis added). The resolution implies that such a role opens up the possibility of reaching both *a post-2012 agreement* and the *2-degree target* (European Parliament 2010c, 2). Surprisingly, the EP talks here about keeping the leading role instead of regaining it. Hence, the EP uses strategeme III, representing the EU as leader. Furthermore, the EP follows strategeme II, splitting the discursive space by distancing the EU from the USA but also from China (European Parliament 2010c, 4), two of the key actors who circumvented a binding and ambitious outcome back in Copenhagen. Using all three core strategemes, the EP tries even post-Copenhagen to fulfil its hegemonic strategy to push for the *leading role* as the empty signifier representing the identity of the EU.

Since the second Barroso cabinet needed to be constituted and inaugurated in early 2010, the period of reflection was only of short duration. The conservative Connie Hedegaard, former Danish Minister for Climate and Energy—and therefore the person leading most of the Copenhagen negotiations—was designated for the newly created position of Commissioner for Climate Action. Unsurprisingly, she argued in the compulsory European Parliament's hearing of nominated Commissioners that the results of Copenhagen were disappointing (Hedegaard 2010a, 2). Referring to the disunity of the EU member states and the overly extensive process of finding a compromise on financing topics in the run-up to COP15, she called for *greater unity* and *earlier formation of common EU positions* in the future (Hedegaard 2010a, 4). Hedegaard equivalised these demands with the demand for a *stronger European voice* (Hedegaard 2010a, 2, 4) but also with *economic success* (Hedegaard 2010a, 2) following *technical solutions* which will create *new jobs* (Hedegaard 2010a, 2, 3). These were of course desperately needed as the EU still faced a tremendous economic crisis in 2010. Equivalising the reconstructed demands, the designated Climate Action Commissioner used the first strategeme, shaping the EU's chain of equivalence from her own perspective. Furthermore, she followed strategeme II, demarcating the EU from other actors, especially the US, which she marked as a state which must now rapidly catch up in terms of energy efficiency and sustainable technology, implying that the EU is already in good shape in this regard even though it also needs to move forward (Hedegaard

2010a, 2). Although Hedegaard forms the two focal points of sustainable economic success and a stronger future role, I am not able in this case to reconstruct the use of strategeme III (representation), so the question of which hegemonic project she is pursuing must remain open.

While the new Commission had to be formed and elected, the rotating Presidency of the Council passed to Spain. In this role, the Spanish Prime Minister José Luis Rodríguez followed strategeme I equivalising *sustainable industry* which is in his eyes "a priority in the context of combating climate change" (Rodríguez Zapatero 2010) with *technical innovation* and therewith also *better conditions in higher education,* indicating the need for *economic regeneration* regarding the economic crises (Rodríguez Zapatero 2010). Seeing that Zapatero deals only peripherally with the challenge of climate change in his speech, no further strategemes can be reconstructed. Nevertheless, it must be noted that he links the EU's climate change policy almost exclusively to economic aspects, which are culturally assigned to the masculinised sphere (cf. Sect. 2.2.3, 2.3.3).

After the Commission was elected, it undertook an attempt in March 2010 to reframe the EU's climate policy in its communication "International climate policy post-Copenhagen: Acting now to reinvigorate global action on climate change". Here, the Commission is pursuing the hegemonic project of a *leading role*, but less offensively than the EP and in a specific form. Rather, the Commission recommends setting an example by transforming the EU into "the most climate friendly region of the world" (European Commission 2010, 2) and trying to motivate others to follow, as this is "the most convincing leadership that the EU can show" (European Commission 2010, 8). All three core strategemes are used by the Commission in this discourse contribution reproducing a specific form of a leader as the EU's identity—the role of a *leader by example.* Whether these invitations succeed is questionable, since the EU already tried through ambitious actions to acquire followership in Copenhagen, and failed, due to, among other things, insufficient diplomatic efforts in the run-up to the summit.

One part of the answer intended to solve this problem was the creation of the new DG for Climate and the new Commissioner post solely in charge of climate action. These performative articulations show the improved approach to thinking of climate change policy more as a diplomatic rather than solely as a scientific field (Çelik 2022, 1030), trying to mainstream climate change policies "across all Community policies" (Herrero and Knaepen 2014, 4). Certainly, these actions alone are not proof that the Commission is pushing for a leadership role, but it should be seen as an additional building block.

A comprehensive impression of the EU's climate change discourse as the Commissioner on Climate Action wants to shape it is provided by Connie Hedegaard in a detailed contribution. In the framework of the Climate Lecture at the Harvard College (USA), she gave a speech entitled "Europe's view on International Climate Policy" in September 2010. Hedegaard outlined the cornerstones of the EU's climate policy and explained why the EU is advancing targets such as the 20/20/20 package. Describing her vision as "shifting from our present model to a low-carbon future" (Hedegaard 2010b, 3), Hedegaard argued that the mitigation measures must address several goals: not only climate protection, but also new impulses to overcome the economic crisis. Hence, she follows strategeme X, occasional but deliberate breach of a specific discourse's limits (cf. Sect. 3.2.8), importing the demand to overcome the economic crisis into the EU's climate discourse. To this end, she rhetorically opens the borders of the discourse, which on the one hand poses a great risk for the stability of the discourse, and on the other hand leads to the assumption that Hedegaard regards the discourse as relatively stable. Following strategeme X, she is suggesting that climate change policy should not only serve this specific policy field but also the interests of the future economy. On another note, the added value of analysing two case studies is also evident here. Strategeme X, which I have newly identified, is used in both the Indian and EU discourse. This is probably a specific feature of climate discourses, as the policy field of climate protection must in many cases be viewed as a cross-cutting issue.

Further demands of the chain of equivalence pursued by Hedegaard are the *low-carbon economy*, which is to be achieved in particular with *low-carbon technology* and the associated *energy efficiency* (Hedegaard 2010b, 8, 3). Hedegaard also equivalises the demand for *renewable energies* with these concerns, as well as *energy security* (Hedegaard 2010b, 6, 8) and *CBDR* (Hedegaard 2010b, 7), using strategeme I. Pointing to the US, she demanded no less than they become part of the "low-carbon revolution" (Hedegaard 2010b, 6) regarding which she underlined that "the European Union intends to stay at the forefront" (Hedegaard 2010b, 4). Arguing this way, Hedegaard uses strategeme II, splitting the discursive space by creating the forward-heading leader EU, actively forcing a "low-carbon revolution" on the one side and the US, which seems to hold back, on the other side of the antagonistic line of division. At the end of her speech, Hedegaard's concept became more concrete, following strategeme III. Pointing again to the EU's planned mitigation actions, she concludes that "we are doing this because we believe the fight against climate change needs developed countries to lead by example" (Hedegaard 2010b, 8).

Concerning Hedegaard's attempt to anchor a specific reading of the EU's identity, she followed the three core strategemes as well as strategeme X, which can all be clearly reconstructed. Although the disappointment in Copenhagen had occurred only a few months ago, the entire line of argumentation ends up running in the direction of representing the EU as a *leader by example.* At the beginning of the speech, strategeme III is used more cautiously, by pointing out the aspects in which the EU is already preceding others and where the USA could therefore follow the EU's example. Towards the end of Hedegaard's lecture she articulated the demand *to lead by example* in a more offensive manner and used the illustration of a revolution. The demand *to lead by example* fulfils the condition that it encompasses the other articulated demands and goes hand-in-hand with them. Final proof that it is the empty signifier is given when it embodies the radical other of the opposite chain of equivalence. If we assume that the opposite chain—which is most likely that of the US (Krzyżanowski 2015, 126)—includes, for example, restraint in mitigation pledges or *lower climate protection regulations* (compared to those of the EU), then it is conceivable that *leading by example,* which means in this case inter alia to follow stricter climate protection rules, is diametrically opposed to all demands of the US.

However, a potential empty signifier also needs support from other politico-societal forces. Preparing for the upcoming COP16 in Cancún, the different EU institutions, NGOs and further actors debated and adopted their positions. Although not all contributions will be analysed in detail here, it is worth reproducing the most convincingly recurring patterns shaping the discourse, which can be inter alia reconstructed in a specific EP debate. The EP discussed its position for COP16 with the Climate Action Commissioner on November 29, 2010. Remarkably, there are some similarities between the political parties as well as relevant differing nuances in the EP members' reasoning, which help us to further characterise the EU chain of equivalence.[11]

The Irish MEP Bairbre de Brún (European United Left/Nordic Green Left; GUE/NGL) stated: "Real leadership and political will are needed to keep the temperature limit to 1.5°C" (Brún 2010) and linked the demand for an *EU leadership role* with further demands such as *stronger mitigation efforts* of the "other developed countries", *more ambitious EU targets* and a united EU *speaking with*

[11] The following analysis does not address the articulations of the right-wing populist MPs, some of whom completely question climate change or reject all EU measures. Since their contributions have not shaped the climate policy discourse at this point, for instance evidenced by the resolution text, the reproduction of their contributions is not helpful as we search for the offensive hegemonic strategy used to install the hegemonic project.

one voice. These demands were equivalised following strategeme I, while the reference to the other developed states furthermore separates the EU's discursive space from the opposing chain's space. Hence, de Brún also uses strategeme II. The terse outlined remarks of the parliamentarian suggest that the leadership role needs to be pushed as the empty signifier to achieve the 1.5-degree target as well as the other demands. This is in line with strategeme III showing that *leadership* represents the EU's identity in the climate change discourse. By implying that all other demands are fulfilled if the EU has "real leadership", de Brún uses it as an encompassing demand, which is one of the necessary conditions of a demand serving as the empty signifier.

Further contributions of other MEPs support these findings. The green MEP Bas Eickhout also followed the first strategeme, equivalising for example, the demands for *more ambitious mitigation efforts by the EU*, pointing to the need for a higher EU goal than the actual 20% target, *financial support for developing states*, and ideally the *1.5-degree objective*. He points out that all these demands have to be met, as well as a *better negotiating performance by the EU*, since "only then can we truly make this leading role a reality" (Eickhout 2010). Following the third strategeme, his line of reasoning also culminates in the *leading role* as empty signifier. Accordingly, these members of parliament agreed—across party lines—to construct the EU as a leader, representing the identity of the Union in global climate change policy. In addition to these MEPs of the Left and Green groups, members of the S&D parliamentary group constructed the chain of equivalence already shown in an almost similar way.

The Hungarian MEP Csaba Sándor Tabajdi (S&D) used all three core strategemes to construct the demand for a *leading role* of the EU as the empty signifier, although he applies different subtleties. He starts his statement by dividing the discursive space clearly distinguishing the EU from the main emitters USA, China and India, which in his opinion needed to be influenced (Tabajdi 2010) probably to head for a more ambitious international way of fighting climate change. Hence, Tabajdi conveys the image that only the EU knows how to react satisfactorily and ambitiously, though he insists that "a 30% unilateral commitment makes sense only if there is real action behind it" (Tabajdi 2010) calling for a stronger EU commitment in the areas of *research* and *development* in order to generate *green jobs* and *sustainable green growth* (Tabajdi 2010). Thus, Tabajdi expands the portfolio of different demands which he equivalised using the first strategeme, but follows the same thrust as the other parliamentarians, as shown by supporting the *leading role* as a common demand. Interestingly, in his speech Tabajdi outlines that the EU can "regain" (Tabajdi 2010) the leading role by complying with his demands, but not that it could currently be considered a leader without question.

Rovana Plumb (S&D) argues similarly to her colleague that the EU must "re-establish" its "leadership position" within the international climate negotiations, fulfilling different demands (Plumb 2010). Using strategeme I, she introduces further demands that have not yet been expressed in detail, such as *climate justice* and *equity* (Plumb 2010). Adding a new perspective to the discussion, she ends her statement with the words: "Let us also not forget the gender dimension in climate change policies", which she equivalises with the demands *green growth* and *a good life for citizens* (Plumb 2010). Insofar as these newly-formulated demands become nodal points, which means that they are taken up and reproduced by other actors, one can speak of an extension of the chain of equivalence. Irrespective of this, we can identify feminised discourse patterns here.

How did the largest group in the EP, the conservative European's People Party (EPP) try to shape the EU's chain of equivalence? Overall, the demands articulated by EPP members are very similar to those of the other parties already presented. José Manuel Fernandes (EPP) equivalised the demands for a *legally binding future agreement* and *mitigation action by the major global emitters*, concentrating on *scientific research* and related *technological innovation*, to enhance *energy efficiency*, *renewable energies* and *competitiveness* (Fernandes 2010). Using strategeme I, Fernandes equivalised these demands with the EU's *leadership* role even though he stressed that the EU should not forget to give weight to the efforts already made by the industrial sector and to protect its competitiveness (Fernandes 2010). Hence, what distinguishes his line of reasoning from that of other EP members is that he does not demand an increase in the EU's 20% target. Nevertheless, he does also demand future ambition, but with a greater emphasis on innovation and research.

This example shows that the conflicting demands for (firstly) sticking to the 20% target particularly articulated by the EPP on the one side and increasing the target to 30% by 2020 as demanded by S&D, ALDE, the Greens and the Left on the other side, are both part of the demand for *ambitious mitigation action by the EU*. This discourse-theoretical finding can best be illustrated with reference to the spider's web depicted in Sect. 2.2.3, which shows that parts of a demand can be conflicting insofar as these contradicting elements can be equivalised in relation to the radical other—which is most likely one of the major emitters. When Fernandes refers to other either not sufficiently or not at all engaged emitters, he explicitly names the USA, India and China, thus splitting the discursive space using the second strategeme. Those MEPs also following strategeme II in their brief EP debate statements seem to agree—across party lines—on the division of the discursive space distancing the EU from the USA (Del Castillo Vera 2010; Hibner 2010; Jazłowiecka 2010; Seeber 2010; Tabajdi 2010; Wils 2010), as Hedegaard also did in her lecture analysed above. Interestingly, some MEPs reason in the same way as Fernandes, additionally delimiting the EU's chain of equivalence from China's and in some cases also from India's, marking them as further major emitters without enough ambitions or who must be motivated to join future action (Gardini 2010; Hibner 2010; Seeber 2010; Tabajdi 2010). Danuta Jazłowiecka (EPP) condenses the underlying argumentation: "…since rising economic powers want to play a greater international role, they must also accept greater commitments" (Jazłowiecka 2010). Distancing oneself from the 'radical other', i.e. in most cases from the USA and often also from China or India, serves to reaffirm the EU's identity as a climate leader who takes the lead, for example, by means of more ambitious targets than some of the parties call for, or simply by being more committed to international climate protection, as the EPP members suggest.

Compared to the USA, China or India, it is irrelevant whether the EU commits to a 20% or 30% target for 2020, as all options are more ambitious than the absent legally binding goals of the other major emitters back in 2010. Hence, it seems that if this is already enough to define oneself as a leader in global climate change policy and to be recognised as such, the EP members can certainly stick to the empty signifier *leading role* even after Copenhagen. Combining various nuances, the different MEPs followed strategeme V "gradual constitution of greater leeway for interpreting the symbolic equivalent of the universal". As discussed elsewhere (cf. Sect. 2.2.4), this strategeme can only be demonstrated at the discourse level, for which this EP discussion provides great insights.

However, the European Parliament's resolution, adopted the next day, commits to the 30% target for 2020 compared to 1990, linking the more ambitious target with "the interests of the future economic growth of the European Union" (European Parliament 2010b, 4) which seems to be the necessary compromising formula and is highly interesting regarding the subsequent years when the world discussed the green economy due to the Rio + 20 summit in 2012. Nonetheless, the unanimously constructed chain of equivalence still leaves room for interpretation. This is illustrated by the diverse understandings of the empty signifier *leading role*, which can partly be described as an economical forerunner trying to generate followers, and also as adhering to the 20% while also reaching for the 30% goal. Furthermore, we can state that the EP particularly sticks to androcentric demands mostly focusing on the economy and technical answers fighting climate change.

How do other actors outline the EU's chain of equivalence? In its annual report, the WWF underlined that it "argued for European leadership such as a move to 30% emissions reductions by 2020 unilaterally" (WWF European Policy Office 2011, 7) agreeing with the basic line of argumentation of some EP members also demanding the 30% target and linking it with the EU leadership role. Interestingly, in the foreword, Tony Long, then Director of the WWF European Policy Office (WWF EPO) articulates another demand that was mentioned in the post-Copenhagen EP resolution: *green economy*. Whereas green economy was not mentioned in the EP discussion and resolution or the Council's Conclusions regarding Cancún, Tony Long attaches high importance, stressing that "A newly emerging political and potentially progressive agenda [is] built around green economy" and therefore promising regarding the awareness to "safeguard natural capital" (WWF European Policy Office 2011, 3). Since the commitment to the *green economy* is described as new and freshly emerging at this time, further analysis must show whether and in what form the demand is pushed by other actors and whether it becomes more important for the EU's strategy to answer climate change.

The WWF only briefly touched on Cancún, emphasising that the conference "ended on a positive note" (WWF European Policy Office 2011, 3). Indeed, along with founding the Green Climate Fund (GCF) and reaffirming the two-degree target, leaving the option open to increase it due to possible newer scientific findings was not insignificant. Even so, the results did not represent a breakthrough. Reflecting the outcome achieved in Cancún, Connie Hedegaard and most EP members agreed that COP16 was a cautious success, while the then President-in-Office of the Council Joke Schauvliege even attested an EU leading role discussing the Cancún outcome with the EP (European Parliament 2010a). In line with the previous findings, the appraisal of the MEPs differed regarding the empty signifier of that time, *leading role*, along the party groups elaborated above. Hence, the EPP members continued to stand by the 20% target (Jordan Cizelj 2010) and the other democratic parties demand an increase to 30% (Jørgensen 2010) while both groups attested the EU a *leading role* back in Cancùn and equivalised it with their specific mitigation goal. After COP16, parts of the press also called for the 30% target, arguing that the EU would otherwise lose credibility in continuing to push for reaching the two-degree target internationally (Politico 2010). However, Hedegaard stuck to the 20% target at the time and dismissed the argument that the EU was making itself untrustworthy with the remark "Everyone knows that you cannot find a region in the world that has made as ambitious targets as Europe" (Hedegaard in Rankin 2011a) using strategeme II splitting the discursive space, which separates the EU und the US.

Regardless of the level of the EU's mitigation target, various actors shaping the EU climate change discourse seem to have agreed that the EU's identity either as a leader or as a leader by example should be (re-)established, even though the measures taken leave a gap concerning the "fair share"[12] of the Union which may not fit into the picture of a leading entity.

This initial analysis of the discourse after Copenhagen is astonishing, as different scholars agree that after Cancún the EU was not able to find its way back to the leader role it had before Copenhagen, despite sharing a similar *idea* of how to being a leader and despite of being able to achieve its—compared to previous summits quite modest—goals (Groen, Niemann, and Oberthür 2012, 183). That's why Lisanne Groen and her colleagues (2012, 187) called the EU rather a "bridge builder" than a leader. This is not surprising as the EU lacked a common agenda, shown by the different demands regarding the EU's climate goals, which were not clarified even on the ministerial level (Groen, Niemann, and Oberthür 2012, 183). However, since establishing an identity requires more than just one year of discursive effort to implement a specific empty signifier, the following years of the EU climate discourse and further potentially used strategemes for strengthening and temporarily fixing the gendered identity must also be reconstructed.

4.2.2 The EU and the Green Economy—an Idea whose Time Has Come?

The years after Cancún were marked by both the deal to swap a second Kyoto Protocol period against a concrete track negotiating a universal successor treaty to be reached in 2015 and a broader discussion of the future economic system stimulated by the upcoming Rio + 20 summit, in this specific context debated under the buzzwords 'green economy'.

Formulating the climate policy part of the EU's 10-year strategy for "smart, sustainable and inclusive growth", the Commission published the "Roadmap for moving to a competitive low carbon economy in 2050" in March 2011. The Roadmap discusses different scenarios modelled by the Commission, which should enable the EU to reach its "objective of reducing greenhouse gas emissions by 80–95% by 2050 compared to 1990", a target which "the European

---

[12] "Fair share" is defined by the Climate Action Tracker as "a country's total contribution [that] would [be] need[ed] ... to make a fair contribution" to implement the global targets such as the former 2 degree or the Paris 1.5 degree goal (Climate Action Tracker 2022).

Council reconfirmed in February 2011" (European Commission 2011c, 3). Strikingly, within this discourse contribution the Commission explicitly emphasised that it is not a matter of formulating new 2020 climate goals or concrete measures in the Roadmap, but rather a basis for dialogue and future action—for other European institutions, the member states as well as international partners (European Commission 2011c, 15). Since the Commission stressed that "the EU is taking a new initiative to stimulate international negotiations in the run-up to Durban" (European Commission 2011c, 13, 14), the Roadmap could also be understood as an articulation trying to portray the EU as a leader in international climate change debates.

Formulating different demands and equivalising them, the Commission used strategeme I to shape the EU's chain of equivalence. Indeed, the Roadmap focuses particularly on the economic aspects regarding mitigation efforts in the sense of technical innovation targeting "power producers and energy-intensive sectors" (Voßwinkel and Reichert 2011, 1), equivalising *low carbon economy* and its co-benefits such as building *new jobs* and generate *technical innovation* (European Commission 2011c, 12). Demanding more *ambitious mitigation action* by the EU (European Commission 2011c, 4), the Commission points out that it "gives ranges for emissions reductions for 2030 and 2050 for key sectors" (European Commission 2011c, 14), trying to win over other EU entities and further actors to favour higher mitigation goals in the long term, which would be a vital part of a future low carbon economy. As the Roadmap "is framed almost solely from a strictly economic point of view" (Krzyżanowski 2015, 124), the Commission predominantly rearticulates demands that can be assigned to the culturally masculinised sphere.

Using strategeme II, the Commission demarcates the EU from India, Brazil and Korea (European Commission 2011c, 11–13), despite calling them "partners" (European Commission 2011c, 13). As the US is not named, it can be assumed that the Commission does not see the US as a partner in tackling climate change and pushing for a low carbon economy—but more as an actor inhibiting engagement in these fields (Krzyżanowski 2015, 124). However, the Commission does not forget to mention that a partner in climate protection also means competitor on the global green market (European Commission 2011c, 13). Regarding the third strategeme, it seems that the Commission tries to install the *low carbon economy* as the empty signifier, promoting it as the universal answer to the climate policy challenges of that time. Accordingly, the Commission and the EP are pursuing different hegemonic projects at this moment, as the EP's hegemonic project was the EU's leading role as reconstructed above. To clarify whose hegemonic project finds more support and is therefore probably most likely to succeed, the discourse must be further analysed. Before turning to other contributions, I will discuss several responses to the new stimulus of the Commission that were prompted by the Roadmap.

The Commission generated controversial discussions. These particularly concerned the conclusions drawn from its broad analysis which showed that reaching the decarbonisation goal cost-efficiently by 2050 would require a 40% GHG mitigation goal for 2030 (European Commission 2011c, 14). Having in mind that a 2030 goal did not exist at that time, most business groups such as the pan-European employers' association BusinessEurope opposed the Commission's suggestions, whereas those belonging to the renewable energy sector like the European Wind Energy Association responded positively and called for even more ambitious goals (Rankin 2011b). The Commission's ideas also found support among environmental NGOs. Pointing to the Roadmap, the Climate Action Network Europe (CAN Europe) underlines the Commission's findings regarding the need of ambitious future mitigation goals and action (Clima Action Network Europe 2012, 1).

Whereas some of the member states opposed higher EU targets which could be a consequence of the Commission's model calculations (Rankin 2011b), they discussed and largely welcomed the Roadmap in the Council's Presidency Conclusions published in June 2011. Although the informal Presidency Conclusions reconstructed below are not binding, they serve as a summary of the discussion within the Council. The Council is not convinced about medium-term mitigation goals, but at the same time recognises the Commission's concept "as the basis for further work on the action needed to make the transition in a gradual, cost-effective way" (Council of the European Union 2011b, 2; emphasis in original). Inviting the Commission to develop plans in the different policy areas elaborating "cost-efficient reduction pathway[s] to a competitive low carbon economy in 2050" (Council of the European Union 2011b, 3), the Council supports the ongoing drafting process to find out how to reach a low carbon economy by 2050. Practically, this means that the Council shows its general interest in decarbonising the EU, without deciding for more ambitious goals or efforts at this point. Indeed, this is a well-known process of the Council, trying to balance the different member states' interests, not least as they are traditionally divided regarding their climate change protection ambitions (cf. 4.1.3).

But which hegemonic project does the Council follow and do the Presidency Conclusions provide a glimpse into the Council's discursive strategy? While strategeme I can be reconstructed, as several demands are equivalised, it remains to be discussed how the Council tries to write the EU's identity collectively.

Both strategeme II and III are hard to track within this contribution. In the Presidency Conclusions it only becomes clear that the economy, in this case the *low carbon economy*, is supported widely, and the co-benefits such as *employment opportunities* (Council of the European Union 2011b, 2) are welcomed and appreciated, but the demand to fulfil a leading role is not articulated—not even in a convoluted way, as it was by the Commission in its Roadmap. Hence, we can assume that the Council may work more towards the *low carbon economy* as the empty signifier rather than towards a somewhat shaped leading EU role, not least because the Council follows strategeme VI offering business actors subject positions regarding upcoming investments implementing a low carbon economy (Council of the European Union 2011b, 2). Nevertheless, it must remain open as to whether the Council is pushing for *low carbon economy* as the EU's identity at that time, as we need to reconstruct at least the three core strategemes to determine a hegemonic strategy (cf. Sect. 2.2.4). Hence, further Council contributions must be analysed, interpreting them on the discourse level. Irrespective of this, the possibility remains that the Council does not share a common hegemonic project due to the existence of sub-groups of member states which may follow different hegemonic projects, but these cannot necessarily be reconstructed from the Council Conclusions. Thus, we must leave it open as to whether or how the member states' ministers meeting in the Council want to collectively paint the EU's identity in global climate change politics.

The EU's discussion of the future economy model continued in 2011 as part of the preparations for the Rio + 20 Summit in 2012. Within the Commission's Communication "Rio + 20: towards the green economy and better governance" (2011a) the Commission builds on its 2011 Roadmap (2011c). However, the Commission is now no longer speaking about the concept of the *low carbon economy* but the *green economy*. Is this a new, more socially and ecologically inclusive understanding of a future economy concept which serves the fight against climate change, or is the Commission merely hiding the old under the cloak of the new by using the more progressive wording and concepts promoted by the UN for Rio + 20?[13]

[13] UNEP (2011, 2) defines green economy as "one that results in improved human well-being and social equity, while significantly reducing environmental risks and ecological scarcities. In its simplest expression, a green economy can be thought of as one which is low carbon, resource efficient and socially inclusive". This definition reflects a more progressive understanding of a sustainable economy, both socially and environmentally, than the low carbon economy presented by the Commission. Further international organisations were also working on the concept of a sustainable economy at this time, for instance the OSCE under the term green growth (Zeigermann 2020, 97).

In its Communication, the Commission equivalised various demands such as *growth while preserving the natural capital, development* (European Commission 2011a, 5), *energy security* and *renewable energies* (European Commission 2011a, 11) using strategeme I. Interestingly, the Commission also equivalised demands rarely encountered so far[14] such as *human well-being, decent jobs* and the *reduction of inequalities* (European Commission 2011a, 5), which are described as welcome side effects of future growth. This impression is also conveyed by the definition of green economy articulated in the Communication as "an economy that generates growth, creates jobs and eradicates poverty by investing in and preserving the natural capital offers upon which the long-term survival of our planet depends" (European Commission 2011a, 2).

Although the Commission expands its idea of a sustainable economy to include social and ecological aspects as compared to the previously promoted low carbon economy, it appears that the Commission sees the green economy primarily as an instrument helping the EU and its member states out of the economic crisis, mentioning growth and jobs first and highlighting the social and ecological co-benefits second. Nevertheless, the Commission is now also following feminised discourse patterns, although they play only a subordinate role—while the masculinised demands such as growth remain in the foreground. However, further analysis will show whether or not these new demands become a permanent aspect of the EU climate discourse and thereby whether they expand the EU's chain of equivalence.

Overall, the Commission is constructing *green economy* as an encompassing demand, implying that it goes hand-in-hand with all other equivalised demands (European Commission 2011a, 5), which is why the third strategeme can be attested. Stressing that "Rio + 20 should encourage countries, in particular industrialised and emerging economies, to develop domestic and regional carbon *emission trading schemes*" (European Commission 2011a, 12; emphasis in original), the second strategeme splitting the discursive space can also be reconstructed, even though it is only used implicitly. From the backdrop that the EU was almost the only actor running an emission trading scheme at that time, we can understand this articulation as an attempt to split the discursive space into a 'we the ETS operator' and 'the others not running an ETS yet'.[15] Marking the

[14] The EP member Rovana Plumb (S&D) has already articulated similar demands in the climate discourse, although she did not represent the majority of the EP back in 2010 (cf. Sect. 4.2.1).

[15] From 2005 to 2008, the EU was the only entity with an ETS (International Carbon Action Partnership 2022, 36). This was followed in 2008 by New Zealand, which of course only has a very small share of emissions, as well as further initiatives by individual US cities and

EU as the ETS forerunner can furthermore mean that the Commission wants to show its ability to lead by example, not least as the EU ETS is often used as a blueprint for other schemes (cf. Sect. 4.1.2). However, the demand to be recognised as a leader is not explicitly articulated anywhere in the text, so this can only be an indication. Hence, it seems that the Commission tried again to change the focus in the EU's climate change discourse from a *leader by example* towards a *low carbon economy*, now denoted as *green economy*. The further analysis will show whether the Commission's new hegemonic project succeeds, and further actors follow its attempt to anchor the *green economy* as the empty signifier of the whole discourse.

In the light of the accelerating discussion of the Rio + 20 Summit, the annual COP took place in Durban at the end of 2011. There, the second Kyoto commitment was agreed and combined with the negotiations on a roadmap elaborating a global binding agreement to be concluded in 2015, known as the Durban Platform. As Connie Hedegaard "was hailed as a hero because of her critical role in Durban" (Bäckstrand and Elgström 2013, 1380), COP17 ended on a very positive note for the EU. Successfully "attaining its goals" (Groen and Niemann 2013, 319), which were more realistic than in previous years, the EU gained a firmer foothold on the international climate stage, and was back with "structural leadership" (Bäckstrand and Elgström 2013, 1380). As this was possible in particular due to the fruitful cooperation with AOSIS and some of the G77 states, Bäckstrand and Elgström stress that (2013, 1381) "the Union combined directional leadership with bridge-building activities, that is, it acted as a leadiator". Thus, the demand to lead in global climate change policy, which is often articulated by the EP or some of the NGOs as well as the Climate Action Commissioner back in 2010 (cf. Sect. 4.2.1), was performatively rearticulated by the EU negotiators in Durban—with a specific note on the EU's mediation abilities. Indeed, other demands made by the EU were also rearticulated in Durban, particularly demands for cooperation with other ambitious actors as well as for stronger mitigation action. However, the evaluation of the results achieved in Durban suggests the representation of the EU's identity by the possible empty signifier leader

Tokyo. Only in 2013 did a noticeable increase in emissions trading schemes follow, with Switzerland, California and the large Chinese cities called "Chinese Pilots" (International Carbon Action Partnership 2022, 36). Of course, the schemes which were introduced in 2013 were still adopted when the Commission outlined the EU's forerunner role, but nevertheless the great emitters were still not a part of these global efforts. Since 2021, the share of global GHG emissions covered by emission trading schemes has been 17% (International Carbon Action Partnership 2022, 37), compared to only 5% when the EU ETS was launched.

or leadiator. In her COP17 speech, Connie Hedegaard focused on the urgently-needed roadmap and the indirectly-linked second commitment period, success of which secured the EU's leading role—yet didn't mention sustainable economy or even green economy (Hedegaard 2011). Does this picture also correspond to the way in which other EU institutions positioned themselves in the run-up to Durban?

The Council's stand on COP17 was formalised in the adoption of its Conclusions: In addition to already familiar demands such as the rapid adoption of a *legally binding agreement*, *mitigation by "all major economies"* (Council of the European Union 2011a, 3; emphasis added) and *keeping the increase of global temperature below 2°C* (Council of the European Union 2011a, 1, 3, 4), it furthermore calls for a future *low-carbon economy* (Council of the European Union 2011a, 4) and the *transition to environment-friendly growth* (Council of the European Union 2011a, 6). Equivalising these demands, the Council broadened its climate change policy focus from concentrating particularly on the need for global action and a legally binding agreement to targeting more concrete or—as Feyyaz Çelik (2022, 1028) puts it—"rational" aspects "such as the use of low-carbon technologies as instruments of minimising the costs of transition into environmentally sustainable methods of operation" In this context, the Council again welcomes the Commission's Roadmap looking forward to working on the topic and recognises "the importance of transitioning to a low-carbon economy and society" (Council of the European Union 2011a, 4). Although the Council focuses more on the economic future of the EU, the promotion of a low carbon/green economy cannot be clearly reconstructed as its hegemonic project due to missing core strategemes II and III.

The EP largely adhered to the demands already articulated in 2010, such as a *legally binding agreement*, a *second Kyoto commitment period* (European Parliament 2011, 87) and *ambitious mitigation efforts by the EU* (European Parliament 2011, 89). In other parts of the EP's resolution, it calls for the global *below 2-degree target* (European Parliament 2011, 85,87,88) and furthermore refers to the *1.5-degree goal* (European Parliament 2011, 90). As the Council also adhered to the *below 2-degree target* (Council of the European Union 2011a, 1), but does not take up the 1.5-degree goal, the EP is more progressive in its positioning. The EP is also more clearly committed to *CBDR* (European Parliament 2011, 85, 87, 88, 97) than the Council (Council of the European Union 2011a, 3) while both agree that the main commitment must come from the industrialised countries/major economies stabilising *CBDR* as a demand of the EU's chain of equivalence. The EP expands the chain additionally when it calls for gender-sensitive climate policies for instance regarding the fair participation of women in the upcoming

Green Climate Fund (European Parliament 2011, 92). Pointing to the need to "effectively address[ing] gender-specific impacts of climate change" in manifold adaptation efforts (European Parliament 2011, 92), the EP follows strategeme X, introducing several demands of the gender equality discourse into the EU's climate policy discourse. In addition, the *protection of particularly vulnerable states* (European Parliament 2011, 90, 92) and active support for developing states—both *technically* (European Parliament 2011, 94) and *financially*—is demanded in the EP-resolution (European Parliament 2011, 90). Following the first strategeme, the EP equivalises these outlined demands with each other and also with *climate justice* (European Parliament 2011, 88) as well as the *transformation of the EU's economy* (European Parliament 2011, 93) welcoming the Commission's Roadmap (European Parliament 2011, 88). In this way, the EP creates a more diverse chain of equivalence, partly in lockstep with the Council and the Commission, which also drew greater attention to sustainable economy demands than previously. The trend to focus more on socially inclusive aspects of the green economy, which Çelik (2022) has already identified with respect to the Council, can also be reconstructed for the EP and the Commission's communications. This goes along with the observation that the concept of a green economy "has generally been subsumed within the meta-discourse of sustainable development" (Ferguson 2015, 17), whereby James Goodman and Ariel Salleh (2013, 417) state in their ecofeminist critique of the green economy that it is some kind of a "Trojan Horse, to be incorporated into the sustainable development agenda but then outgrow it". This aspect is to be taken up again later (cf. Sect. 4.3.1).

Coming back to the question of whether the demand *green economy* or the *leading role* served at this point as the hegemonic project of the EU's entities, I can state for the EP that it still adheres to the leading role. Besides actively demanding an EU leading role in global climate change policy in order to attract other states to follow (European Parliament 2012a, 89), the EP used strategeme II and III to paint the EU as a leading entity, emphasising that other actors such as the US and China are working towards a green economy and that the Commission should therefore encourage the EU to also follow this economic path for the sake of the EU's leadership (European Parliament 2011, 92). Hence, the EP welcomes green economy efforts but understands them as one piece of the mosaic of a leading EU in climate change policy—which is therefore still its hegemonic project.

In summary, the EU's performance in Doha continues to speak for the hegemonic project of the leading role, which the EP also supported. The Commission's behaviour seems ambiguous, as evidenced by the fact that its delegation in Doha rearticulated in a performative way the demand for the *leading role*, but buttressed the *green economy* and earlier also the *low carbon economy* in different communications discussing the EU's future climate change policy.

How did other actors position themselves in this fragmented field? Some of the NGOs have been won as socio-politico forces supporting one of the potential empty signifiers. Building on its annual report of the previous year, the WWF was now particularly supporting the demand for the *green economy*. Under the headline "From Economic Crisis to Green Economy" (WWF European Policy Office 2012, 8) the WWF dedicated the whole first chapter of its annual report 2011 to this future economic concept. In saying that it had launched "a programme to provide input to the debate on how greening the economy could contribute to easing the financial crisis", the NGO linked a possible greener way out of the crisis with "a more optimistic future" (WWF European Policy Office 2012, 8), thereby becoming a prominent advocate of the *green economy*. Following strategeme I, the WWF equivalised *green economy* inter alia with *sustainable development* and *economic innovation* (WWF European Policy Office 2012, 8). Reporting that the WWF called on the EU in its 2050 Energy Report "to strengthen its *global leadership position* in realising a *carbon-free economy*", (WWF European Policy Office 2012, 16; emphasis added), these demands were also equivalised. Here, it is important to note that even though the WWF demands the EU's leadership, it does so solely for the sake of the implementation of the green economy. Moreover, to push the green economy further to succeed, the NGO follows strategeme VI creating an alliance together with 72 European companies calling for the 30% goal for 2020 (WWF European Policy Office 2012, 16) to bring the EU closer to the green economy.

The WWF's reasoning in its climate policy section points exclusively in the direction of a green economy, which is also announced as one of the top priorities for the upcoming year. Implying that the other outlined demands go hand-in-hand with a green economy, which is able to fill the lack of the common universal, the WWF constructs the green economy as the empty signifier. Since the WWF is consulted by MEPs or Commission officials at this time (WWF European Policy Office 2012, 8) particularly for its expertise regarding green economy, and might be influential if they succeed in shaping the EU's understanding of a well-established green economy, supporting this hegemonic project could be a temporary strategic choice. However, it looks as if it would work for the WWF to envision the EU as a future green economy trying in parallel to achieve the other WWF goals such as more ambitious mitigation efforts or environmental protection. Hence, on the one hand WWF pushes for the green economy as the empty signifier and on the other hand it fills it with more ecological aspects than does, for instance, the Commission.

Friends of the Earth Europe (FotEE) reproduced some of the demands articulated by the WWF. Interestingly, the two NGOs seem to totally disagree when it comes to the green economy. Trying to shape the concept to fulfil its own interests, the WWF followed strategeme V, generating greater leeway. On the contrary, FotEE articulates its critique on the green economy by rejecting it completely. While Rio took place in 2012, various protests were held by the NGO, for example in Belgium or in Paris, publicly criticising the concept as greenwashing of the prevailing economic system based solely on the exploitation of nature (Friends of the Earth Europe 2013, 12, 13). Understanding green economy merely as the strengthening of corporate groups that neither serves the climate nor the environment is another reason for FotEE's resistance (Friends of the Earth Europe 2012). This example shows that the green economy is not able to unite all European NGOs behind it because it is not encompassing different genuinely climate-protecting demands in the same way. Nevertheless, even if the NGOs do not equally commit to this demand representing the identity of the EU, there is still a chance that the green economy will—at least temporarily—become the empty signifier, since some NGOs, such as the WWF or—in part—CAN Europe (2012, 1), do support it. However, the reconstruction of the discourse vis-à-vis Durban has shown that the green economy dominated the discourse there only marginally. In light of the findings regarding the NGO voices, it is therefore necessary to reproduce the further discourse to find out whether the hegemonic project of some to replace the *leading role* by the *green economy* succeeded—or whether green economy is just a flash in the pan due to the Rio summit's agenda.

After years of preparations and discussions, which were indeed part of the EU's climate change discourse, the Rio Earth Summit 2012 took place with roughly 44,000 participants. The decisions in Rio were not as far-reaching as the global debate in the run-up to the summit might have suggested; Biermann (2013, 1100) even speaks only of the "least common denominator". Nevertheless, it is worthwhile to reconstruct then Commission President Barroso's speech at the opening session to gain further insight into the Commission's understanding of green economy.

In Rio, Barroso equivalised a huge variety of demands such as the *fight against poverty* and *better development cooperation* which focuses on the aspect of *sustainability*, as well as *sustainable economy* (Barroso 2012a, 2). Even though I am not able to reconstruct the use of a hegemonic strategy in the speech, it is interesting to note that Barroso equivalised these outlined demands with each other saying that: "For the European Union, this is what green economy is all about" (Barroso 2012a, 2). Combining this quote with my former analysis of the

Commission's discourse contributions, it seems that Barroso widened the Commission's understanding of the green economy, incorporating additional social and ecological aspects without making them into a hierarchy. In the Commission's Communication the economic aspects had dominated the ecological and social dimensions which had (only) been described as welcome co-benefits of sustainable growth.

Besides discussing the green economy, the year 2012 saw several initiatives to bring gender-related aspects more into the EU's climate change discourse. Following the appeal of the Danish Council Presidency (Council of the European Union 2012, 5), EIGE, the European Institute for Gender Equality, published "the first EU-wide report on gender equality and climate change" (Langbakk 2012, 3) providing data and examining various correlations.[16] In conclusion to the research report, EIGE formulated some policy recommendations, such as to increase "women's participation in climate change decision-making", as it "is an important factor for more gender-responsive and efficient climate change policies" (European Institute for Gender Equality 2012, 73). Furthermore, EIGE demands an increase in women's "enrolment in science and technology-related educational fields leading to careers in the energy and transport sectors [...] obtaining access to institutions and power structures that support and control climate change policymaking" (European Institute for Gender Equality 2012, 73). Although these articulations are a vital part of the EU's climate change discourse, they do not yet count as nodal points. Even though some of the demands have already been sporadically articulated by others, for instance the EP, these gender-specific demands need to be rearticulated and therefore stabilised by a critical number of actors if they were to shape the EU's chain of equivalence.

However, it is important to note that EIGE particularly articulates gendered demands serving the first dimension of gendered climate change discourses (cf. Sect. 2.3.3). While the first dimension is rare within the EU climate change discourse, EIGE interpellates women on the one hand as particularly affected by climate change (European Institute for Gender Equality 2012, 19–20), but on the other hand also as actors who urgently need to be more involved as politicians (European Institute for Gender Equality 2012, 36) or as engineers and planners (European Institute for Gender Equality 2012, 38–43), because otherwise considerable parts of the reality of many people's lives would not be taken into account

[16] These are, for instance, the different ways in which women are affected by the consequences of climate change due to gender structures, the importance of women's participation in improving various policy fields linked with climate protection, as well as the finding that emission levels differ due to specific socially-created gender roles (European Institute for Gender Equality 2012).

when formulating solutions to fight climate change impacts and to mitigate GHG emissions. Indeed, if climate change policy is made by a homogenous group it is very likely that the policies particularly address their specific realities. Keeping that instance in mind, EIGE's report implicitly shows how gendered discourse patterns of the first and the second dimension interact, as EIGE points out that the use of public transport or climate-friendly behaviour are gendered insofar as both are more often used by women, while policies in these respects have at the same time been underutilised. Whether other actors rearticulate the subject positions interpellated by EIGE and thus consolidate them in the discourse must be shown in the further analysis. A final interesting aspect of the report is that EIGE rearticulated the well-established demand of *EU leadership* in global climate change negotiations (European Institute for Gender Equality 2012, 73). Demanding the use of this role for gender improvements in the climate field, it seems that EIGE follows two strategemes. First, it uses strategeme III, representing the EU as a *leader*, and second, EIGE tries to bring the gender-equality discourse or at least parts of it into the climate change discourse, following the example of the EP (2011, 92) using strategeme X.

The Council took up some of EIGE's recommendations in its Conclusions "gender equality and the environment" and integrated some of the report's specific demands into the EU's chain of equivalence. Demands such as that of increasing the employability of women and particularly the *participation of women* in scientific research, and in climate change politics (Council of the European Union 2012, 3) were rearticulated by the Council and therefore stabilised as a probably new nodal point of the EU climate change discourse offering women the subject position as agents. The Council followed strategeme I equivalising these demands with demands such as *mitigation action by the EU* and *integrating a gender equality perspective* into the different fields of mitigation such as transport or energy (Council of the European Union 2012, 4). Interestingly, it seems that the Council is not interested in women's integration into the labour market solely for equity reasons, but more for reasons of future competitiveness and economic success pointing to "the emerging green economy, a potentially job-rich area" (Council of the European Union 2012, 4). Although the Council is actively extending the EU's chain of equivalence here, it is again not possible to detect the pursuance of a specific hegemonic project.

Besides the Council, the EP in two different documents also discussed and adopted its position on gender and climate change. Before the EIGE report was published, the EP adopted a Resolution in April of the same year. Demanding gender-sensitive initiatives, the EP uses surprisingly clear language in asking both the Commission and the Council to *gender-mainstream* future climate change policy (European Parliament 2012b, 4). Since the EP has a decades-long tradition of using a gender-sensitive approach in its development policy (Debusscher 2021, 293), especially in its Development Committee (Allwood 2020a, 2020b), it is able to build on existing awareness and expertise if it decided interlink gender and climate policy more strongly.

Later in September 2012 the Parliament approved a resolution on women in the green economy. In its September Resolution, the EP mentions the EIGE report and equivalises different gender-related demands such as the *increase of female participation* in different sectors of environmental-related policies as well as research and climate-friendly green jobs (European Parliament 2012c, 7). Interestingly, the EP does provide its definition of green economy in this Resolution, pointing out, that

> *"green economy is defined as a sustainable economy, which means social and ecological sustainability; whereas social sustainability involves a social order permeated by gender and social equality regardless of gender, ethnicity, colour, religion, sexual orientation, disability or political opinion" (European Parliament* 2012c, *3).*

Accordingly, the EP uses a broad green economy definition, incorporating ecological and social aspects with a particular focus on gender, and seems to extend the Commission's understanding. Even though I am not able to reconstruct a hegemonic strategy in this resolution, the EP's green economy definition is particularly important if it becomes hegemonic, representing the EU's climate identity.

At this point, we can state in principle that all actors considered so far clearly advocate for integrating various aspects of the EU gender discourse into the EU's climate discourse. Consequently, they use strategeme X (again), the "Occasional but deliberate breach of a specific discourse's limits" to integrate different gender-related demands into the climate change discourse of the EU. As theoretically derived in Sect. 3.2.8, this endeavour does carry a certain risk, as opening up the EU climate discourse's borders could destabilise the chain of equivalence if too many "external" demands are able to enter the discourse. Therefore, those EU actors using strategeme X assume added value, additionally stabilising the hegemonic project by integrating specific demands of the gender-discourse. The motive for integrating gender aspects is to strengthen the social acceptance and support of the empty signifier. Indeed, integrating gender aspects into the EU's climate change discourse could serve both empty signifier candidates—the *leading role* as well as the *green economy*—if this suggested that they stand not only for an ambitious climate policy, but for an ambitious and gender-sensitive climate policy.

While gender-related demands became nodal points in the EU's climate change discourse, the timing of the actor's articulations of these demands might have been triggered by both the Rio + 20 summit[17] and the UNFCCC-initiated greater gender debate in Doha at COP18. In Qatar at the end of 2012, the so-called Doha Miracle was achieved, namely the official agreement that all UNFCCC member states commit themselves to gender equality, particularly in their delegations (UNFCCC 2012, 48). The key decisions in Doha were the final acceptance of the second Kyoto commitment period, the binding schedule to reach an agreement in 2015 and to increase the climate protection ambitions before 2020 (United Nations Framework Convention on Climate Change 2023b). The EU actors' speeches at the conference and the internal EU discussions prior to COP18 will only be briefly touched on here in trying to answer the question of which hegemonic project was followed both in and prior to Doha.

The EP positioned itself as usual in the run-up to the COP by releasing a Resolution. In this framework, the Parliament constructs the EU identity as represented by the demand for leadership—most likely through *leading by example.* Concretely, it demands to *lead by example* concerning the implementation of the Durban platform for action (European Parliament 2012a, 5) and the sustainable use of raw materials (European Parliament 2012a, 11) but also regarding the financing of much-needed future research (European Parliament 2012a, 12). Besides these cases of the EP demanding to *lead by example*, it also demands to lead more generally in climate protection (European Parliament 2012a, 4). It is also important to note that the EP is committed to transformation leading towards a sustainable economy (European Parliament 2012a, 10, 11), but neither in the sense of an identity representing the EU nor using the term green economy. In the framework of the EP debate prior to the adoption of the Resolution, a few MEPs also perpetuated the demand to *lead by example* (Jørgensen 2012) or of the *EU as a leader* in climate change policy (Seeber 2012; Jadot 2012; Panayotov 2012). The EP discussion gives the same impression as the Resolution, since *green economy* is demanded just once in the debate (Estrela 2012). It seems that the EP had lost its careful interest or hope in the green economy, perhaps because the concept was only a vaguely formulated component of the Rio Declaration 2012 (Beisheim, Lode, and Simon 2012, 2) and did not seem to spark any new climate policy momentum.

---

[17] The Earth Summits look back to a decades-old advocacy on the part of feminist activists to adopt a gender perspective in sustainable development and related debates (Cornwall and Edwards 2014). However, the nexus of development, climate and women/gender, was only really taken into account in the final document of 1992—but not of 2012 (see Foster 2017 for a detailed critical discussion of both outcome-documents).

Back in Doha at COP18, Hedegaard formulated her demands, very cautiously calling for *global cooperation* and underlining the importance of a *binding agreement in 2015* (Hedegaard 2012a, 1–2). She however underlined that the EU had already delivered, indirectly portraying the EU as a leading actor trying to bridge globally differing interests (Hedegaard 2012a, 2). Hedegaard fulfilled this balancing role in particular within the framework of the informal group of the Cartagena Dialogue, which she co-created. In the Dialogue's framework, high level debates between ambitious countries from different negotiating groups were held during the year to build common understanding and trust (Schunz 2015, 185), which was an important piece of the mosaic for the subsequent Paris success.

The Council Conclusions before Doha are again ambiguous as to which hegemonic project the Council follows—or whether it follows a hegemonic project at all. Hence, the hegemonic project can once again only be clearly named as the *leading role* for the EP, while the Commission was apparently undecided whether to still stick to *green economy* as hegemonic project or to follow Hedegaard's articulations pushing the EU's *leading role* again. For the NGOs we can draw almost the same conclusion. In the meantime and only for a short period, the chain of equivalence of the EU—informed by the question of how to get out of the financial crisis, and by the global dynamics due to Rio + 20—developed towards a *green economy* as a new hegemonic project. However, this hegemonic project was not successfully established as a hegemony because it didn't find the support by a 'critical mass'. Thus, the green economy's time did come, but its heyday was short-lived and focused primarily on economic aspects. Hence, it can be assumed that at least by the Commission the *green economy* was rather followed as a greenwashed idea for recovering economically. For this purpose, both masculinised and feminised discourse patterns were used, but with a strong focus on masculinised demands. We will return to them later, but first the wider EU climate discourse will be reconstructed.

4.2.3 From Warsaw to Paris: Future EU Goals in Limbo?

The period from 2013 onwards was marked by negotiating and preparing for the Paris Summit in 2015 and thus also for a broad discussion of the future EU climate policy and its 2030 targets.

In February 2013, the EP met to look back at the Doha COP and discuss the EU's climate policy in general as well as future challenges. Compared to former EP debates, the MEP's reasoning differs only slightly. Hence, solely new patterns and aspects necessary for gaining an insight into the EP's idea of the EU's identity are reconstructed. Then Chairman of ENVI, Matthias Groote (S&D), stressed that the Council had too often needed clarification and recalibration of its position back in Doha. Therefore, he renewed the demand for a united EU *speaking with one voice* (Groote 2013), giving the chance for the EU to be recognised for the role in which he would want them to be seen by third countries. Insisting that "we as Europeans should not wait for others, but should continue on a path—towards a low-carbon society, towards a sustainable society—and get the 194 states on board" (Groote 2013), Groote clearly shows that he wants the EU to *lead by example*.

Lucinda Creighton, representing the then Irish Council Presidency, also emphasised the demand that the EU *lead* in the international climate change negotiations in her two statements during the EP debate. Stressing that "There are expectations and opportunities for the EU and its Member States to play a key role in facilitating agreement and in continuing to lead an ambitious agenda to keep global temperature increases below the 2°C threshold" (Creighton 2013) she used strategeme I equivalising a *leading role of the EU*, the *below 2-degrees goal* as well as *opportunities for the EU and its member states*. Interestingly, in her second speech she becomes clearer and stresses that the EU institutions and the individual member states "can take the lead" in drafting and shaping the post-2020 agreement and that this "is the objective of the Irish Presidency (…)" (Creighton 2013), constructing the demand to *lead* as the empty signifier representing the EU's identity. As the Council hadn't claimed the EU to be a leader in its contributions analysed so far, it seemed fragmented regarding the question of how it wanted the EU to be seen in global climate change policy, while some pushed for the leading role in the same way as Creighton. Climate Action Commissioner Hedegaard once again did not articulate the demand of a *leading role* in this EP debate, even though she did so in 2011 (cf. Sect. 4.2.1). Rather, she drew fresh attention to the urgency of *speaking with a common voice* (Hedegaard 2013a).

Pressing for unity was necessary not least as the already existing conflict with the Visegrád group concerning the EU's climate policy had become tighter in 2013, particularly with Poland. Without calling them out, S&D MEP Dan Jørgensen underlined in his statement that potential failure of EU climate policy lay in his view notably in Poland's opposition regarding more ambitious climate protection measures and targets (Jørgensen 2013)—which is why he called on the

Commissioner to push for the use of majority voting.[18] MEP Gerben-Jan Gerbrandy (ALDE) was even more explicit about the Polish brake block[19] and the related opacity of the Council when asking "why can we not in Parliament invite the Polish Minister to explain his position on climate change?" (Gerbrandy 2013). Equivalising the demands for *stronger climate targets* and higher *transparency by the Council*, Gerbrandy pointed his finger to a major weakness of the Council's climate change policy, as the discussions of the member state's environment ministers are often not accessible—and rather based on national interests. Therefore, the demand *to speak with one voice* was unsurprisingly articulated across party lines in this EP debate. It seems that a major problem has been identified which could inhibit the discursive production of the EU's identity as a leader in climate change politics. Indeed, a leading entity who does not speak with one voice can easily be weakened, as the different demands such as ambitious mitigation action or a higher share in renewable energies need to be rearticulated again and again as the hegemonic project is otherwise not able to succeed.

All parties under the UNFCCC protocol must adopt its Intended Nationally Determined Contributions (INDCs) prior to Paris. Due to new general conditions, such as the ongoing economic crisis, the lack of investment capacities in economically weakened countries and major changes in the energy sector, the Commission pushed for a broad process debating the EU's future climate change policy to begin. Hence, the Commission published its Green Paper "A 2030 framework for climate and energy policies" in March 2013, inviting all interested parties to contribute to following digital consultations (European Commission 2013). In the Green Paper, the Commission primarily gives an overview of future challenges and numbers outlining only a few demands which are mostly derived from its 2011 Roadmap (cf. Sect. 4.2.1). Following strategeme I, the Commission equivalises several demands, such as the concrete *40% target by 2030*, arguing that it would pave the way for the necessary *80–95% target by 2050* with the *increase of renewable energies* by a share of 30% until 2030 (European Commission 2013, 3). Furthermore, the Commission articulates the demand for *fair burden-sharing* among the member states (European Commission 2013, 12) while no further concrete targets—such as an energy efficiency- or energy saving goal—are proposed.

---

[18] Although the EU's climate policy was already subject to majority voting at the time, the focus continued to be on finding consensus.

[19] One of the central disputes at this time was the revision of the ETS and the planned backloading of emission certificates (Creutzig et al. 2014, 10). Poland was the only opponent, not least against the background of its energy mix as well as energy-intensive economic model.

One of the actors replying to the Green Paper was CAN Europe. In its extensive response, the NGO reproduces several of its previously known demands, such as *increasing the EU's mitigation efforts* and *taking its historical responsibility* in climate politics (Clima Action Network Europe 2013b, 5). Stressing that "European targets and policies have in fact shaped policies worldwide by providing political leadership, by helping new technologies go down the learning curve faster, and by helping other (in particular developing) countries" (Clima Action Network Europe 2013b, 2) CAN clearly uses strategeme III representing the *EU as a leader* for various reasons. Hence, the NGO network constructs the demand for a *leading role* of the EU as the empty signifier encompassing the other articulated demands. Another emblematic illustration of CAN pushing its hegemonic project of the EU as a leader is its argumentation for higher EU climate targets "maintaining its global leadership on climate action" (Clima Action Network Europe 2013b, 2). Bailing up more than 120 European ENOG's views, CAN represented a large share of the tenor of the NGOs at that point.

Eurochambres, the Association of European Chambers of Commerce and Industry, represents the view of some of the business stakeholders. In its position paper, Eurochambres demands a strong role for the EU "in international efforts to combat global warming" (Eurochambres 2013, 1), calling the Union a "pace setter" (Eurochambres 2013, 4) for global emission targets. Equivalising the demands *technical innovation*, *renewable energy* and *energy efficiency* (Eurochambres 2013, 1), Eurochambres uses strategeme I. Strikingly, the chain of equivalence of Eurochambres furthermore demanding (voluntary) European *mitigation goals* (Eurochambres 2013, 5) and welcoming the EU's climate policy efforts is very similar to the chain shaped by CAN. Since these demands articulated by Eurochambres serve the purpose of ensuring cost-effective and "more flexible and innovation-based approach[es]" (Eurochambres 2013, 1) and competitiveness, we can attest that the same demands are articulated with differing motives by the two different actors. Interestingly, Eurochambres follows strategeme III representing the EU as an actor keeping a *strong role* in global climate change policy. Even though *a strong EU role* is not the same as a leading role, the hegemonic projects of Eurochambres and CAN point in a similar direction. Following strategeme II, Eurochambres also clearly distinguishes the EU from other major emitters by insisting on global rather than unilateral action. Arguing that unilateral commitments instead of a globally binding agreement only shift emissions to third countries, the demands for an *international legally binding treaty* and indirectly for *climate justice* are rearticulated (Eurochambres 2013, 3). Indeed, calling for GHG emissions not to be externalised particularly protects the region's economy, which might be the real motive behind the demand for a *strong EU role* in global climate politics.

The reconstruction of the discursive strategy of Eurochambres and CAN shows that constructing the same chain of equivalence does not automatically mean that the actors have the same idea of the shared demands. Rather, the examples illustrate the effect of the logic of equivalence; very different actors follow the same—at least superficially so—demands which are equivalent in relation to the radical other's chain of equivalence. While both entities demand *future mitigation action*, CAN means ambitious and binding mitigation- and energy-related targets, whereas Eurochambres only supports voluntary goals. But they share the common demand for *mitigation action* picturing the radical other as less actively engaged in climate protection efforts.

In September 2013, in his last year in office, the President of the European Commission, Manuel Barroso, addressed the European Parliament with his annual State of the Union address. Taking stock of "the past achievements, the most pressing challenges (…) and set[ting] the Union's priorities for the years to come" the State of the Union address is a "public occasion for reflection on the identity and the future of the European Union" (Pansardi and Battegazzorre 2018, 854) with a broad audience, because in addition to the EP, these speeches are intended to reach EU citizens in particular. The EU's climate change policy takes up very little space in Borroso's multifaceted 2013 speech. Discussing the EU's economy, he links future economic success closely with the Union's climate change policy (Barroso 2013a, 5). Barroso's bottom line is that the EU should become an *economic leader* with the help of *stronger research* and the resulting *technical innovation* (Barroso 2013a, 5)—aspects he also expects to decisively advance the EU's climate protection campaign. Without demanding an EU leadership role in climate change policy, the Commission President paints the picture of a future economically-leading EU serving the fight against the consequences of climate change. Back in 2012, Barroso had formulated that demand very similarly in his State of the Union speech, stressing that "The world needs an EU that keeps its leadership at the forefront of development and humanitarian assistance. That stands by open economies and fights protectionism" and continuing that this EU also "leads the fight against climate change" (Barroso 2012b, 11). The President of the Commission again constructed a tight connection between climate change leadership and economic issues, demanding EU leadership in global climate policy even more clearly under the caption "Political Union" (Barroso 2012b, 9). In 2012, Barroso had pushed very clearly for the EU's leading role in climate policy—representing the EU's identity in the sense of strategeme III—and although he was more implicit about it in 2013, he nevertheless maintained his stance for the second year running. Furthermore, both State of the Union speeches in 2012 and 2013 provide insight in Barroso's central answer to the challenges posed by the climate crisis; technical innovation and economic success for which an EU leadership role in climate change seems to be needed. Regarding his favoured solutions for fighting climate change, we can attest that his—of course only briefly outlined—ideas are androcentristic, because the demands *technical innovation, leading role* and *economic success* are clearly part of the socially constructed sphere of masculinity (MacGregor 2010; Bruff and Wöhl 2016, 96).

In the subsequent EP discussion following his 2013 address, Barroso responded to accusations from the party whip of the Greens, Rebecca Harms, who argued that the Commission had neglected climate protection efforts in recent years (Harms 2013). Responding to this allegation, Barroso became surprisingly clear and rearticulated the demand for a *leading role*: "Ms Harms, when you say that Europe has done nothing or that on climate change there is zero change, come on! Let us be real! Europe is leading the world in terms of the objective for climate change" (Barroso 2013b). In doing so, Barroso openly constructs the identity of the EU in climate policy as a *leader*, using strategeme III. This rare moment of the Commission President clearly pursuing his hegemonic project confirms the above-mentioned impression that his demand for a *low carbon economy* and associated *innovations* were part of Barroso's vision of EU leadership in climate change policy. Hence, he tried to shape the character of his hegemonic project, the leading role of the EU, putting economic aspects into the empty signifier using strategeme V.

Parallel to these engaging debates about climate policy on the EU level were discussions on the global scene which were supposed to result in another agreement at COP19 in Warsaw at the end of 2013. What demands were formulated and what strategemes were used prior to the COP hosted by EU member state Poland? The Council Conclusions in preparation for the upcoming COP will provide a first impression. Equivalising different demands such as *more ambitious international mitigation efforts* (Council of the European Union 2013, 2, 3, 4), *transparency* and *MRV* (Measurement, Reporting, Verification) (Council of the European Union 2013, 3) as well as future *stocktaking* (Council of the European Union 2013, 5), the Council clearly tries to shape the future climate agreement, again without openly articulating its hegemonic project. Against the backdrop that, besides the emerging powers, the USA in particular resisted any MRV mechanisms enabling the global community to compare the planned INDCs ex-ante for Paris,[20] the Council pursued strategeme II, splitting the discursive space pointing to 'the others' who are not committed to reviews and are therefore less encouraged than 'us', the EU. It is also noticeable that the Council attaches more importance to adaptation than in previous years, although the focus remains on mitigation. When the Council demands *adaptation by all,* the EU institution equivalises it with *loss and damage,* using strategeme I. But the demands are given different weights. Stressing that the Council "REITERATES its commitment to continue constructive discussions on approaches to address loss and damage at the Warsaw Conference" (Council of the European Union 2013, 3; emphasis in original), it chose an extremely noncommittal wording. While adaptation is described as playing "a key role", the Council demands higher commitment by all partners trying to really enhance global action in the framework of the Paris Agreement. This would also include emerging powers; accordingly, the Council tries to burden them with the responsibility the industrial states in fact have when it comes to loss and damage.

Also differentiating the EU from the US and the emerging powers, the EP used strategeme II in its resolution on the upcoming COP19 in Warsaw, in the same way as the Council did. Welcoming the recent US government's signals to take climate action and to play a "greater role in global efforts" (European Parliament 2013, 9), the EP indicated the differences between the two, portraying the EU as an already highly-engaged entity. The EP points out that the EU acts as a "first mover" (European Parliament 2013, 9) insinuating that the USA stays as the laggard on the other side of the discursive border. Aiming to be a first mover, the EP furthermore uses strategeme V, as it shows that a leading role means both being a first mover in the international negotiations (European Parliament 2013, 9) but also to profit from possible economic advantages generated by the development of "climate-related innovation in the energy and industry sectors", because such innovation creates future jobs and stimulates "economic growth" (European Parliament 2013, 4).

Furthermore, the chain of equivalence as (re-)produced by the EP includes the demands for *technical innovation* (European Parliament 2013, 9) and the promotion of *energy efficiency* and *renewable energies* (European Parliament 2013, 6) but also *CBDRRC (European* Parliament 2013, *5).* The trend of not mentioning the green economy, which was already observed in the period after the Rio conference, continues. On the contrary, the EP mentions the concept of a future "sustainable economy" (European Parliament 2013, 10) twice. Although

[20] For the Indian debate on MRV/ex-ante review process and how it was seen as encroaching on India's sovereignty, cf. Sect. 3.2.3.

the term is not fully defined by the EP, it stresses that climate protection as well as room for development are an inherent part of sustainable economy. Since similar aspects had been addressed when the EP had discussed the green economy in former years, it is worthwhile to observe that only the wording changed so that the green economy was still pursued by the EP—but now called sustainable economy. Although this distinction may seem sophistical, it is important to know which future economic concept is being pushed by the actors to be able to identify and discuss their inherent gendered discourse patterns in the subsequent ecofeminist examination of the results.

To sum up, the EP still focused on the *leading role* as the empty signifier encompassing the other demands to represent the EU's identity in climate change policy. Painting EU climate policy as a cross-cutting issue interwoven with several other policy fields, such as "sustainable development, industrial policy and resource policy" (European Parliament 2013, 10), the EP understood the EU climate discourse as incorporating parts of other policy fields. Accordingly, the use of strategeme X, *occasional but deliberate breach of a specific discourse's limits,* can be identified. This will develop as one of the hallmarks of the EU climate discourse, as will be shown later.

This finding is in line with the hegemonic strategy of further EU actors, as Connie Hedegaard's speech at the opening of the high-level segment at COP19 shows. Commenting on the development that climate policy is increasingly attracting the interest of business actors, she stated that:

> *"If we want to move faster we need to mainstream climate action into the economy, into our key political choices and into our development strategies. We need a paradigm shift" (Hedegaard* 2013b, *1).*

By arguing for exporting climate-related demands into other discourses such as the EU economic discourse, Hedegaard is taking a certain risk, namely to open the borders of the climate discourse and therewith jeopardise its stability which might trigger a change of the EU's chain of equivalence in an unexpected way, probably to the detriment of the hegemonic project.[21] Hedegaard uses strategeme X here in a new form, not importing individual demands from another discourse, but as an attempt to export climate-related demands into other discourses. Further investigation must show whether she or other actors were using strategeme X successfully at this point. This form of climate mainstreaming will be discussed

[21] For an in-depth discussion of the risks using strategeme X and its theoretical development see Sect. 3.2.8.

in more detail in Sect. 4.2.7, when I will be able to potentially attest the broader use of this specific form of strategeme X with the new European Green Deal.

What was the outcome in Warsaw and did the EU have the chance to position itself as leading in the framework of COP19? Although hopes were expressed that the EU could gain an additional positive profile through Poland hosting COP19, the fears voiced in advance that the Polish Presidency would actually harm climate protection for example through NGOs, proved to be true (Keating 2013). The COP presidency was recognised as unambitious and disappointing. It was accompanied by the Polish blockade of higher ambitions within the the EU (Creutzig et al. 2014, 11) and public demands for shale-gas extraction. The fact that the Polish Minister of Environment, who chaired the COP negotiations, was removed from his ministerial office in the middle of COP19 by then Prime Minister Donald Tusk is also a discourse contribution underlining Poland's lack of priority for climate protection (Marcinkiewicz and Tosun 2015, 14) as the chief negotiator's authority was severely undermined (Fuhr et al. 2013). Nevertheless, some results were achieved in Warsaw since particularly European states salvaged the international adaptation fund that help developing countries to deal with the changing climate (Boyle 2013, 5). Another success was the introduction of the loss and damage mechanism which was indeed only a very first step down the road towards more climate justice, not least because the newly established Warsaw International Mechanism for Loss and Damage (WIM) was only provided with a mandate "focus[sing] on knowledge and expertise sharing activities" (Calliari 2018, 726). Remembering the Council's highly cautious statement on Loss and Damage, WIM was more a success for the developing countries (Calliari 2018) than for the EU. Besides finding a compromise regarding WIM, these COP negotiations almost failed, illustrated, inter alia, by the premature departure of many well-known NGOs due to "the slow speed and lack of ambition" in Warsaw (Vidal and Harvey 2013). The EU-demanded schedule en route to Paris was adopted only with some cutbacks (Herrero and Knaepen 2014, 6), as MRV measures and an earlier submission date for the INDCs were not agreed upon. Hence, actual leadership of the EU can only be poorly attested regarding COP19. However, it was the USA, among others, that put the brakes on MRV prior to the Paris summit, but also on an early pledging deadline. Thus, the EU still formulated more far-reaching demands; announcing its INDCs earlier than the other major emitters as well as providing transparency harboured the chance to establish itself as a leader.

Reflecting its actions and policies in 2013 in its annual report, WWF Europe also calls for the EU to play a leading role. Using strategeme I, the WWF continues to equivalise its already formulated demands such as higher share of *renewable energy* and *energy efficiency* (WWF European Policy Office 2014, 22) but also *low carbon economy, more sustainable lifestyles* and *investments in green technology* (WWF European Policy Office 2014, 22). Furthermore, the organisation follows strategeme VI offering other politico-societal forces subject positions shaping the EU's chain of equivalence to support the WWF's demands. As in previous years, these are business actors (WWF European Policy Office 2014, 12, 32), other ENGOs, for instance in the framework of Green 10 (WWF European Policy Office 2014, 32), and citizens (WWF European Policy Office 2014, 12). Regarding the hegemonic project, it is surprisingly clear that in 2013 the WWF no longer chose the green economy as its hegemonic project in the climate policy field, as it had back in 2011, and then slowly phased out in 2012. Instead, it seems that the 'old' *leading role* serves as the 'new' hegemonic project of the WWF, as illustrated by the demand that the EU should lead regarding the phase-out of investments in fossil fuels reasoning that the EU has a globally "key role to play" which is why the WWF is "pushing for Europe's leadership on global issues" (WWF European Policy Office 2014, 13). But, these are only first indications, as the WWF does not explicitly demand the EU's leadership in climate change policy. However, it is nevertheless obvious that the WWF changed its focus.

The following year 2014 was formative for the EU climate discourse in several respects. The election campaign for the EP and the resulting new Commission, but also mainly the preparations for Paris, ensured lively debates and decisions on EU climate policy. In February, the Commission presented a communication that was supposed to shape the debate initiated with the Green Paper into a proposal for the EU's climate and energy policy for 2020–2030 and to lay the groundwork for the EU's INDCs. Equivalising the demands for *more ambitious mitigation action* with a *low carbon economy, growth, jobs* and *energy security* (European Commission 2014b, 3) the Commission follows strategeme I. To this end, the Commission recommends adopting three concise targets: a 40% mitigation target by 2030 compared to 1990 (European Commission 2014b, 5), an increase in renewable energies to 27% of the EU-wide energy-mix by 2030 (European Commission 2014b, 5) and increased energy savings, which should be specified at a later date, although approximately 25% (European Commission 2014b, 8) are envisaged in the Communication. Alexander Bürgin (2015) gained insight into the internal Commission negotiations when drafting the Communication by conducting ten semi-structured interviews. These revealed that then Energy Commissioner Günther Oettinger tried to reduce the 40% target and to prevent additional national renewable energy targets due to concerns regarding future energy prices and a loss of European competitiveness (Bürgin 2015). Barroso found a compromise formula to keep the 40% goal—which was of course fought for by Hedegaard—while cancelling the individual targets at member state level and instead introducing a general EU-wide 27% goal regarding both the share of renewables and the level of energy savings (Bürgin 2015, 703). Bürgin even describes this compromise as a "policy shift" because his interviews showed that without Oettinger's intervention, the individual member state targets for renewables would very likely have been proposed, securing the member states' commitment (Bürgin 2015, 703). The compromise that was found reveals both the power structures in the EU of that time (Fuchs and Feldhoff 2016) and underlying political conflict between the demands of competitiveness (which includes cheap energy prices) and ambitious mitigation action. Nevertheless, the demands which seemingly stand diametrically opposed to each other are equivalised in their contrariety to the outside. In concrete terms: *Competitiveness* and *ambitious mitigation action* are articulated as going hand-in-hand with one another, in contrast to the USA, which is portrayed as inactive.

Besides proposing future goals, the Commission explains in detail that it is now a matter of introducing a low carbon economy and implementing it in the long term to preserve the EU's "first mover advantage" in these fields (European Commission 2014b, 18). The Commission thus follows other EU institutions in no longer referring to the green economy. Instead, it equivalises *low carbon economy* with other economic-related demands such as *growth, renewable energies, technical innovation* and *jobs* (European Commission 2014b, 4) emphasising the advantages of this specific economic system. Even though these advantages might remind us of the green economy concept, the latter placed a stronger focus on the interlinking of social, ecological and economic concerns. Therefore, it can be argued that the Commission is now attaching more importance to the economic aspects again. Since the economy is located in the culturally masculinised sphere, while social issues (understood as being part of reproductive practices) (MacGregor 2010) and the ecology are associated with the feminised sphere (Plumwood 1993, 43), a change can be observed regarding the use of gendered structures in the Commission's line of argumentation.

Following strategeme II, the Commission distinguishes the EU from those states that no longer participate in the second commitment period of the Kyoto Protocol and other major emitters who are not part of a legally binding agreement (European Commission 2014b, 17). Although strategemes I and II can be clearly reconstructed, it is not apparent which demand is positioned as the empty signifier, as both possibilities, *leading by example* metaphorically described by the 'first mover' or the *low carbon economy*, do have the potential to encompass the other demands of the EU's chain of equivalence. Hence, further documents must be analysed.

Reactions to the Commission's Communication were not long in coming, as the EP adopted its Resolution answering the Green Paper on February 5, 2014. The EP stressed that it "regrets that the Commission's communication (...) is short-sighted and unambitious on a number of levels, specifically as regards the lack of national targets for renewable energy and of any meaningful new action to incentivise energy efficiency (...)" (European Parliament 2014c, 11). As in the past, the EP calls for more far-reaching targets than the other EU institutions, introducing the addition of demanding "at least" a 40% mitigation target by 2030 (European Parliament 2014c, 11), furthermore demanding a 40% energy efficiency target (European Parliament 2014c, 12) and recalling the goal of decarbonisation by 2050 (European Parliament 2014c, 13).[22] On the contrary, it is of course not surprising that Eurochambres (2014, 1) wants to partly prevent these goals, or at least keep them low enough as their priority is not to jeopardise growth.

Before we shed light on EUCO's reaction, it is interesting to note that the EU member states were divided over the level of the targets proposed by the Commission. Consisting of 13 member states,[23] the so-called Green Growth Group (GGG) was built in early March 2014. The formation of GGG itself must be understood as a performative articulation in the EU's climate change discourse demanding high ambitions and backing the Commission's vision. In its position paper, GGG articulates the demand for *higher mitigation ambitions by the EU* calling for an "at least 40%" target (Green Growth Group 2014), following the EP's penchant for adding "at least" to the targets opening the necessary space to increase them at a later stage. Using strategeme I, further demands were equivalised, such as the *increase of the renewable energy share*, *fulfilling the below-2-degree target*, the *use of carbon market mechanisms*, *international cooperation* and *energy security* (Green Growth Group 2014). As the paper is only a short

[22] Although the Commission's 2011 Roadmap to a low-carbon economy by 2050 "was never formally adopted due to resistance by some Member States" (Wachsmuth et al. 2022, 10), its long-time goal of reducing GHG emissions by 80–95% by 2050 (from 1990 levels) was later committed to by both the Parliament and the Council (European Parliament 2014c, 13). In 2009 EUCO agreed the first time to the 80–95% by 2050 goal (European Council 2009, 3) "as the EU's contribution to a collective 2 degrees trajectory" (Dröge et al. 2015, 5) referring to the IPPC's findings.

[23] The Green Growth Group consists of Belgium, Denmark, Estonia, Finland, France, Germany, Italy, the Netherlands, Portugal, Slovenia, Spain, Sweden and the United Kingdom (Green Growth Group 2014).

catalogue with nine briefly formulated points, it does not become clear which hegemonic project GGG pursues, if any. Nevertheless, it can be stated that GGG fully supports the demands of the Commission and in some cases even goes further, calling for "at least" 40% mitigation and referring to the energy efficiency target which should be "at least 27%" (Green Growth Group 2014), while the Commission suggested approximately 25%. It is therefore interesting to see how EUCO reacted, as the EU's heads of state and government negotiated the general climate policy guidelines and therefore took the process over from the Council.

In the Conclusions of March 21, 2014, EUCO did not adopt any concrete targets for 2030, but articulated various demands which were equivalised following strategeme I. Linking *energy security*, *economic competitiveness* and *mitigation actions* (European Council 2014a, 7) at the beginning of the Climate and Energy section of the Conclusions shows that the negotiation of future goals must be understood in the context of the slowly beginning economic recovery of the EU, and should also serve—besides the fight against climate change—other goals in the economic and energy sectors, not least regarding the Crimean crisis. Confirming that "the European Union will submit its contribution at the latest by the first quarter of 2015" (European Council 2014a, 7), which is also demanded by the other major economies, EUCO uses strategeme II, splitting the discursive space. Stressing that the EU will be part of the early group submitting its INDCs, EUCO differentiates the EU by indirectly showing that a timely announcement of targets is not expected from the other major emitters, which can certainly be substantiated by the reluctance of the US in the past. In the articulation of this claim, the use of strategeme III, representation, also becomes clear, because it is simultaneously insinuated that the EU will again *lead the way* with an early announcement. Although EUCO merely defines the framework of the 2030 targets, it explicitly spells out in which areas these goals will be located: *mitigation*, *energy efficiency* and *renewable energy* (European Council 2014a, 8). This message is particularly important against the background that Barroso had to put his foot down within the Commission, finalising the proposed framework in early 2014, when then Energy Commissioner Oettinger—trying to protect the fossil fuel energy industry and dependent countries—sought to prevent both a nominally-fixed energy efficiency target and also a renewable energy target.

Besides negotiating the 2030 targets, climate policy was also a subject of the election campaigns of the 2014 European elections, albeit not one of the main foci. Still, as a special Eurobarometer on Climate Change in spring 2014 found, "Half (50%) of all Europeans think that climate change is one of the world's most serious problems and around one in six Europeans (16%) think it is the single most serious problem" (European Commission 2014a, 5). Hence, making climate policy an election issue was a wise decision by at least some parties. For the first time the campaign was enhanced by nominated *Spitzenkandidaten* of the parties running for the Commission's Presidency. Observers therefore attested the creation of a more stable European public sphere as well as partially unified campaigns across national borders for the first time of EP elections (Optenhögel, Humuza, and Thalhofer 2014, 3). Although it is difficult to speak of a unanimous election campaign among the European parties because topics and political disputes differed depending on the national context (Hrbek 2014, 212), the TV discussions still offered insights into which political fields were most discussed. These debates were dominated by various issues, in particular how to answer the effects of the economic crises, migration and asylum policy, as well as the annexation of Crimea by Russia and the related debate on the enormous dependency of European countries on Russian energy sources. Thus, if climate issues were debated, they were discussed in the context of energy security and the economy or in specific national contexts. Nevertheless, the manifestos of the major political groups give a deeper insight into their climate change discourse contributions. These manifestos inform both the party's voting campaigns prior to the elections and the goals they set themselves for the 2014–2019 legislative term (Hrbek 2014, 208).

Unsurprisingly, the Greens' manifesto is particularly dedicated to the EU's environment and climate change policy. Nevertheless, various demands as well as the use of strategemes and in some instances also hegemonic projects can be reconstructed in the other party manifestos analysed. Here, the manifestos of ALDE, Greens, EPP and PSE are reproduced and discussed.[24] ALDE already outlines the role of the EU in the international fight against climate change on the very first page of its manifesto and calls for "a Europe that takes the lead in fighting climate change" (Alliance of Liberals and Democrats for Europe 2013, 1). The party positions its other climate-related demands under the heading "Creating jobs and opportunities". Accordingly, ALDE links climate policy concerns very closely with economic ones; one could even say that climate action is viewed through an economic lens—which of course corresponds to the character of this party family. Following strategeme I, the demands of *sustainable development*, *low-carbon economy*, *renewable energy* and *jobs* are equivalised (Alliance of Liberals and Democrats for Europe 2013, 3). Furthermore, the demands for

[24] It was planned to analyse the six major groups of the EP of that time. However, neither the fourth largest group ECR (European Conservatives and Reformists), which is considered to be eurocritical and conservative to open right, nor the sixth largest group at the time, the Left, published a manifesto.

renewable energy, *carbon storage* and to *end fossil fuel subsidies* are also articulated (Alliance of Liberals and Democrats for Europe 2013, 3) in an effort to weave them into the EU's chain of equivalence. Although ALDE relies partly on feminised demands such as sustainability, the focus on economics leads to strongly masculinised discourse patterns. Regarding ALDE's hegemonic project, the party makes the leading role of the EU a cornerstone to fight climate change. Nevertheless, it cannot be clearly identified as ALDE's hegemonic project when considering only the manifesto. The party comes back to different terms describing a future economy model at different places in the document and calls for "sustainable development" (Alliance of Liberals and Democrats for Europe 2013, 3), "low-carbon economy" (Alliance of Liberals and Democrats for Europe 2013, 3) and "sustainable economy" (Alliance of Liberals and Democrats for Europe 2013, 4)—equivalised in every case with different demands. It seems that one of these demands—which the party probably considers synonymous—could also serve as ALDE's hegemonic project, as they could all go hand-in-hand with a less climate-intense future economic model as well as with a leading EU role.

The conservative party family EPP, on the contrary, only calls for a "stronger" role of the EU in climate policy and devotes itself in particular to energy policy issues when it comes to the EU's future climate policy (European People's Party 2014, 8). The EPP also refers to sustainable growth (European People's Party 2014, 7, 9), even though other concepts such as green economy are not pushed further.

The only party calling for the implementation of a green economy is S&D. The Social Democrats use strategeme I equivalising *green economy*, *renewable energy* and *technology* and the promotion of *future green technologies* (Party of European Socialists 2014, 6). Interestingly, S&D begins its brief "Green Europe" chapter with the words "the EU must regain global leadership on the protection of nature and natural resources and the fight against pollution and climate change" (Party of European Socialists 2014, 6), which is why we can assume that S&D wants to construct the EU's identity in climate change politics as a *leader*. This also corresponds with the articulations that have been reconstructed so far from the EP debates. Since S&D emphasises that "this requires both close cooperation with our global partners and leading by example" (Party of European Socialists 2014, 6), the use of strategeme V can be traced here, giving the empty signifier various meanings.

The Greens' election manifesto offers greater insight into their positioning in the EU climate discourse as compared to the other parties. Equivalising different demands, strategeme I is followed. These demands are, inter alia, a *legally binding treaty*, *climate protection* (European Green Party 2014, 20), *sustainable agriculture* (European Green Party 2014, 14) and the expansion of *renewable energies* and *energy efficiency* (European Green Party 2014, 9). In contrast to the other parties, the Greens furthermore call for *support for developing countries,* both technically and financially, to adapt and mitigate (European Green Party 2014, 30). Even though the Greens demand *innovation* to transform the economy (European Green Party 2014, 13), they are less focused than the other parties on technological innovation as a response to the climate crisis—and thus also less focused on mastering nature. Instead, emphasising the need for saving energy by demanding the goal of a 40% reduction by 2030 (European Green Party 2014, 20) alongside mitigation action (European Green Party 2014, 9) and underscoring a savings target of at least 55% for 2030 (European Green Party 2014, 20), the Greens give greater weight to feminised aspects of climate change policy. Insisting several times that the EU must play a leading role in global climate change policy (European Green Party 2014, 14, 20, 30) and even describing taking the "lead on climate change" as "one of our prime Green foreign policy concerns" (European Green Party 2014, 30), the Greens do indeed follow their hegemonic project and try to shape the EU's identity as a *climate change policy leader*.

A final important aspect of the Greens' election programme is the central demand for a "Green New Deal" (European Green Party 2014, 13) to renew the EU and make it more sustainable by bringing together "three coherent avenues to sustainability: fighting unemployment, poverty and all forms of social injustice; transforming our economies with innovation and eco-efficient solutions to tackle climate change and environmental degradation" (European Green Party 2014, 13). On the one hand, the Greens show their understanding of sustainability which includes social, economic and ecological components in equal measure. On the other hand, this part of their manifesto confirms the above-mentioned finding that the Greens also focus on innovation to combat the climate change but see it only as one mosaic piece of the response to the climate crisis. We can again observe the use of feminised discourse patterns as the Greens give special attention to social and ecological aspects in the framework of a future economy model. Having in mind that the EU launched its Green New Deal in 2019, it is furthermore worth noting that this wording was already used by the Greens in 2014 and was later incorporated into the EU climate discourse as an important demand. Hence, history teaches us that a demand expressed for the first time in a specific context[25] can become a weighty nodal point at a later and broader stage.

[25] Indeed, the concept of a "Green New Deal" has already been discussed in various contexts. UNEP, for example, has referred to its intention to make future financial programmes more sustainable and furthermore situated the idea of a green economy within their concept of a Green New Deal in 2008 (Unmüßig 2012, 4).

After the elections in late May 2014 which were won by EPP,[26] the EU climate discourse was shaped by various aspects. On the one hand, the discussion of the 2030 targets continued in view of the approaching Paris negotiations. On the other hand, the process of composing the future Juncker Commission generated various articulations shaping the climate discourse. Although these discussions took place in parallel, the formation of the Commission will be discussed first, not least since different NGOs and MEPs feared that the EU's future climate policy would be weakened, both because of the Commission's newly tailored portfolios and the politicians elected (Robert 2014b).

Juncker nominated Miguel Arias Cañete from the conservative Christian Democratic party of Spain as the future Climate and Energy Commissioner in charge of both DG Climate and DG Energy (cf. Sect. 4.1.3) which had often been at odds with each other. Indeed, it was already apparent in the reconstructed discourse that climate and energy are becoming increasingly intertwined. Now this development was also reflected in the Commissioner's portfolio which must be understood as a discourse articulation intended to link energy security and climate action and to underscore the demand for energy security in the EU's climate change discourse. NGOs such as Greenpeace Europe (in Robert 2014b), WWF Europe (2015, 13) and Green 10 (2014, 1) as well as MEP Claude Turmes (in Robert 2014b) criticised this move as they feared that climate change policy could be sidelined. At the same time, some MEPs welcomed the idea of merging energy and climate change protection, even though they seriously questioned the personnel choice in the subsequent EP hearing of the nominated future Commissioners (Berès in Robert 2014b). Indeed, Cañete's nomination faced strong opposition in the EP, on the one hand because of a past misogynist statement made after a lost TV debate and on the other hand because he was closely linked to the fossil-based energy industry (Wandler 2014). As a former manager and shareholder of an oil company, Cañete was summoned for questions by the EP for a second time, which was of course a sign of disapproval. Although Cañete passed the EP confirmation in the subsequent vote, his closeness to the oil industry remained an issue. Cañete's former engagement in the industry combined with his reluctance in selling his shares must be understood as discourse contributions in the sense of performative support of fossil energies.

Against this background, the mission letter Juncker addressed to Cañete is interesting to analyse. Equivalising inter alia the demands *energy security*, *green*

[26] EPP finished with 29.43%, closely followed by S&D with 25.43% of the votes cast (European Parliament 2014b). Implementing the results, Juncker was elected Commission President.

growth and *renewable energy* (Juncker 2014, 4), Juncker follows strategeme I. Two findings are striking: first, the various demands around energy are constructed in particular as an urgently-needed investment in future growth and energy security against the unverbalized backdrop of the annexation of Crimea by Russia only a few months ago. Hence, the fear that energy is at the forefront of Cañete's tasks seems at first glance to be well-founded. However, Juncker formulates at the end of his task description the demand "that the EU plays a leading role in international climate policy" (Juncker 2014, 4). Of course, further contributions need to be reconstructed and analysed to see how the new Commissioner or the Commission as a whole tries to shape the EU's climate change discourse and whether they follow a more energy-related hegemonic project. Another aspect of the political reconfiguration after the elections is also likely to influence the EU's climate change discourse. Donald Tusk, then Prime Minister of Poland, became the new Council President. Given that he followed a highly critical policy towards the bulk of ambitious climate actions of the EU, the journalist Aline Robert feared another attempt to water down the EU's climate agenda (Robert 2014a).

While Brussels was repositioning itself, the acting Commission President Barroso was able to speak quite freely in New York at the UN Summit for Climate Change in September 2014. He optimistically assumed that EUCO will follow the Commission's proposal of "an ambitious reduction target of 40% of domestic emissions by 2030, compared to 1990 levels, plus a renewables target of at least 27% and energy savings of 30%" (Barroso 2014, 1), stressing that "in the critical run-up to Paris, the European Union is leading by example". Differentiating the EU from other major economies Barroso emphasised that the EU will pledge its INCs well in advance of Paris, splitting the discursive space using strategeme II. In this speech it becomes crystal clear that Barroso wants the EU to be recognised as leading by example in global climate change policy. Now at the end of his term, it seems that he no longer needs to take too much notice of EU-internal blockers such as single member states or lobbyists from the energy-intensive branches. Furthermore, he underlines that the EU is willing to mainstream climate change policy into appropriate policy fields (Barroso 2014, 2), thus following strategeme X.

Although Barroso was certain in New York that EUCO would follow the Commission's proposal, EUCO adopted the common framework for the EU's climate policy until 2030 at its 24th October meeting in 2014 with slightly lower targets. In summary, the heads of state and government agreed on the formula of at least 40/27/27 by 2030: the reduction of EU GHG emissions by at least 40% by 2030 compared to the base year 1990 (European Council 2014b, 1), the increase of energy efficiency by at least 27% and a renewable energy share of at least 27% by 2030 (European Council 2014b, 5). Analysing these targets, it is important to note that "at least" is added to the goals—inter alia successfully lobbied for by European NGOs (Clima Action Network Europe 2015a, 5), the EP and GGG. Hence, to some extent the goals are more ambitious than proposed in the Commission's Green Book, but at the same time the energy-saving goal is lower than the 30% predicted by Barroso.[27] However, adding "at least" EUCO lays the foundation for the EU's own demand to regularly review the targets and, if necessary, to increase them (in the Paris agreement jargon known as stocktaking). In several places in the text, EUCO emphasises that no individual targets will be imposed at member state level, although this option is open to them. Knowing that the EU had developed precisely such MS targets to achieve the Kyoto goals in the sense of internal burden-sharing, a tactical change is emerging here and will be detected when the EU translates the Paris goals into EU legislation. Concerning the hegemonic strategy, this document provides little insight into the strategy of EUCO, although the discourse pattern already used in March to distinguish itself from the other major emitters is again pursued. Hence EUCO used strategemes II and I, while III is not reconstructable at this point. Even though the target formula of at least 40/27/27 by 2030 forms the heart of the future goals, the announcement also includes the EUCO's demand to the Commission to initiate an Energy Union. This should ensure closer cooperation between the member states in the energy sector and reduce possible future risks, for example through unilateral dependencies in the supply of fossil fuels. The discussion and implementation of the Energy Union will not be analysed in depth here. Not surprisingly, however, it is also partly negotiated in the EU climate discourse—not least because energy security achieved by an increase in renewable energies brings the co-benefit of GHG mitigation.

The EU's climate and energy targets for 2030 faced different responses; the demands of some actors were met, while other actors were disappointed. Greenpeace, for example, was upset: "The global fight against climate change needs radical shock treatment, but what the EU is offering is at best a whiff of smelling salts" (Sideridou in Darby 2014). The NGO demanded at least 55% mitigation by the EU and a renewables share of at least 45% by 2030 (Clima Action Network Europe 2013b, 2–3). Thomas Becker, from the European Wind Energy

[27] It is almost ludicrous that Juncker wrote in Cañete's mission letter that "A binding 30% objective for energy efficiency by 2030 is to me the minimum if we want to be credible" (Juncker 2014, 4).

Association, expressed a similar view: "The 27% [renewable energy] target is disappointing and is contrary to the incoming Commission's plans to make Europe the world leader in renewables" (Becker in Darby 2014). Other actors who feared more far-reaching targets, however, welcomed the adopted goals (Darby 2014). Nevertheless, CAN reflected in its annual report 2014 that the EU "could have ended up with considerably weaker targets" and pointed to the EU's challenges in these times such as euro-scepticism, backward-looking member states and the consequences of the economic crises, but at the same time to the tireless engagement by several NGOs(Clima Action Network Europe 2015a, 5).

Preparing for COP20 in Lima, we can attest almost the same strategy by the different actors that I have reconstructed to this point. While the Council once again provides no insight into its hegemonic project, the EP does so in its Resolution prior to the COP. The EP emphasises both the first mover position of the EU which "could encourage international partners to raise their ambitions accordingly" (European Parliament 2014a, 33) and that the EU should *lead by example* in the field of *research and innovation* (European Parliament 2014a, 38). Although the EP emphasises it calls for mitigation targets higher than those adopted by EUCO, the comparison of the EP Resolution with the Council Conclusions in the run-up to Lima shows that both actors nevertheless reproduce the same demand for *ambitious mitigation action by the EU* (European Parliament 2014a, 7; Council of the European Union 2014, 2). Another demand articulated by both institutions is *gender sensitivity* in climate change policy (European Parliament 2014a, 36; Council of the European Union 2014, 4). Up till the preparation for Lima, only the EP had formulated this demand in official statements prior to COPs; the Council now follows suit, thus strengthening the nodal point *gender sensitivity* in the EU's chain of equivalence. Of course, the new sensitivity might be traced back to the international discussions prior to Lima, where the UNFCCC Work Programme on Gender was adopted. Further analysis will show whether gender sensitivity becomes a long-term demand in the EU's climate change discourse in case other socio-political forces are won for its support. Besides Lima, the actors also took up strands of discourse from the EIGE Report 2012 or the green economy debate, which at least partially touched on gender sensitivity.

The Lima Summit was disappointing for some observers, but given its role as a precursor summit to Paris, it served mainly to prepare the 2015 agreement while greater announcements were placed on hold for 2015. In the end, a draft text for Paris was adopted. Even though various EU demands such as MRV were not appropriately taken into account, the EU saw itself as a well-regarded actor in 2014. Not least as it was the first entity to publish its targets for 2015, Cañete ended his speech by saying "The EU is ready and urges others to join"

(Cañete 2014, 2). Though China and the USA followed the EU's example in November of the same year (cf. Sect. 3.2.4), their climate goals fell far short of their responsibilities—but nevertheless created pressure on other major emitters. Still, the pressure to build appropriate diplomatic alliances to help ensure success in Paris remained. This aspect and the future shape of the Paris Agreement were the central topics of the EU climate discourse in 2015, the course of which, and contribution to, the (re)production of the EU's identity in global climate policy will be deconstructed and discussed in the next chapter.

4.2.4 The EU towards and in Paris: Co-Building Alliances to Lead the Way

The EU climate discourse in 2015 was completely dominated by the upcoming Paris summit. In February, the Commission addressed the Council and the EP with the communication "The Paris Protocol—A blueprint for tackling global climate change beyond 2020". It shapes the EU's climate discourse as it contains both ideas to improve the Paris Agreement (PA) to "strengthen[ing] ambitions over time and recognise[ing] the need for more financial support for mitigation and adaptation" (Dröge et al. 2015, 3) as well as the future climate and energy policy architecture of the EU.

The Commission uses all three core strategemes, trying to anchor its empty signifier according to its hegemonic strategy. Regarding the future climate agreement, demands which are already well-known, such as *mitigation action by all*, *MRV* and a *legally binding protocol* but also more specific demands such as *stocktaking in a five year rhythm* (European Commission 2015b, 5) and *increased climate-diplomacy* particularly due to the upcoming summit (European Commission 2015b, 11, 13) are equivalised, following the first strategeme. The Commission also splits the discursive space, insisting that the other major emitters should follow the EU and also *submit their INDCs to the UNFCCC* in the first quarter of 2015 (European Commission 2015b, 7). The agreement to communicate the INDCs "well in advance" of COP21, ideally "by the first quarter of 2015 by those Parties ready to do so" (United Nations Framework Convention on Climate Change 2015b, 3), was finally reached in Lima, although the regulation was not binding. Pointing to the fact that the US and China announced their "indicative post-2020 targets" in November 2014 some weeks after the EU did so (European Commission 2015b, 7), the Commission used two strategemes. First, it followed strategeme II again creating an 'us'—the forerunner EU—and a 'them' of other major emitters as the followers. In doing so, the Commission represented the EU as a *leader*, using strategeme III. At the same time it suggested that none of the demands incorporated by the EU in its chain of equivalence are represented by the radical other and are thus represented by the empty signifier *leading role*. Hence, the Commission tailors the various demands in the style of a leader, implying for instance that achieving *MRV*, a *legally binding protocol*, but also *technical and financial support for developing countries* in the framework of an upcoming agreement means to *lead*. Furthermore, the Commission locates additional multifaceted demands such as upcoming *diplomatic initiatives* as well as *early submissions* within the empty signifier *leading role*. Therefore, the Commission follows strategeme V, generating greater leeway.

Two further strategemes can be identified as being used to help the hegemonic project to succeed. On the one hand, the Commission—by now routinely—relies on strategeme X, locating European climate policy as a cross-cutting issue emphasising that it should be "mainstream[ed]" into the EU's "economic and development cooperation" (European Commission 2015b, 14). Strikingly, the Commission uses strategeme X in reverse, not importing a specific demand of another discourse, but planning to export parts of the climate change discourse. This further variation of strategeme X—which has only been detected once so far (cf. Sect. 4.2.3)—will be considered in more detail at a later stage, when it is used more broadly in the context of the European Green Deal (cf. Sect. 4.2.7). Moreover, strategeme VII (occasional but deliberate breach of the antagonistic line of division) is used to extend the outreach of the hegemonic project, as the Commission tries to inscribe the demand to *announce the INDCs early* in the opposite chain of equivalence, which is to be assigned to the USA. Later, we will be able to identify this specific strategeme as being used by other EU entities as well.

Fulfilling its leadership ambitions, the European Union officially submitted its INDCs to the UNFCCC (United Nations Framework Convention on Climate Change 2023a) very early (Torney 2019, 174) as just the second actor to do so. In an unusually blunt statement, the Council stressed that it "will send a strong signal to [its] international partners" submitting its INDCs in early March "presenting an ambitious common EU position" which is "crucial for the EU's credibility as leader in the fight against climate change" (Council of the European Union 2015b, 3). Interestingly, in its brief report of its meeting on the sixth of March, the same day the INDCs were officially submitted, the Council used the three core strategemes to discursively consolidate the EU's identity as a leader. Besides being the second submitter, the form of chosen INDCs should also be understood as an expression of the EU's leadership identity as it again leads by example following the non-binding technical presentation guidelines of the INDCs adopted in Lima (Tobin and Schmidt 2021, 147). Stating within its INDCs that "the EU and its Member States urge all other Parties, in particular major economies, to communicate their INDCs by the end of March 2015" (Latvia and the European Commission 2015, 5) is an additional rearticulation of the demand to lead (Tobin and Schmidt 2021, 148), and another try at using strategeme VII to pressure others to submit.

The concrete goals, summarised in the concise formula of at least 40/27/27 by 2030, their evaluation by different actors and also their development process have already been reconstructed (cf. Sect. 4.2.3). However, the now only five-page long officially-tabled INDCs provoked different reactions. Although being the second submitter attracted positive attention from observers, the EU's INDCs were also criticised. WWF Europe for instance censured the neither precise nor committing nature of the adaptation and finance sections (Anderson in Keating 2015). CAN Europe noted that the EU's INDCs were not sufficiently comprehensible and not ambitious enough and called for a mitigation goal of at least 55% (Clima Action Network Europe 2015b, 2). This criticism, however, did not change the shape of the EU's chain of equivalence, as the ENGOs agreed with the underlying general demands. Furthermore, the support of European society was overwhelming, as found in a representative Eurobarometer survey conducted in May and June 2015. 93% of the respondents believed that climate change can only be fought in a global manner (European Commission 2015a, 45), articulating and stabilising the demand for *global mitigation action*. In addition, more than 90% of the respondents shared the assessment of the EU institutions that there is a need for action particularly in the areas of energy efficiency and renewable energy (European Commission 2015a, 52, 56), rearticulating and consolidating these demands within the EU's chain of equivalence.

Besides broad support by different politico-societal forces, a chain of equivalence is also stabilised in regard to the (not necessarily that radical) 'radical other'. Since the US serves as "the EU's identity constitutive other" when it comes to global climate change policy (Krzyżanowski 2015, 126) the US submission of its INDCs could have weakened the EU's leadership aspirations and its underlying hegemonic strategy. But as outlined above, the EU and its institutions managed at the different COPs and in several texts to discursively implant the demand for early pledging into the US' chain of equivalence using strategeme VII. Hence, one can read the US engagement as part of the EU's successful discursive strategy in painting themselves as a leader, followed by other actors. However, another reason that the demarcation from the 'radical other' still works is that the US goals continued to be less ambitious than the EU's (Biedenkopf and Walker 2016, 31).[28]

Nevertheless, the USA still had an interest in gaining a climate policy success as part of Obama's future legacy, and this is why the country also engaged diplomatically in the run-up to the COP ensuring a positive Paris outcome. Katja Biedenkopf and Hayhley Walker (2016, 31) identify an implicit form of US-EU division of labour which partly shaped the EU's climate change discourse. While the US focused on other major emitters in bilateral talks, Cañete, French host Laurent Fabius and French Special Ambassador for Climate Change Laurence Tubiana embarked on an enormously extensive diplomatic tour addressing developing and small island states as well as various multilateral bodies and developed states (Biedenkopf and Walker 2016, 26–29). Thus, the demand for *strong diplomatic engagement* articulated by various sides shaping the EU discourse was both followed and performatively reproduced by the Commissioner and others. In this way, they wanted to ensure that the agreement was both highly ambitious and universal—both important demands stabilising the EU's hegemonic project.

Following the idea of Connie Hedegaard back in 2011 to overcome the differences between ambitious developing and developed states (Brun 2016, 121), and building on the former Cartagena Dialogue as well as an informal female climate negotiators' network (Yamin 2021, 226), the foundation was laid for an informal alliance symbolizing ambition and universality. The alliance was created to avoid the mistakes made in Copenhagen when the EU went ahead with motivated targets but without the necessary compact. As the co-founder of the later so-called High Ambition Coalition (HAC), particularly created by the Marshall Islands, the EU was instrumental in backing HAC financially and also in convincing other countries to join them (Parker, Karlsson, and Hjerpe 2017, 248). Some of the central HAC goals coincided with the demands of the EU's chain of equivalence, such as the *five-year frequency of stocktaking*, *financial support for developing states* and future *accountability and transparency rules* (Parker, Karlsson, and Hjerpe 2017, 248). Cooperating with the small island states within HAC the EU then also fought for the long-term 1.5-degree target (Högl 2019, 374) which was one of the Island state's central demands (Yamin 2021).

[28] Probably the most concrete evidence of the EU's higher ambition compared to the US is provided by the Climate Action Tracker. While rating both actors' goals as "medium" and stressing that both INDCs are not living up to their historical responsibilities, the EU's INDCs are characterised as more promising since the US reduction target of 26–28% alludes to the reference year 2005, where emissions were higher compared to the reference year of 1990 chosen by the EU (Climate Action Tracker 2015, 2016).

While the alliance-building process was in full swing in the background, Commission President Juncker gave his first State of the Union speech in September 2015. He delivered the annual address to the EP on September 9, barely three months before the Paris negotiations began. Juncker's choice of how to introduce the topic of climate change was remarkable. The chapter is called "United in Leadership in addressing Climate Change" and his first sentence "one example of where Europe is already leading is in our action on climate change" (Juncker 2015a, 22) clearly demonstrates that he is pursuing the hegemonic project of the EU's identity as a leader. Following the first strategeme, he equivalises the demands of *ambition*, *legally binding agreement*, and *greater diplomatic engagement*, but also the *end of fossil fuels* (Juncker 2015a, 22). Although Juncker demands the *end of fossil fuels* together with then German-chaired G7 and only for the end of the century, he is a new supporter of this demand. So far, the end of fossil fuels or subsidies has been demanded particularly by MEPs from the Greens, Left and ALDE as well as NGOs, but not by the Commission itself.[29] One theory to explain this new support for this demand is that the HAC opted for the 1.5-degree target, which actually means decarbonising by the end of the century. This and his desire to position the EU as a forerunner are the most likely explanations and probably part of his hegemonic strategy.

Interestingly, Juncker repeatedly refers to a leadership role for the EU in his address, stressing that "my Commission will work to ensure Europe keeps leading in the fight against climate change" (Juncker 2015a, 23). This quote illustrates how Juncker follows strategeme III representing the EU as a leader, not least by using the word "keeps". In a subsequent paragraph, Juncker also makes clear to what he attributes the still prevailing identity of the EU equivalising his demand for *renewable energies*, in which he sees the EU as a leader, with *green technology* and *technical innovation* (Juncker 2015a, 23). These masculinised aspects (cf. Sect. 2.3.3) are also Juncker's main foci formulating answers to the worsening climate crisis. It is furthermore interesting to note that in his view, the EU's leading role is capable of overcoming the supposed contradiction between reducing emissions and economic interests, pointing to the power of renewable energies and related technical achievements (Juncker 2015a, 23). Using strategeme II, Juncker says:

[29] While the Commission stated in 2011 in preparation for Rio + 20 that "Rio + 20 should launch a coordinated set of actions by countries to identify and phase out environmentally harmful subsidies, accompanied by targets and deadlines", it did not further clarify the demand or formulate a concrete goal.

> *"Let me be very clear to our international partners: the EU will not sign just any deal. My priority, Europe's priority, is to adopt an ambitious, robust and binding global climate deal." (Juncker* 2015a, *23)*

He thereby implies that the EU as a reliably ambitious player on the one side of the discursive border is able to convince the still hesitant partners on the other side. Overall, Juncker's position towards Paris is refreshingly clear, pushing the hegemonic project of the EU's leading role in global climate change policy. Is this hegemonic project, already reconstructed for the Commission and the Council, supported by other EU actors shaping the EU climate change discourse vis-à-vis Paris?

The chorus of different voices expressing their respective positions in trying to shape the EU's climate change discourse in the run-up to the Paris Summit was loud and diverse. Before we turn again to the actors of the Brussels institutions, the contributions of non-governmental entities should be reconstructed. Using strategeme III, the ETUC—the umbrella organisation of trade unions in the EU—also pursues the hegemonic project of a leading role for the EU, stressing in its pre-COP21 position that the Union "must continue to play a leading role in the negotiations" (European Trade Union Confederation 2015, 1). The ETUC furthermore follows strategeme I, equivalising the demands for a *below 2-degree goal*, *global net-zero in 2100*, a *legally binding agreement*, *mitigation by all* as well as *CBDR* (European Trade Union Confederation 2015, 1) constructing almost the same chain of equivalence as for instance the EP. Beyond that, the ETUC also articulates demands that are not so widespread, such as *decent work* which the ETUC equivalised with the outlined demands and a *just transition* (European Trade Union Confederation 2015, 1). These demands can be traced back to the nature of the actor, representing the interests of workers. However, since different actors bring new demands into the EU's chain and try to anchor them, the ETUC has to find the support of other socio-politico forces to make this labour-friendly version of an EU leading role successful. Overall, the ETUC uses more feminised patterns since both decent work and a just transition centre on the wellbeing of people and nature instead of the economy.

A counterpart of the ETUC is BusinessEurope, unifying more than 30 European national business federations. Even though no hegemonic project can be extracted from their COP21 positioning it should be noted that BusinessEurope—unsurprisingly—focuses on economic aspects of a possible agreement. The demand for *competitiveness* (BusinessEurope 2015, 2) seems to form the heart of its reasoning, although other broadly supported demands such as a *legally binding agreement* (BusinessEurope 2015, 1), *mitigation and adaptation action by all* and binding *MRV* measures (BusinessEurope 2015, 3) are also rearticulated. It is very likely that BusinessEurope supports these demands because they aim to attain the same legal conditions for competitors around the globe. This impression is reinforced as BusinessEurope clearly distinguishes the EU from other major emitters, using strategeme II (BusinessEurope 2015, 1). In this case it follows exactly the same strategy as the ETUC; even if their interests are different, they correspond on the general demands rearticulating them particularly regarding the "radical other". While the ETUC is more concerned with globally-secured labour rights and BusinessEurope with competitiveness, both see the common demands for a legally binding agreement or mitigation action as keys to safeguard their own interests. It can be stated that BusinessEurope in particular uses masculinised discourse patterns pointing to economic success and competitiveness. As I am not able to distinctly reconstruct strategeme III, it remains open as to which hegemonic project BusinessEurope follows.

Likewise, the question of the hegemonic project of the European ENGOs in Paris also remains open. While in previous years the WWF or CAN pushed for the leading role of the EU as their hegemonic project (Clima Action Network Europe 2013b, 2; WWF European Policy Office 2012, 16, 2014, 13), it is not clear whether these actors still wanted the EU to be represented as a leading actor in Paris. CAN Europe, for instance, supported most of the demands of the EU's institutions—although it located these on the more ambitious end of the scale (Clima Action Network Europe 2015b, 2). However, neither the demand for a leading role in international climate change policy nor an empty signifier can be traced. Even though the WWF demands that EU leadership achieves the 2015 adopted Sustainable Development Goals target (WWF European Policy Office 2016a, 30), for which the EU was an important policy entrepreneur (Zeigermann 2020), the NGO's annual report does not call for the same EU role when it comes to global climate change policy. Rather, it seems that WWF concentrated more on specific fields of European climate policy such as the *100% renewables* demand (WWF European Policy Office 2016a, 8) instead of working through a greater EU identity. Michael Jacobs (2016, 318) supports this finding when he notes that the NGOs generally changed their strategy for Paris and focused more on concrete areas of action such as the fight against fossil fuels to mobilise large numbers of people. The enormous success of the ENGOs in organising very broad civil protest in the run-up to Paris (Morgan 2021, 259; Jacobs 2016, 322) proved the ENGOs right on this point.

In direct preparation for the negotiations in Paris, both the Council and the EP have positioned themselves on the climate change map. The Council's Conclusions for Paris stand out from the respective Conclusions of previous years.

Underlining "the critical importance of the 2015 Paris Conference as a historic milestone for enhancing global collective action and accelerating the global transformation to a low-carbon and climate-resilient society" (Council of the European Union 2015a), the Council put more pathos and pressure into its words. Furthermore, the Council makes a—quite rare—hint in the direction of a global transformation. In the other documents examined, the Council only made similar statements in 2010 and 2011 calling for a global transformation towards a low-carbon economy (Council of the European Union 2010, 7) or a low-carbon society (Council of the European Union 2011a, 4). The latter demand is rather uncommon in the EU's climate change discourse so far. There are multiple options for how to interpret the demand for a low-carbon society, as it could mean either to privatise climate change efforts, locating them in the people's private sphere, or to strive for a more comprehensive idea of future climate action, addressing all GHG-emitting sectors. If the demand finds further support, we will come back to it at a later stage. Without reconstructing the Council's chain of equivalence shaped prior to Paris, it is important to note that it followed its 2014 Conclusions regarding the Lima COP20 demanding *gender equality* and a *gender-sensitive approach* (Council of the European Union 2015a). In addition, the Council also calls for *human rights* as an integral element of future global climate policy (Council of the European Union 2015a). These findings lead to a twofold bottom line: The Council once again does not position itself towards a concrete hegemonic project, even though it clearly called for an EU leadership role back in March when submitting the EU's INDCs. Furthermore, the Council seems to articulate more feminised demands than in the years before.

The EP has passed a resolution that contains pithy sentences similar to those of the Council's Conclusions. However, it clearly follows a hegemonic strategy when stressing that

> *"the EU has thus far played a leading role in the efforts to mitigate global warming and will continue to do so in the run-up to a new international climate agreement in Paris at the end of 2015; demands that this ambition be matched by other big emitters;" (European Parliament* 2015, *70).*

The EP evidently positions the EU as a leader here—both in the immediate past, albeit aiming in particular at mitigation efforts made by the EU itself, and in the future. Interestingly, the forward-looking demand is not only semantically related to the ductus of leading by example. The EP uses strategeme II in the last half-sentence by demanding a similarly ambitious commitment from the other major emitters. At a later point in the Resolution the EP split the discursive space again when addressing the USA and calling for stronger future efforts, (European Parliament 2015, 79).

The comprehensive document calls for the already variously reconstructed demands such as a *legally binding agreement*, *mitigation action* but also *adaptation action* (European Parliament 2015, 71), *technical support* (European Parliament 2015, 76) and the *recognition of and compliance with scientific findings* (European Parliament 2015, 71). Calling on the EU to continue to live up to its role as a leader (European Parliament 2015, 78, 79), the EP once again strengthened its hegemonic project. Moreover, the EP shapes up to its reputation of being the most ambitious EU institution (Burns 2021) by continuously demanding the 1.5-degree target be set (European Parliament 2015, 70). Furthermore, strategeme X can be traced once again, as the EP, as well as the Council, demands the incorporation of a *human rights perspective* as well as a *gender perspective* into global climate policy (European Parliament 2015, 79). In this way, the basis of the climate discourse can be broadened, which may help the EP's hegemonic project to succeed.

Although the hegemonic strategy of the EP can be reconstructed in the Resolution, such outcome documents are not suitable for depicting the actual underlying discussion, which is why we should reconstruct the conflicts bargained in the debate held prior to the adoption of the Resolution. The central lines of conflict within the debate run between the conservatives of the EPP and the other democratic parties. Basically, the reasoning of the conservatives does not differ that much, as they rearticulate the demands we already know from EP debates. Only the decarbonisation goal of 80–95% by 2050, which had already been accepted elsewhere (cf. footnote 22) and which was also adopted by the EP later in its Paris Resolution, is clearly contradicted by the EPP members. Most of them argue that the 2050 goal is not realistic (Ayuzo 2015; Florenz 2015; Liese 2015) and additionally point to the often articulated demand of *competitiveness* (Ayuzo 2015; Belet 2015). Even though no EPP member is saying that the 2050 targets are too far-reaching, most of them fear a future loss of competitiveness if the EU were to stand alone with highly ambitious climate goals in Paris. It may seem striking that the different parties with their partly varying preferences agreed nevertheless on a common empty signifier, the leading role. Of course this does not apply to some MEPs from populist or right-wing parties and factions. Since most of them question climate change as a whole or downplay the human-induced share, it is hardly surprising that the parliamentarians from Europe of Freedom and Direct Democracy (EFDD) and Europe of Nations and Freedom (ENF) question EU climate action altogether in this debate (Stuger 2015; Helmer 2015). However, since their demands were neither able to prevail in the EP's Resolution, nor made up a strong percentage, we will now continue to turn to the partially successful hegemonic projects of the other parties.

In this debate, various strategemes were used in trying to reach the hegemonic project. The discursive space was split inter alia by Krišjānis Kariņš (EPP) (2015) and Jo Leinen (S&D) (2015) delimiting the EU from the other major emitters and specifically from the USA. Many demands of the chain of equivalence were equivalised following the first strategeme. Representing the EU as a leader—through almost all political groups—strategeme III was also used. Kathleen Van Brempt (S&D) (2015) for instance said "above all, we need to maintain our leading role", Giovanni La Vie (EPP) (2015) points to the EU's leading role in the past demanding the same for the future. On a further note, Cañete, who also spoke during this EP debate, stressed the need for a leadership role several times and ended his speech with the words:

> *"I would like to thank you once again for your support and enthusiasm in strengthening climate policy in the European Union and our global leadership in this area" (Cañete* 2015a*).*

His primary demand vis-à-vis Paris—besides *stocktaking*, *MRV* and overall *a unilateral legally binding treaty*—is obviously the *EU's leading role,* insinuating that a leading role encompasses all the other common demands of the EU's chain of equivalence. Referring to himself and to his extensive diplomatic tour of recent months with the words "my job will be to ensure that the EU leads the negotiations" (Cañete 2015a), Cañete illustrates how to achieve leadership: through enormous alliance-building and global recognition for such an endeavour.

The start of COP21 was accompanied by massive demonstrations with more than half a million people attending worldwide (BBC 2015). In Paris, the protests were smaller than planned, due to increased security measures restricting some demonstrations following the deadly Islamist attack in Paris just a few weeks prior to the summit (France24 2015). ENGOs and civil society mobilised throughout the whole year leading up to the negotiations, reinforcing their respective demands, particularly the *1.5-degree target* and *immediate action.*

The key EU speech was given by the President of the Commission in the high-level panel at the beginning of the summit, which underlined the importance of Paris with 150 heads of state and government attending (European Council 2015). In his speech, Juncker equivalised the four central demands of the EU for the summit in just one sentence:

> *"We need a binding, robust and lasting agreement which reflects a long-term vision of a carbon-free economy, the implementation of which will be periodically assessed and adapted to raise the level of ambition necessary." (Juncker* 2015b*; emphasis added)*

Interestingly, he pointed to the EU's long-term *below two-degree goal*, while the EU lobbied for the 1.5-degree target being part of the High Ambition Coalition (HAC), and he demanded the same in his State of the Union address. However, as he stressed the need for a *carbon free economy*, he indirectly also supported the *1.5-degree target* as it is a more concise expression of the long-term goal of a carbon free economy by 2100. Hence, Juncker's (2015b) words that "we must do more over the years to come" are not surprising. Since he adds that "others need to follow our example" (Juncker 2015b) he uses strategeme II, splitting the discursive space, and the audience gets a glimpse into his hegemonic project. Although less explicit at this point, it can still be said that Juncker used strategeme III, representing the EU as a leader who sets different examples for the other major economies to follow.

Analysing the EU's hegemonic strategy at COP21 we need to reconstruct the EU's performative articulations. As a major driver of the HAC, the EU was able to lend even more weight to its negotiation targets through further alliance-building. Of course, HAC was not a one-way street, finding other actors to support the EU's will. As the EU, for example, adopted the demand of the small island states for a long-term 1.5-degree target as common ground, it also had to walk down the road to meet with the developing states' demands for more adaptation and more finance, as well as greater loss and damage efforts (Oberthür and Groen 2017, 5). This compromising formula paid off; today, scholars are certain that the HAC was an important factor in ensuring the success of Paris by breaking the former firewall of developing and developed states (Savaresi 2016, 19; Streck, Keenlyside, and Unger 2016, 4)—alongside public pressure and the fact that none of the actors who were blamed for the Copenhagen outcome wanted to be the fall guy again. Tony de Brum, then Foreign Minister of the Marshall Islands, was the face of the loose HAC which grew from day to day in Paris as months of shuttle diplomacy bore fruit. The breakthrough came when the USA joined HAC and Brazil followed a little later (Yamin 2021, 237). Brazil's support meant breaking down the closed resistance of the BASIC states and thus generating a majority for the more far-reaching demands such as the 1.5-degree target and the five-year stocktaking. Thus, big players like India or China could hardly contradict the PA without being considered the obstructionists of Paris. All these factors led to the historical PA, which indeed achieved more than was generally expected.

The decisions of Paris are well known and will be revisited later when it comes to their successive implementation.[30] To understand the construction of the EU identity as a leader in the context of Paris, it is important to know that the old Kyoto logic of a top-down agreement has now been reversed into a bottom-up mechanism of pledging and—within a later elaborated but binding form—reviews, setting a common frame. This new logic must be seen as another key to the success in Paris (Falkner 2016, 1114). Furthermore, in this way the EU had the chance to lead, pledging its INDCs as the second entity and inviting others to join. Additionally, the EU acted as a "leadiator" (Bäckstrand and Elgström 2013) or bridge-builder, being an important part of the HAC in following the hegemonic project of discursively constructing its identity as a leader. Indeed, it also depends on whether and how an identity is—at least to some extent—recognised from the outside. Charles Parker and Christer Karlsson (2018, 529) conducted a representative long-time survey at the COPs from 2008 to 2015 and found that the EU was consistently recognised as a leader by at least 41% by all respondents, albeit to varying degrees.[31]

How did the EU's chief negotiator assess the performance immediately after Paris? Celebrating a "historic climate deal", Cañete points out:

> *"Paris is much more than the deal. Paris is also about diplomacy, geopolitics and influence. And in this context, Paris is a major win for Europe and its allies" (Cañete* 2015b, *1).*

He uses strategeme III by making clear from the beginning of his brief speech at the press conference on the day after the adoption that Paris is a story about "how Europe and its allies shaped the Paris deal" (Cañete 2015b, 1). Showing that it was also the alliance of the EU and its allies that was able to put pressure on the big emitters, he used strategeme II to split the discursive space. Pointing out that the HAC was the "master plan of Europe" (Cañete 2015b, 1), he clearly stylises the EU as a leader that succeeded in doing no less than helping the PA to come to fruition. Interestingly, Cañete stressed that he visited besides many other states the key actor Brazil. This is another indirect articulation of the EU's importance within the HAC and thus the EU's leadership role, as the different

[30] For a good overview see Falkner 2016.

[31] The level of recognition of the EU as a leader in the context of the COPs varied considerably, peaking in 2008 with 62% and reaching a low of 41% in 2015. It is worth reading the authors' analysis in its entirety, because they emphasise on the one hand that it is not possible to speak of a single leader, especially in Paris, and on the other hand that the same recognition as leader cannot be expected for all forms of leadership (Parker and Karlsson 2018).

EU institutions made clear in advance: An important demand encompassed by the *leading role* is *multilateral climate diplomacy*. Whether and how the hegemonic project, the leading role of the EU, is gendered will be discussed in more detail later (cf. Sect. 4.3.2). At this point, however, it should be emphasised that the form of leadership, which relied in particular on bridge-building and leading by example, certainly employs feminised patterns and to a lesser extent the aspect of a powerful strong role, the kind of strong role India reproduced in its climate discourse. As I also reconstructed demands which are culturally situated in the masculinised sphere (particularly regarding the economy and technical innovations), a broader discussion will follow after reconstructing the further EU climate change discourse post-Paris and until 2021.

4.2.5 The EU after Paris: A Changing Global Context—A Changing EU Identity?

The importance of the PA for the EU climate change policy and its discourse can hardly be overestimated. As it seems that the EU chose the right strategy in Paris by relying on ambitious pledges and even more on smart climate diplomacy, it now had to follow up with action in both respects to maintain the picture of a leading EU. In 2016, therefore, the focus was on laying the much-needed political and legal groundwork to reach the NDCs on the one hand and on rapid ratification on the other hand. But the unexpected pro-Brexit vote in June 2016 and the election of US President Donald Trump, which coincided with COP22 in Marrakesh, overshadowed these priorities. What significance do these developments have for the EU's climate discourse and the hegemonic strategy followed by the actors so far?

After Paris, the Climate and Energy Commissioner initially stuck to the empty signifier *leader*. Cañete gave a speech at a Bruegel think tank event in February 2016 entitled "How will the Paris agreement impact EU climate and energy policies?". In his talk, Cañete builds on the identity of the *EU as a leader* perpetuated in Paris by referring to its historic success (Cañete 2016b, 1) and now calling for its *fast ratification* (Cañete 2016b, 2). Following Strategeme I, he equivalises these demands with concrete implementation-oriented demands such as *sustainable agriculture and land use*, *sustainable transport* and the *improvement of the ETS* (Cañete 2016b, 2, 3). Laying his focus on the demands of *renewable energy* and *energy efficiency,* which should lead to *energy security* (Cañete 2016b, 4), Cañete outlines one of the foci of the Juncker Commission: the transformation of the energy system (Juncker 2016, 27) serving both energy security and mitigation needs. He made these points referring to the upcoming Commission Communication "The Road from Paris" tabled just a few days later. This is intended to initiate the *fast ratification*, which Cañete rearticulated in his contribution insinuating that a quickly-confirmed EU signature induces a leading EU. In summary, the three core strategemes can be reconstructed in the speech, continuing to consolidate the empty signifier of the *leader* as still representing the EU's identity in climate change policy.

The Communication "The Road from Paris" presented by the Commission pursued a slightly different goal: It rearticulates and stabilises the empty signifier of the *EU as a leader by example*. Hence, the Commission has slightly changed the direction of its thrust by referring strongly to the idea of *leading by example* to reach the equivalised demands. These are inter alia the *transition to a low carbon economy*, *energy transition*, *innovation*, and *competitiveness* but also *circular economy* (European Commission 2016a, 4). The fact that leading by example is conceived and treated as an encompassing demand is underlined by the Commission stressing that "To reap those benefits, the EU will need to continue to lead by example" using its "first mover advantages" particularly in the renewable energy sector and low carbon technology (European Commission 2016a, 4). The Commission follows the same pattern in its later Communication "Clean Energy for all Europeans" of November 2016 (European Commission 2016b). Demanding again the EU's global leadership in renewable energies (European Commission 2016b, 7), the Commission equivalises its empty signifier to *lead by example* with the *implementation of the PA*, *economic growth*, *new jobs* and *enhancing citizens' welfare (*European Commission 2016b, *7)*. Likewise, in DG Climate's eyes, the ideas described in this and further Commission Communications "cement[s] the European Union's leadership role in the clean energy transition" (DG Climate Action 2017, 9). The estimation of DG Climate is interesting here because it focuses the leadership on one specific field and thus insinuates that the EU can set a good example in the area of energy transition, leaving it open as to whether leading by example encompasses further areas of action.

The Commission's wide-ranging Communication furthermore outlines the way forward implementing the EU's NDCs. The basic pattern running through the document is that it demands *innovation and technical solutions* for the necessary transformation of the economy, referring to some good examples of EU action already taken and stressing the importance of *future measures* (European Commission 2016b). Following the three core strategemes, this document is the expression of a slightly adapted hegemonic Commission strategy, rearticulating demands that can be attributed to the masculinised sphere, in particular to fill the empty signifier *leading by example* with the various necessities for transforming the energy system. Whether or not this adaptation of the hegemonic project is also apparent among other actors who shape the EU climate change discourse remains to be seen in further reconstruction.

In its outline of future climate diplomacy in February 2016, the Council—in this case in its formation of the foreign affairs ministers—looked back at Paris and the preparatory phase and described the EU as a *leader* (Council of the European Union 2016a, 2). The Council urges that climate diplomacy must be pursued with similar vigour as back in 2015 to achieve the pledged goals of the PA—inter alia through the EU's foreign policy and its various forms of cooperation with third countries to generate followership (Council of the European Union 2016a, 5, 6, 8). Hence, the Council shares the idea of *leading by example* at least as an important nodal point of the EU's chain of equivalence. The EU becomes even clearer in its Global Strategy published in June 2016, emphasising that it "will lead by example by implementing its commitments on sustainable development and climate change" (European Union 2016, 40). Although both the foreign minister's Council statement and the Global Strategy cannot be used to reconstruct the actor's hegemonic strategy in its entirety, these contributions support the Commission's hegemonic project to install the demand to *lead by example* as the empty signifier representing the EU's identity in climate change policy.

The subsequent parallel discussions on the implementation of the PA at EU level and its ratification were dwarfed by the UK's Brexit vote at the end of June 2016. While the consequences of the vote did not have an immediate impact on the ratification process or the EU's climate policy in 2016, it was clear that losing an ambitious climate change policy advocate (Schreurs 2016, 221) would change the future dynamics (Fischer and Geden 2016). After all, the UK had so far "made an overall positive contribution to the EU's ability to act internally and internationally on climate change" (Bocse 2020, 267). Thus, the question was virulent which form a *leader* identity will take when a major member state leaves the EU.

This question can only be dealt with during the analysis of the following years. Mentioning Brexit only intermittently, and referring to the resulting need for unity within the EU (Juncker 2016, 24), Juncker addresses the EU climate policy in his State of the Union speech 2016 irrespective of the Brexit vote. Even though I am not able to trace a hegemonic strategy, it can be noted that Juncker wants the EU to *lead* in global climate policy. Designating this role could be a first cautious reaction to Brexit. Stressing that it was the EU who brokered the PA, Juncker furthermore linked the demand of *early ratification* of the agreement with the EU's global role (Juncker 2016, 20). Thus, he implies that leading means to ratify at an early stage, giving *leader* the specific meaning of acting as an example. That's why the Commission President calls on the member states to support and not to stand in the way of a rapid ratification (Juncker 2016, 20).[32]

The NGO WWF followed the same strategy, calling for the *fast ratification* of the agreement as it is one of the keys "if the EU wants to remain a climate champion" (WWF European Policy Office 2016b). However, due to a special legal construct, the USA was able to announce its ratification together with China at the G20 meeting in September 2016 before the EU had the chance (Biedenkopf and Walker 2016, 32); the various EU institutions, then a fortiori, insisted on their demand to ratify the PA quickly for a couple of reasons: First, some of them defined it as an element of the empty signifier to *lead by example*, second, the EU had great interest in shaping the future framework conditions of the PA that were still to be negotiated, whereby ratification of course reinforces the negotiating position.

Long debates in the Council, where Poland in particular opposed a fast-track ratification without clarifying the binding national obligations in advance (Fischer and Geden 2016, 4), were followed at the end of September 2016 by the decision of the Council to ratify the PA in accordance with only some member states convincing the still-critical states to postpone the outstanding discussions. After the EP ratified the PA on October 3, 2016 and the Council followed suit a few days later (Council of the European Union 2016b), the PA was able to enter into force on November 4, 2016—shortly before COP22 started and 30 days after the threshold values of 55% of the states with 55% of global GHG emissions were reached with the EU's signature. Although the EU was therefore perceived as the key actor in lifting the PA above the 55% threshold, this cannot hide the fact that at this point of time six member states had ratified the PA and that therefore only the corresponding 4% of global GHG emissions instead of the EU's then share of 12.8% had been bindingly submitted to the UNFCCC (Becker 2016). Furthermore, the fly in the ointment persisted that the EU was not one of the first ratifiers.

[32] The ratification of international treaties that affect the sphere of member state competences—as the PA does—usually takes a long time: Besides the Council and the Parliament, the individual member states also have to go through their respective processes to ratify these "mixed agreements" (Müller-Brandeck-Bocquet and Rüger 2015, 212–13). In addition to long processes, the positioning of some member states was also a problem, as Poland, for example, made ratification (initially also at the EU level) dependent on the future design of the ETS.

Before COP22 began, the EP and the Council formulated various demands. As in previous years, only the EP pursued a hegemonic strategy in its positioning in the lead-up, which aimed to anchor the empty signifier of *leading by example*. In its resolution, the EP tried to shape the EU's identity, equivalising some almost typical EP demands, for instance to *increase the share of renewable energy*, *energy efficiency* and *energy security* (European Parliament 2016a, 8, 9) and the demands for *technical innovation* and *research*. Furthermore, the EP articulated those demands which were most likely to help implement the EU's obligations confirmed in the PA, such as the demands to support the 2018 *facilitative dialogue* (European Parliament 2016a, 4, 5), *increase the ambition level of the EU NDCs* (European Parliament 2016a, 3, 4), but also the demands to *protect human rights* (European Parliament 2016a, 6, 11) and *increase gender equality* in global climate change policy (European Parliament 2016a, 6). Interestingly, those demands culturally associated with the masculinised sphere of domination, for instance those with energy-related aspects, are painted as the main fields in which the EU already *leads by example* and in which they should continue to take this role (European Parliament 2016a, 9). Splitting the discursive space, the EP uses the twist of welcoming China's and also some other competitors' introduction of "carbon trading or other pricing mechanisms" (European Parliament 2016a, 8), marking the EU as a pioneer on the one hand and China as a follower on the other. In this specific case we can reconstruct the use of both the second and third strategemes by the EP. Accordingly, the EP acts in the same way as the Commission did, relying on the empty signifier *lead by example*, as supposedly representing the identity of the EU. In terms of the breadth of demands, however, the EP again stands out from the Commission, calling for more social aspects such as a *just transition* or *gender equality*, while *technical and market-based solutions* are still the focus when formulating answers to the climate crisis.

In the previous EP debate, three EPP MEPs (La Via 2016b; Kelly 2016; Christoforou 2016) and one S&D MEP (Bonafè 2016) articulated the need for the EU to take a classic *leadership role*, partly building on the pre-Paris discourse. On the contrary, EPP member Giovanni La Via (2016b) stressed the need to *lead by example* as did S&D member Jytte Guteland (2016). Interestingly, while the EP resolution seems to be part of the adaptation of the hegemonic project in the sense of leading by example, in the EP debate the strategy of the factions and single MEPs is less clear. The EPP members seem to still push for the *leading role of the EU* as the empty signifier while the other MEPs focus more on the implementation side of the PA, which speaks for the empty signifier of *leading by example*. Therefore, it remains to be seen whether or not the parliamentary groups will unite behind the common hegemonic project of *leading by example* as adopted in the EP Resolution.

While the rapid ratification of the PA gave COP22 in Marrakesh a boost, the important framework conditions for implementing the PA, such as the rulebook, the design of the stocktaking mechanism and particularly the 2018 facilitative dialogue which was planned to ensure a first round of stocktaking and would ideally lead to higher mitigation-ambitions, were still to be negotiated. The aim of the EU for Marrakesh was to generate followers for its political preferences within these fields. This is also suggested by the frequently formulated demand for continued *strong climate diplomacy* articulated by different MEPs, the Council, and the Commission. Building on the HAC network[33] (Obergassel et al. 2016, 5) and committing to close cooperation with allies, the EU rearticulated the demand to *lead by example* in a performative way in Marrakesh. However, the COP22 negotiations were "overshadowed" by the US elections (Mathiesen 2016); in the second week of the conference, Donald Trump was elected as the future US president. During the campaign, the Republican had already announced that he "would 'cancel' [the] Paris climate deal" (BBC 2016). Hence, in Marrakesh, it was not only the US delegation that seemed to be shocked to learn of the election results, but indeed all those in attendance seemed to be taken aback (Obergassel et al. 2016, 2). The future change in the chain of equivalence of the 'radical other'—the USA—became instantly apparent, which is highly significant for the EU and its hegemonic project of leading by example. Did the prospective change of the counterpart in the chain of equivalence have immediate implications for the EU discourse?

In the EP discussion reflecting the outcome of COP22, MEPs across party lines agreed that the EU must take an even stronger leadership role in global climate policy in the future due to the new global context (La Via 2016a; Leinen 2016; Gerbrandy 2016; Affronte 2016; van Brempt 2016). It is important to note that besides Green MEP Claude Turmes it was Climate and Energy Commissioner Cañete who demanded an EU leading role in renewables in this EP discussion (Turmes 2016; Cañete 2016a). Demanding that the EU *lead by example*, Cañete equivalised his empty signifier particularly with demands located in the masculinised sphere, stressing that climate change should be addressed with *energy efficiency*, *renewables* and *technical innovation* (Cañete 2016a). Hence, in addition to the core strategemes, Cañete uses strategeme V, expanding the leeway of his empty signifier. MEP Giovanni La Via (EPP) also uses strategeme V, but goes further by putting "the way towards a sustainable and competitive economy"

[33] On the EU's mixed record on mobilising the HAC network after Paris, see Earsom 2023.

(La Via 2016a) into the empty signifier of *leading by example.* We can therefore reconstruct two simultaneous developments in this EP discussion: On the one hand, Cañete and La Via, who fill the empty signifier *leading by example* with far-reaching demands trying to already compensate for the new context including future US President Trump, and, on the other hand, MEPs of different parties who stick to the former empty signifier *leader,* for the same reason. Whether one of these strategies will fill the vacuum that is likely to arise, will be an exciting question that will also accompany my assessment of the year 2017. This new strategic question is particularly interesting because before COP22, parts of the pan-European press were still talking about the G2 consisting of China and the USA, which the EU would only be able to contest if it *leads by example* alongside these actors (Stefanini 2016); indeed, in the new situation, this development seemed to be very unlikely.

Only six months later, the EU climate discourse was once again stimulated by events on the other side of the Atlantic. The announcement of the USA's withdrawal from the PA on June 1 was promptly followed by reactions from different countries that both regretted the US exit from the PA and emphasised the now even more important need for ambitious global climate action. While observers feared a domino effect (Jotzo, Depledge, and Winkler 2018, 814; Pickering et al. 2018), most major emitters reacted quickly and assured the world that they will not leave Paris. Stating that "Today is a sad day", Cañete (2017) pursued the hegemonic strategy of constructing the *EU as a leader* in climate policy in his statement reacting on the US announcement. Stressing that "The world can continue to count on Europe for global leadership in the fight against climate change", the Commissioner follows strategeme III. Interestingly, he uses strategeme V to draw a specific form of leadership which builds on cooperation with other actors. Splitting the discursive space, he also uses strategeme II, saying that "We are on the right side of history" (Cañete 2017). He thus constructs the US as the—in this case literally—radical other staying on the wrong side of history. Following strategeme I, Cañete creates the chain of equivalence by equivalising the demands for *leadership* and *increased cooperation with partners* as well as *financial support for developing countries* and *low-carbon transition,* which he even describes as an engine of growth (Cañete 2017). At this point, it is too early to conclude that Trump's announcement of the US exit changed the Commission's hegemonic project from *leading by example* to *leading together with partners*, or just to *being a leader*—but first tendencies could certainly be identified.

The EU heads of state and government did not publish a common statement in reaction to Trump's withdrawal announcement. It is obvious that Theresa May, then Prime Minister of the UK, did not want to offend her US partner who would become ever-more important post-Brexit (Müller-Brandeck-Bocquet, Pietzko, and Zürn 2018, 14). Only Italy, France and Germany issued a joint statement where they emphasised the enormous importance of the PA—not least as an economic engine—and stressed that the agreement could not be renegotiated (France, Germany, Italy 2017). Although no hegemonic strategy can be reconstructed in this rather brief statement, the demands articulated for *sustainable growth* and *economic success* show that those leaders radically questioned the dichotomy, constructed and perpetuated by Trump, of climate protection on the one side and economic success on the other.

Various EU-ENGOs also reacted to the announced US withdrawal. Interestingly, this included the then Chair of CAN Europe, Wendel Trio, who underscored in accordance with the three heads of state and government the possibility of future growth through ambitious climate protection. Trio also called on the EU to "step up its game now, to trigger more climate action and smooth the way for scaling up the Paris climate pledges" (Trio 2017). Susann Scherbarth (2017) of Friends of the Earth Europe argued in the same way and additionally called for an end to the EU's fossil fuel dependency by 2030. The European Environmental Bureau even called on the EU to double its mitigation efforts in trying to fill the gap that the US will inevitably leave (European Environmental Bureau 2017). The reactions of the ENGOs, however, do not indicate which hegemonic project they follow, if any. On the contrary, the European Climate Foundation (ECF) with its well-known CEO and PA co-architect Laurence Tubiana took the announced withdrawal of the USA as a moment to clearly construct the EU as a *leader* in global climate change policy. In its 2017 annual report, the ECF, a foundation with a large scope and recognition, stressed that "European leadership and innovation, both politically and financially, became more necessary than ever in 2017" (European Climate Foundation 2017, 16)—thereby demanding the recognition of the EU as a leader and pushing it to act as such. Using the three core strategemes, the ECF also pursued a reconstructable hegemonic strategy.

The reiterated response that the dwindling commitment of the USA must lead to increased efforts by the EU was also followed by some parts of the EP. On June 14, the EP used the visit of Hilda Heine, President of the Marshall Islands, a co-creator of the High Ambition Coalition, as an opportunity to discuss US withdrawal. After Heine's address, which was particularly dedicated to the challenges of the climate crisis, Helena Dalli, who represented the then Maltese Council Presidency, openly demanded the *EU's leadership role* in international climate policy as a reaction (Dalli 2017). Interestingly, all three core strategemes can be reconstructed in Dalli's speech, so that she was demonstrably pursuing a hegemonic strategy. In the context of the Council, this is a rare open moment. The later position adopted by the whole Council must show whether Dalli is speaking here primarily to represent the Council Presidency or for the whole Council. Commission President Juncker also spoke in the EP on the same occasion, but his remarks allow for fewer conclusions about the Commission's hegemonic project. At the very least, he rearticulated the frequently-discussed demand *stronger cooperation*, especially with China (Juncker 2017a). We should understand this contribution as a hint that Juncker shares Cañete's concern that the demand of cooperation should be given greater weight within the empty signifier of a leader post-Paris.

In the later EP debate, Kathleen Van Brempt and Jo Leinen, both part of the S&D group, supported the approach of Juncker in demanding *more intense cooperation* with other major emitters such as China. Both MEPs placed this demand into the empty signifier of the *leader* EU, following strategeme V (van Brempt 2017; Leinen 2017). The Greens pursued a more progressive direction in this EP discussion. Ska Keller used all three core strategemes, splitting the discursive space and representing the EU also as a *leader*. Interestingly, she uses strategeme V differently, putting three aspects into the empty signifier: leading by example, leading for the sake of leading and leading by increasing the EU's NDCs. Emphasising that "we have the possibilities and we have to go forward. We can really show leadership here" (Keller 2017), she stressed that the Greens see the demand to show higher ambitions through higher NDCs now as a part of the reaction to the US withdrawal. Indeed, the demand to increase the EU's NDCs is not new, particularly from the Greens' side, but it has a new quality: Keller now also cited the gap the USA will leave in the fragile NDC architecture of the PA, and did not merely appeal to the EU's (historical) responsibility.

While Trump and his withdrawal from the PA did have a small following in the ranks of the right-wing populist MEPs, it is interesting to note that the democratic parties of the EP stood together in sticking with the *leader* EU as the identity-representing empty signifier. Nevertheless, the parties kept their differences. When EPP MEP Giovanni La Via (2017) for instance stated that the EU is "ready to play a leading role", he closely linked the demand to *lead* with *competitiveness*—in clear contrast to the Greens. Hence, we see that trying to install a hegemony is an ongoing struggle between different actors (Diez 2013,

204–5), although such a decisive event as the US withdrawal can bring them closer together. This is also evident in Juncker's State of the Union speech in September 2017 where he said "I want Europe to be the leader when it comes to the fight against climate change", adding that "Europe must ensure we make our planet great again" (Juncker 2017b, 3). This brief overview of different reactions has shown how the actors now firmly demanded that the EU act as a *leader* in global climate change policy, with a special emphasis on the feminised demand *cooperation.*

This conclusion is also illustrated later in the EP's pre-COP23 debate. Apart from the Eurosceptic or right-wing populist MEPs, the EP again agreed that the EU must act as a *leader* in climate policy. The roles are similarly assigned as in the debate on the US withdrawal from the PA. The EPP MEPs stress the need to achieve the EU's existing goals, linking the achievement with the *leading role,* which is indeed pointing in the direction of a *leader by example* (Spyraki 2017; Gambús 2017). The Greens, on the other hand, insist that too little is being done to meet the current targets and that the EU's NDCs must be increased, making these two aspects a part of their understanding of a *leader* (Jadot 2017). Besides the three core strategemes and strategeme V, the green MEP Jadot also uses strategeme VI. Stressing that being a leader in climate change policy in his understanding also means to *create jobs* and *renewable energy* sources, he offers subject positions for politico-societal forces calling on no fewer than "all Europeans" to support this project (Jadot 2017).

The COP23 in Bonn, which was chaired by the Fiji Islands, was supposed to discuss the implementation of the Paris Agreement and to launch the Facilitative Dialogue 2018 (also known as Talanoa Dialogue)—but was almost completely dominated by Trump's announced withdrawal from the PA. The EP's resolution of the common position vis-à-vis Bonn must also be understood in this context. First of all, the EP emphasises that the broad commitment to the PA after Trump's declaration is emphatically welcomed (European Parliament 2017a, 4). This statement is accompanied by the demand that the EU must now assume the *leadership role,* not least to defend the PA. Interestingly, the chain of equivalence constructed within the EP's resolution includes very diverse demands, both in the field of energy (European Parliament 2017a, 9) and support for developing countries (European Parliament 2017a, 11), as well as in social aspects of climate policy. For instance, the EP calls for the *strengthening of human rights* in climate policy, "the promotion of fair workforce transition" and better *gender equality* (European Parliament 2017a, 7). Overall, it is clear from this document that the EP is demanding an even *higher EU-engagement* against the backdrop of the US withdrawal. The EP also calls for *greater involvement of local non-state actors* in the process of implementing the PA (European Parliament 2017a, 8). This probably addresses the movement in the USA called "We Are Still In" whose members—including cities, counties and companies—pledged to continue adhering to the US climate goals.

The Council's Conclusions dedicated to COP23 generate an insight into its hegemonic project. First of all, the Council acts like the EP and calls for *greater involvement of non-state actors* such as sub-national actors and civil society and the private sector, in order to be able to achieve the commitments of the PA through other levels (Council of the European Union 2017, 2). I would even dare to suggest that in point six of its Conclusions, the Council is waking up in its rhetoric and also in the clarity of its climate policy positioning—precisely because of Trump's announcement. The Council uses strategeme I, equivalising the demands for *new alliances*, the *further implementation of the PA*, and *EU leadership* with the words that it "REAFFIRMS that the EU will continue to lead global efforts to combat climate change" (Council of the European Union 2017, 3; emphasis in original). The Council had only once before—when it submitted the EU-INDCs—positioned the EU that clearly as a leader. Using all three core strategemes, these Council Conclusions not only unambiguously portray the EU as a *leader*, but also actively pursue a hegemonic strategy. Interestingly, the Council chooses the same understanding of the EU as leader as the Commission did at that time, focusing on alliance-building.

As expected, Bonn turned out to be a mixture of the self-assurance necessary in order to continue to adhere to the PA, and the technical workings of a traditional COP. The demands formulated by the Council and the EP in the run-up, for example around the Facilitative Dialogue, were largely fulfilled. The so-called Talanoa Dialogue ensured a first round of stocktaking so that the time until 2020 could already be used productively—not least to discuss the eagerly-awaited 2018 1.5-degree IPCC report and to reflect the already pledged NDCs (Zürn 2017).

Other outcomes in Bonn were the gender action plan which aims in particular to increase the share of women participating in global climate policy and in climate-protecting measures. Regarding the share of women in the EU's COP delegations, Fig. 4.2 shows that there is still a way to go in reaching a gender balance. Nevertheless, the 30% mark is reliably achieved within the EU delegation so that it can be assumed that the perspective of women is generally taken into account (Ergas and York 2012, 968; Burk and Hartmann 2017, 143). Of course, this is not a statement about the quality of participation, which would need further research.

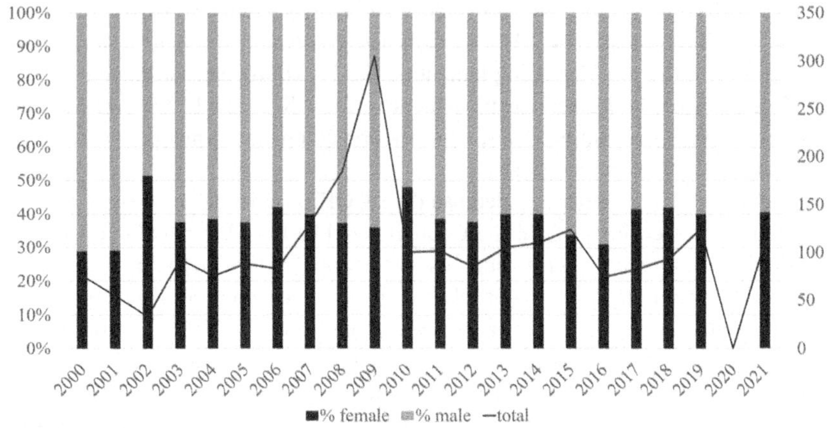

Fig. 4.2 Members of EU COP delegations by sex, 2000–2021. (Own illustration with data from UNFCCC 2000, 2001, 2002, 2003, 2004, 2005, 2006, 2007, 2008, 2009, 2010, 2011, 2012, 2013, 2014, 2015a, 2016, 2017, 2018, 2019, 2021b)

Further agreements were reached in Bonn to increasingly involve indigenous peoples in the negotiations, a demand that the EP, for example, had voiced for some time. For the rulebook, which is an important aspect for the EU institutions, a draft with more than 260 pages was accepted—its adoption would take place at COP24 in Poland, thus remaining open as a major construction site for the next year.

Overall, the analysis of 2016 and 2017 shows that the various actors responded to Trump's announcement of the US withdrawal from the PA in large part by changing their hegemonic project from *leader by example* to *EU as a leader*. However, not all leaders are the same. Rather, the analysis revealed that some of the actors have given the empty signifier a new meaning—for example as compared to the understanding of *leader* during Paris. Now, actors such as the Commission and the Council were concerned that the *leader* should focus on cooperation in order to have a chance of achieving the climate goals and at the same time pursuing the well-known triad of economic success, innovation and mitigation. In contrast, some NGOs and some MEPs charge the identity of the EU with even more ambitious EU NDCs. However, little has changed with regard to the gendered structure of the discourse as most actors rearticulate especially masculinised demands, even though some feminised demands are gaining more importance, particularly when it comes to cooperation.

4.2.6 Meeting the Growing Public Pressure to Set the Course for the Implementation of the Paris Goals

With 2017 particularly marked by the global dimension of climate policy, the EU in 2018 concentrated more on internal tasks. Developing a common vision and laying future tracks, the Union tried to implement its Paris pledges. The question of which sectors should be involved and to what extent affected not only the EU institutions but also broad sections of the business community and non-governmental actors, which positioned themselves regarding the discussion on the long-term EU GHG emissions reduction. In addition, the Fridays for Future (FFF) protest movement was formed in 2018, with Greta Thunberg school-striking in Sweden, and was met with rapidly increasing (media) interest (Pianta and Sisco 2020, 3) and broad transnational support (Marquardt 2020). Both culminated in an unprecedented pro-climate-protection sentiment among European citizens, as the findings of the 2019 special Eurobarometer show (European Commission 2019a). Compared to previous years, and also in contrast to the subsequent 2021 special Eurobarometer, an all-time high was achieved in 2019 (cf. Fig. 4.1) with more than 93% of respondents agreeing that climate change was a serious problem and 60% agreeing that it was one of the most serious global problems (European Commission 2019a, 3). In this wider context, the course was set for the European Green Deal (EGD)—all the more so when the Commission released its Communication "A Clean Planet for All" in late 2018 (Wendler 2022a, 77).

In parallel, further conceptual questions of the future EU climate policy were also debated in 2018. One aspect to be considered was the future design of EU development cooperation regarding mitigation and adaptation in line with the SDGs, which of course include—besides economic and ecological facets—social prospects of development such as gender equality. Known for its internal EU forerunner role of thinking gender and development together (Allwood 2020b, 178–79), the EP adopted its resolution "Women, gender equality and climate justice" (European Parliament 2018b). Since this is a rare moment within the EU climate discourse of showcasing the use of gendered discourse patterns of the first dimension, and a rare instance of interpellating groups for a specific gendered role, the EP resolution will be reconstructed in detail.

In the introductory chapter, the EP sets the scene of climate change and its interrelation with gender structures. There, the EP interpellates women living in the Global South particularly for their subject position as victims of climate change effects and pointing to gendered social structures and their resulting consequences such as "unequal access to resources" (European Parliament 2018b,

3). As theoretically derived in Sect. 2.3.2, if the women of the Global South (GS) are drawn one-dimensionally as victims of climate change, there is a danger that other subject positions in this concrete discourse become difficult to imagine. This is not because women do not occupy different subject positions at the same time, but because they are primarily anticipated and addressed as victims. In the following chapters, the resolution states that these gendered structures shape different spheres such as "decision-making processes and labour markets" (European Parliament 2018b, 3). Therefore, it demands *women's greater participation* in different decision-making processes (European Parliament 2018b, 8), as they are still insufficiently represented. Stressing the fact that climate policy would profit from women's views, the EP interpellates women of the GS also as "agents of change" (European Parliament 2018b, 3). Accordingly, the EP tends to recognise multifaceted subject positions of women of the GS. But unfortunately, it is not clear whether the EP recognises the simultaneity of the possible roles of women as victims and agents—or whether most women of the GS are primarily interpellated as victims, while the more privileged women, for example those assigned to higher classes, are understood as agents.[34] Nevertheless, these gendered discourse patterns of the first dimension, inducing different women's subject positions, can have a meaning for the second dimension of gendered discourses when it comes to measures to tackle the climate crisis in the Global South, which are to some extent subjects of the EU discourse.

When the resolution briefly comes to the situation in the EU, the EP particularly criticises the insufficient participation of women in the member state's climate policy bodies, trying to interpellate women as decision-makers (European Parliament 2018b, 4). Beyond these subject-positions, the EP introduces an additional role for all women, interpellating them as consumers who make relevant family-wide decisions in daily life due to their high share of care work (European Parliament 2018b, 7). Thus, the EP does not completely fall into the trap Arora-Jonsson (2011) reconstructed in the typical climate change discourses, contrasting a woman's potential roles either as a victim of climate change when she lives in the Global South or as an important actor caring for the climate if she lives in the Global North. Nevertheless, the EP falls into another trap regarding the second dimension of gendered discourse patterns, concerning the question of how to deal with climate change consequences. Discussing gender and climate change and insinuating that everyday consumption is a central component of emission

[34] The danger here is that the attempt to remedy marginalisation on the basis of gender can nevertheless be classist. This in turn can lead to other social problems, as has for instance been studied in the area of post-conflict peacebuilding (Zürn 2020).

reduction without considering energy production or the broader responsibility of the economic system, it seems that the EP is about to privatise responsibility for protecting the climate. The private sphere is associated with femininity, while the public is culturally associated with masculinity (Prügl 2021, 300; Bruff and Wöhl 2016, 95). Hence, caring for the environment is associated with the private and therefore the space in which femininity is placed (MacGregor 2014, 625). The discursively constructed responsibility for fighting the climate catastrophe shows the mutual influence on gendered climate discourses of the second dimension from the first dimension. Whether the EP pushes this idea of gendered responsibility within the EU discourse as part of its hegemonic strategy can only be shown by further analysis.

The EP also dealt with other conceptual issues of the future EU climate policy. In July 2018, it adopted a resolution on the design of EU climate diplomacy, which provides interesting insights into the EP's hegemonic strategy in trying to create the EU's identity as a climate leader. Using the first strategeme, the EP produces a chain of equivalence focusing on demands such as *technological innovation* to reach the PA goals but also to support *human rights* and *gender equality* as a vital part of climate diplomacy. The chain is built around the nodal point *leader* regarding the field of renewables (European Parliament 2018c, 4, 13) and in general in global climate policy (European Parliament 2018c, 6, 12). In doing so, the EP uses two different temporal references, namely the leadership of the EU back in Paris (European Parliament 2018c, 3) and the future vision of the EU as a leader, and thus also the need in terms of its climate diplomacy. Following the second strategeme, the EP renews the distinction between the EU and Trump's USA, emphasising that the announced withdrawal from the PA places the EU under a special obligation to act as a leader in international climate negotiations (European Parliament 2018c, 6). Using the three core strategemes, the EP continues its hegemonic strategy of positioning the EU as a leader.

What further contributions shed light on the discourse and strategy of actors shaping the EU discourse? In March 2018, EUCO invited the Commission to elaborate a proposal regarding how the long-term GHG emissions reduction can be achieved in line with the EU's PA commitments. Thereupon, the Commission laid out different scenarios and pathways, as tabled in its Communication "A Clean Planet for all" in an effort to fulfil the EU's long-term obligations to reach net-zero emissions by 2050. The Communication was presented in November 2018 shortly before COP23. The public consultations in the previous summer generated diverse discursive contributions (Ammann and Boussat 2023, 244).

The association of the various European chambers of commerce and industry, Eurochambres, was one of the economic actors that expressed its views at an early stage. Focusing on the more widely formulated demand of *competitiveness,* Eurochambres insisted that it fears the exodus of companies in the wake of ambitious mitigation measures, which is why it also called for the promotion of global *MRV* commitments (Eurochambres 2018, 2), which would achieve globally comparable framework conditions. It is interesting to note that Eurochambres nevertheless points out that in the long-term positioning of the EU there lies an opportunity for companies, for example in the area of innovations, which will trigger the needed transformation of the entire economy (Eurochambres 2018, 3). Even though this remark seems far-reaching, the paper remains rather vague, concentrating on competitiveness without demanding concrete thresholds or measures.

On the contrary, most involved ENGOs such as Greenpeace Europe and WWF Europe, took a different approach. They demand more ambitious EU mitigation and energy goals in order to meet the long-term goal of net-zero in the framework of the EU's historical responsibility (WWF European Policy Office 2018, 3; Greenpeace Europe 2018). Therefore, both European ENGOs demand *EU's net-zero emissions by 2040* instead of 2050 as envisaged by the EU. ETUC, the confederation of different European trade unions, by contrast still sticks to the demand *EU net-zero in 2050* (European Trade Union Confederation 2018, 2), equivalising it with *low carbon* and *circular economy* (European Trade Union Confederation 2018, 3). Furthermore, aspects of its chain of equivalence are a *just transition* factoring in equity aspects like *gender equality* and *decent work* (European Trade Union Confederation 2018, 2, 3). Attesting the EU current leadership in technologies and beyond, ETUC uses strategeme III representing the EU's climate change policy identity as a *leader* as regards to the broader shift to "a global and rapid move to 'net-zero emissions'" (European Trade Union Confederation 2018, 4). Putting different nuances into this leadership role, ETUC furthermore follows strategeme V, creating leeway in the empty signifier. Doing so, the confederation does not demand any more ambitious goals than the agreed status quo, but at the same time places social aspects such as gender equality and decency in the empty signifier. Hence, ETUC uses both masculinised and feminised discourse patterns in formulating their proposals for tackling climate change.

The Green Growth Group (GGG), the loose alliance of now 16 EU member states which already appeared in the run-up to Paris and advocates a more ambitious EU climate policy, also positioned itself regarding the Commission's Communication. GGG clearly demands that the EU *lead by setting a good example* in preparing the review of the NDCs in 2020. The group equivalises this demand inter alia with *equity* and *competitiveness* in order to increase the EU's NDCs on the one hand and on the other hand to meet the already promised targets such as EU net-zero in the year 2050 and 40% mitigation by 2030 (Green Growth Group 2018, 1). Interestingly, GGG rearticulates several times the demand *to trust in and follow the scientific findings,* referring to the upcoming 1.5-degree report of the IPCC (Green Growth Group 2018, 1). The very concise positioning of the group follows the common hegemonic project discursively creating the EU as a *leader by example.* The distinction from the radical other is implicitly made by stressing the need to set an example that generates followership among others. Hence, GGG uses the three core strategemes in trying to help its hegemonic project to succeed. Moreover, relying particularly on *innovation* and *research* (Green Growth Group 2018, 1) to counter the consequences of the climate crisis, GGG uses foremost masculinised discourse patterns. Demanding "an economically efficient green transition" which is "just for all the European citizens and regions", GGG (2018, 1) also makes careful use of discourse patterns that can be categorised as feminised.

The publication of the IPCC's 1.5-degree special report in October 2018 (IPCC 2018) coincided with the outlined debate on the EU's long-term strategy, but also with its medium-term goals, as in 2018 the NDCs were to be critically reflected in the framework of the Talanoa Dialogue. Therefore, Greenpeace Europe used the IPCC report as an opportunity to remind the EU of its "claims to be a climate leader" (Greenpeace Europe 2018) demanding more ambitious 2030 NDCs in line with the report's shocking findings—but also to fulfil its ambitions to lead by example as the EU was the first actor starting the process to critically reflect and discuss its NDCs. The Commissioners Miguel Arias Cañete (Climate Action and Energy) and Carlos Moedas (Research, Science and Innovation) commented on the 1.5-degree report and underlined the "Juncker Commission's objective to become a global leader in renewables and to put energy efficiency first" as well as their ambition for the EU to continue to be the leader in innovation and research "to tackle climate change" (European Commission 2018b). Hence, we can attest the Commission's hegemonic project to be recognised as global leader in climate change policy, and also to some extent to be seen as leading by example, which is inter alia supported by Greenpeace Europe and GGG. However, the Commission still primarily sticks to masculinised discourse patterns demanding technical solutions such as renewables and energy efficiency, while cooperation or other feminised demands are articulated only on the second glance.

In November 2018, the long awaited Commission's Communication "A Clean Planet for all", outlining "A European Vision for a Modern, Competitive, Prosperous and Climate Neutral Economy" was adopted (European Commission 2018a). Showing that the EU has been at the forefront of global climate action and should continue to *lead by example*, especially in the field of renewable energies (European Commission 2018a, 4–5), the Commission uses the three core strategemes in the second chapter of the proposal to present the EU as a *leader* in global climate policy. The Commission implies that achieving *net-zero by 2050* and *leading in the field of renewables* represent *EU leadership*. Hence, we can trace the use of strategeme V, loading the empty signifier with further meaning. The discursively constructed chain of equivalence includes, among others, the demands for *competitiveness, sustainable growth, jobs* and various *co-benefits* to be generated by climate action (European Commission 2018a, 5). Although the Commission also articulates the demands for a *socially fair transition* (European Commission 2018a, 5), and *the improvement of quality of life* (European Commission 2018a, 6), the chain of equivalence is still characterised by technical and therefore masculinised demands. Following strategeme VI, the Commission offers subject positions for economic actors and the society (European Commission 2018a, 6), trying to generate the necessary support for its hegemonic project.

After the Commission has outlined its more general vision, the third chapter of the document deals more specifically with the eight scenarios the Commission has modelled.[35] It is interesting to note that only two scenarios (7 and 8) are even able to reach net-zero emissions by 2050 (Umweltbundesamt 2018, 3). In the case of scenario 7 it includes the combination of different scenarios of technological optimisation, such as the electrification of several industries (scenario 1) or the innovation of Power-to-X (scenario 3) with carbon capture and storage (CCS). Scenario 8 also combines the technical optimisation scenarios 1–4 with the scenario of an EU-wide circular economy (scenario 5) and drastic lifestyle changes of the society (European Commission 2018a, 7). Bearing in mind that CCS technologies are to this day highly controversial and lack public acceptance due to several—perceived as unrateable—risks (Urbansky 2023, 14; acatech 2018, 49), it seems, that only scenario 8, which additionally requires lifestyle changes, is an appropriate way to achieve net-zero by 2050 on the EU level. However, reflecting the different scenarios, the Commission is still primarily focusing on technological innovation as a response to the climate crisis and thus rearticulating particularly masculinised discourse patterns. Distinctly feminised discourse patterns can only be reconstructed in scenario 8 when lifestyle patterns and consumption decisions are affected. The circular economy cannot be clearly assigned to either feminised or masculinised patterns; this will be discussed later in Sect. 4.2.9.

[35] For a detailed presentation and discussion of the scenarios see Wachsmuth et al. 2022.

As the Commission's Communication already included the increased energy efficiency and renewables goals jointly adopted by the Council and the EP in 2018, it was also considered and discussed outside the EU. With the new targets of a 32% share of renewable energies by 2030 and a 32.5% increase in energy efficiency compared to 1990 (European Commission 2018a, 5), the EU was able to performatively rearticulate its demand to be recognised as a leader.

In the preparation of the Katowice Summit, it is therefore hardly surprising that both the Council and the EP pursued the *EU as a leader* as their hegemonic project using the three core strategemes in their discursive contributions. The Council adopted its Conclusions on October 9 (Council of the European Union 2018), the content of which was controversial regarding the EU's position on the review of its own NDCs. A leak revealed that the Council was divided into different camps: One group expressed the need to "signal readiness to review" the EU's NDCs while another group stuck to a compromising formula outlined by the presidency which "gives a strong political signal and narrative on the EU level of ambition for the upcoming UN negotiations in Katowice" (Stam 2018).

Analysing the subsequently adopted Conclusions, it becomes clear that the latter group was able to prevail in the decisive paragraphs 10 and 23. In paragraph 23, the Council stresses the importance of the results of the Talanoa Dialogue and the need to increase countries' targets globally. Saying that the Council "UNDERLINES that the EU will continue to create a positive momentum among all Parties to enhance global climate ambition" (Council of the European Union 2018, 9; emphasis in original), the Council does not announce any concrete measures or targets, but rearticulates its perception of the EU as an important actor that could lead the way and motivate other states to achieve higher targets. Thus, the demand to *lead* is indirectly articulated here, serving both as a reference point for the EU's identity (strategeme III) and to distinguish the EU from other states (strategeme II) implicitly portrayed as less ambitious. In paragraph 10 the Council stresses that the EU's energy goals will be increased and that the EU and its member states will take stock of these and further sectors (Council of the European Union 2018, 5). However, the need to strengthen mitigation targets is neither discussed nor announced. Although the Conclusions fall short of the leaked demands of the more ambitious member states, the EU is still reproduced as a leader stating that the Council: "RECONFIRMS that the EU will continue to lead the way in the global pursuit of climate action" (Council of the European Union 2018, 3; emphasis in original). Of course, early revised and announced EU mitigation goals would have underscored the demand to be seen as a leader in a performative sense. Nevertheless, all three core strategemes can be reconstructed, and thus the Council continues to pursue the hegemonic strategy of constructing the EU as a leader. Is the EP heading in the same direction?

Typically, the EP goes further than the Council in demanding the EU's future mitigation targets under its NDCs to be increased to 55% by 2030 (European Parliament 2018a, 6). Using strategeme I, the EP equivalises various demands such as *renewable energy* (European Parliament 2018a, 14), *competitiveness* (European Parliament 2018a, 5, 6), *sustainable economic development, jobs (*European Parliament 2018a, *5)* and *gender equality* (European Parliament 2018a, 9) in climate change policy. As it furthermore "Stresses the importance of an ambitious climate policy for the EU to act as a credible and reliable partner globally, of maintaining the EU's global climate leadership" (European Parliament 2018a, 6), the EP integrates the demands for an *ambitious EU climate policy* and a *leadership role* into the chain of equivalence. In doing so, it also follows strategeme III representing the EU as a leader. Furthermore, the EP distinguishes the EU in particular from the USA, namely under Trump (European Parliament 2018a, 6), dividing the discursive space using strategeme II. Although the Council and the EP both pursue the hegemonic strategy of constructing the identity of the EU as a leader, the empty signifier seems to be filled in a different way. The EP places the demand for a significant increase in mitigation targets in the empty signifier, while the Council remains vaguer by particularly relying on generating the allegiance of other states through a broad discussion. However, it is the nature of a well-functioning empty signifier to be able to represent diverse meanings at the same time.

The negotiations at COP24 in Poland in 2018 were supposed to be technical in nature, focusing on the rulebook navigating the future comparability and verification of the NDCs' implementation (Zürn 2019). Bas Eickhout, deputy chairman of the Green group in the EP, complained that the EU negotiators in Katowice shifted the spotlight onto the rulebook as part of a trick instead of focusing more on the first stocktaking process to be concluded at COP24 in the framework of the Talanoa Dialogue (Stam and Simon 2018) and instead of seizing the chance to initiate new dynamics that would increase the EU's NDCs. Actors from civil society and the business community called even more intensely for stronger targets in Katowice. After the first days of COP24, a group of businesses, investors, local and regional authorities as well as civil society organisations grouped in the Coalition for Higher Ambition published an open letter urging the EU leaders to

> *"show leadership and ambition, and commit to increase the EU's 2030 target even beyond the 55% reduction some Member States and the European Parliament are calling for" (Coalition for Higher Ambition* 2018, *1).*

Unsurprisingly, Cañete already declared international EU leadership at the beginning of the negotiations on the basis of the recently adopted long-term strategy, for which "the EU received praise" on the global stage (Stam and Simon 2018). This image was supported by the EU heads of state and government when, just one day before COP24 ended, they officially invited the Council to

> *"work on the elements outlined in the [Commission's] Communication to enable the European Union to submit a long-term strategy by 2020 in line with the Paris Agreement" (Council of the European Union* 2019, *4).*

This is the performative rearticulation to lead, as the EU would undoubtedly be one of the first actors with a long-term plan.

The different stakeholders thus continue to refer to the EU as a leader, with masculinised patterns continuing to dominate, although it was possible to reconstruct more demands to be located in the feminised sphere compared to former years. However, the empty signifier is once more conceptualised more strongly as a leader, partly characterised by a focus on the aspect of leading by example, especially in the field of energy. But the aspect of leading by cooperation—which seemed to develop in the direction of a competing hegemonic project—did not prevail.

4.2.7 The European Green Deal: Making Hay While the Sun Shines

After the EU started to set the course for the implementation of the European contribution to the Paris Agreement, the EU climate discourse in 2019 was shaped by the EP elections, flanked by broad public debate of the climate crisis as well as the constitution of the new Commission and its first climate-related activities, first and foremost the inauguration of the European Green Deal (EGD). The importance of the latter point can hardly be overestimated, as the EGD was also seen as a turning point in EU climate policy (Fischer 2020). Further aspects such as the failure of the *Spitzenkandidaten* system had also an impact on the discourse that should not be underestimated, as the upcoming analysis of the EGD's genesis will show.

In March 2019, the EP dealt with EU climate policy and discussed its position with the invited current Council Presidency of Romania and Arias Cañete, the Climate Commissioner. Both Cañete and Melania Gabriela Ciot, representing the Romanian presidency, discursively constructed the EU as a leader in their lengthy statements. While Cañete rearticulated the demand to *lead by example* (Cañete 2019), Ciot (2019) was much more explicit saying that the EU is "determined to lead the way in the global response to climate change" representing the EU as a *leader* using strategeme III. Following strategeme I, Ciot (2019) links this demand with the demand to *fully implement the PA* and *the well below 2-degree goal.* Interestingly, later in her speech she also articulates the demand to *lead by example,* pointing to the necessary *submission of the EU's long term strategy* to the UNFCCC as required in 2020 and calling on other states *to follow the EU's lead* (Ciot 2019). Hence, she also uses strategeme V creating leeway and putting various meanings into the empty signifier of the leader. Cañete furthermore explains in his speech how the EU should respond to the climate crisis focusing, as in previous contributions, on *technical innovation* and *optimisation in the energy sector* (Cañete 2019), and hence, on masculinised discourse patterns. Repeatedly addressing European society, it seems this is the way Cañete chose to respond to the overall climate in the EU in 2019, when there were increasing and vociferous public calls for an effective climate policy—not only from FFF, but also from further environmental and climate protection alliances, which gained additional traction through the huge FFF movement[36] (Gentes, Löning, and Trapp 2020). At the end of his speech, Cañete stresses that the EU must listen to the will of society, and in particular to the admonitions from the younger generation, and create a transformation to net-zero by 2050 (Cañete 2019). In doing so, Cañete follows strategeme VI by generating different politico-societal forces, hoping for their future support.

Peter Liese (EPP) also follows this strategeme when he turns to the activists sitting in the gallery of the parliament on that day. He insists that the upcoming EP's resolution of the EU's climate neutrality by 2050 is also good news for them—and partly their success (Liese 2019). Whether Liese and the other MEPs following the same strategeme succeed in weaving FFF into their respective chain of equivalence in a supporting role must remain open at this point. While the speeches of the various EU parliamentarians of this day frequently refer to the younger generation and their climate strikes, the MEPs do not detectably follow a hegemonic strategy consisting of the three core strategemes, so that the ecofeminist hegemony analysis reaches its limits here.

The day after this EP debate, the Parliament adopted its long term strategy in response to the Commission's 2018 Communication "A clean Planet for All"

[36] Although FFF is mainly made up of pupils and students, other groups, such as parents and scientists, joined rapidly in 2019. For the socio-cultural structure of FFF see Della Porta and Portos 2023.

by a large majority, which arithmetically must include both S&D and EPP as well as further votes from the other groups such as the Greens, the Left and ALDE (Legislative Observatory European Parliament 2019). It is important to note that the EP continues in its long-term strategy to follow its hegemonic project of constructing the identity of the EU as a *leader*, linking it with the demand for higher NDCs (European Parliament 2019c, 12) of 55% GHG savings by 2030 compared to 1990 levels (European Parliament 2019c, 5). Strikingly, in the previous debate, the EPP group argued for less far-reaching future targets than the subsequently adopted 55% mitigation goal (Liese 2019). Obviously, the EP was able to reach an agreement on these higher goals in the end. This shows that a common hegemonic project, namely the leadership role, can override other conflicts, probably due to the economy-focus adopted in the resolution, which is one of the main interests of the EPP. It is likely that the pressure of more than 1.5 million students who took to the streets on March 15 (Friends of the Earth Europe 2019) has further boosted those parties demanding an (at least) 55% goal in the EP's resolution.

The speech of Greta Thunberg, the then shining star of the global environmental movement FFF, on April 16 at the meeting of the EP's Environment Committee also underlines this impression. Although she did not follow the core strategemes to construct an identity of the EU in her speech, she clearly showed that FFF demands ambitious action with the theme "I want you to panic. I want you to act as if your house was on fire" (Thunberg 2019, 1). She insisted that "the millions of schoolchildren school striking for the right to a future" demand no less than "please wake up and make the required changes possible" (Thunberg 2019, 3). The analysis of the speech reveals two main demands: to *listen to the science* and *climate justice*, which Thunberg increasingly calls for—also regarding the Global South. Overall, she positions the youth as a well-recognised actor trying to shape the EU climate change discourse and is somewhat surprisingly compatible with the position of other actors such as the EP or the Commission. After all, taking adequate account of scientific findings has consistently been a frequently reproduced demand. Of course, the demands of FFF go far beyond the common ground with the European institutions, particularly when it comes to concrete climate protection measures and the pace of implementation.

Thunberg's speech took place one month before the citizens of Europe elected their future European Parliament. It can be understood as a manifestation of the growing importance of climate change as a campaign issue. In the 2019 campaign, climate policy was for the first time a decisive EP election issue (Weyers et al. 2020) and was the most important topic for those citizens who were likely to vote, as a special Eurobarometer showed (European Parliament 2019a, 58). The parties also reacted to this situation in their programmes; in all manifestos of the democratic parties the increasing importance of climate policy can be attested when compared to their 2014 counterparts (Institute for European Environmental Policy 2019, 10). However, as Till Weyers et al. (2020, 292) determined, climate protection plays hardly any role in the 2019 manifestos of the right-wing and right-populist parties, which is why I follow the authors and do not include these parties' programmes in the subsequent analysis.

The Party of European Socialists (S&D) reproduces its empty signifier of the *EU as a leader* in climate change policy right at the beginning of the paragraph of its manifesto dedicated to climate change:

> *"Europe must be a leader in fighting for the environment and tackling climate change, particularly as some of our international partners are falling behind." (Party of European Socialists* 2019, *8)*

Here, S&D indirectly refers to the USA and distances the EU from Trump's PA withdrawal, following the second strategeme. Although the party's programme regarding climate protection is only briefly summarised on one page (as most of the other topics), S&D constructs a complex chain of equivalence. Using strategeme I, S&D equivalises inter alia the demand for an undated *net-zero target for the EU*, with the *protection of biodiversity, just transition, consideration of scientific findings, leading by example in the renewable energy sector*, and a *sustainable transport* while *increasing people's quality of life* (Party of European Socialists 2019, 8). Overall, it is evident that S&D takes a social perspective on the climate crisis while the concrete measures which should be taken are of technical nature in the field of energy. These technical and therefore masculinised demands are combined with feminised patterns such as a just transition or the people's wellbeing, underlying the social democrats' hegemonic strategy to construct the empty signifier of the EU as a leader, also with the help of feminised demands.

The European Left (EL) focuses even more strongly on social demands, which they link with *climate protection measures* and especially with the demand for *sustainable development* (European Left 2018, 14). The Left's articulated social demands, which are culturally associated as feminine, include *lifelong social solidarity* and the *support for disadvantaged regions of the earth* (European Left 2018, 14). Equivalised with *the use of environmentally-friendly technologies,* EL is the only party calling for a transformation of the current economic model (Weyers et al. 2020, 292) based on the *precautionary principle* (European Left 2018, 14) in the sense of strong sustainability.[37] Since the core strategemes II and III of a hegemonic strategy cannot be reconstructed, I am not able to detect the EL's hegemonic project.

The EPP, the then-largest EP group, very clearly pursued the strategy of establishing the EU as a *leader* within the discourse, while remaining rather vague regarding its vision for fighting climate change. As in various EP debates, the EPP views the climate crisis in its manifesto from an economic perspective. Hence, the articulated demands of the manifesto—such as *clean technology*, *economic success* and *a real energy union* serving *energy security*—are often economy-related (European People's Party 2019, 5) and connotated as masculine. Following the first strategeme, the EPP also equivalises other demands, such as *considering scientific knowledge* or *climate protection* (European People's Party 2019, 5), which repeatedly converge with the leading role of the EU as its empty signifier. Stating that the EU should show all other states "how sustainability works" (European People's Party 2019, 5), the EPP divides the discursive space into an 'us' and 'them' in the sense of the second strategeme. Using strategeme III representing the EU as a leader, the EPP completes the necessary set of the three core strategemes of a hegemonic strategy with a clear focal point on masculinised demands.

ALDE also adheres to an economic perspective on climate policy in its manifesto (Weyers et al. 2020, 295) and repeatedly calls for the EU to act as a leader in global climate policy (Alliance of Liberals and Democrats for Europe 2018, 6). Still, the manifesto of ALDE is quite different from that of the EPP, as the Liberals are much more concrete when they call for a future mitigation target of 55% by 2030, making it an important prerequisite for the EU's leadership role (Alliance of Liberals and Democrats for Europe 2018, 6) using strategeme V. Following strategeme I, ALDE equivalises, among other demands, *a sustainable economy*, *competitiveness*, *technical innovation* and *circular economy* with the *leadership role* and also energy-related demands such as *energy efficiency* and *renewable energies* (Alliance of Liberals and Democrats for Europe 2018, 6). The chain of equivalence produced in this way shows, above all, economic-related demands which should particularly be fulfilled by technical and market-based instruments such as the ETS (Alliance of Liberals and Democrats for Europe 2018, 6). Although ALDE represents the EU as a leader and thus strategeme III can be reconstructed, ALDE does not use strategeme II, which would actively delineate the EU from other actors splitting the discursive space. Regarding the

[37] Regarding the distinction between weak and strong sustainability, see Sect. 1.1, 3.3.1.

character of a manifesto addressing potential voters and party members, this finding is not surprising, even though it means that I am not able to attest the use of a hegemonic strategy while ALDE actively supports the picture of the EU as a leader in global climate policy. Hence, ALDE does traceably follow its hegemonic strategy in some EP debates, but not in its 2019 manifesto—building particularly on masculinised demands in both cases.

As was to be expected, the Greens' manifesto stands out from the other party families due to the strength of its engagement with EU climate policy, as it already did in 2014 (cf. Sect. 4.2.3). Like S&D and the Left, the Greens view climate policy from a more social perspective. Outlining very concrete proposals, the Greens call for an increase of the EU's NDCs to "at least 55%" by 2030 (European Green Party 2019, 2), going the furthest of all parties by prefixing the 55% with "at least". Using the three core strategemes, the Greens follow the hegemonic strategy of constructing the identity of the EU as a leader in climate policy and thus join the strategy of most of the other parties analysed. The Greens' understanding of an EU leadership role means further-reaching EU climate targets. Forming the chain of equivalence, the party equivalises the demands for the general *well-being of the people* and their *wealth*, a *net-zero emission economy, improving the ETS* (European Green Party 2019, 2), *moderate consumption, circular economy* (European Green Party 2019, 4), the *end of subsidies for fossil fuels*, a *coal phase-out*, and different energy-related demands such as *100% renewable energy, energy saving and efficiency* as the EU's first future priorities (European Green Party 2019, 3). It is striking that the Greens on the one hand stress social demands and on the other hand also focus strongly on technological solutions to combat the climate crisis. Arguing for their hegemonic project, they thus use both highly masculinised and feminised discourse patterns. Calling for a "sizeable Green New Deal to finance and leverage investments" to make "the necessary transition to a sustainable economy" (European Green Party 2019, 10), the Green's cross-cutting idea of climate policy becomes obvious. Consequently, they position their analysis in the greater context of the economic system and discuss many different policy fields such as transport, agriculture, health and social rights (European Green Party 2019). Furthermore, soliciting for a comprehensive "Green New Deal", their demand anticipates the future Commission's programme to be launched only a few months later, as will be discussed in more detail further on in this chapter.

After analysing the manifestos, it can be said that all the parties studied formulate primarily technological solutions in response to the climate crisis. They all use masculinised patterns, while the Greens, Left and S&D also articulate social and sometimes degrowth demands and weave them into the chain of equivalence.

Thus, they also present feminised discourse patterns as part of the solution. In the cases where a hegemonic strategy could be identified, it is the *leading role of the EU* that was discursively constructed as the empty signifier—in some cases based exclusively on masculinised demands, and in other cases on both masculinised and feminised patterns.

Which demands did other actors articulate in the run-up to the elections? Green 10, the association of the 10 largest ENGOs in Europe, clearly calls for EU leadership in the global climate policy as its first priority for the upcoming elections (Green 10 2019, 2, 3). Although only parts of a hegemonic strategy can be identified within Green 10's EP election manifesto, this is a vibrant indication that the ENGO coalition has been won as a supporter for the hegemonic project portraying the EU's identity as a leader in climate policy. Green 10 probably expects this EU identity to provide the greatest possible support for its own climate policy priorities. ETUC has taken a similar position, also calling for EU leadership (European Trade Union Confederation 2019, 1, 2, 3), closely linked to *more ambitious mitigation targets* and socially oriented demands such as a *just transition* (European Trade Union Confederation 2019, 2). From the business actors, only calls to vote can be reconstructed, but no clear positioning on EU climate policy prior to the EP elections, even though Eurochambres (2019a) calls climate change one of the "big issues (…) which must be addressed effectively at EU level". At least in the election campaign, this demand was met.

The Greens had announced early on that they would concentrate on social, environmental and climate issues (Stam and Schulz 2018). But they were no longer alone in setting this spotlight as *Spitzenkandidat* Frans Timmermans (S&D) also focused on social and climate policy, as was traceable in the last TV debate, where his opponent Manfred Weber (EPP) represented the consensus-oriented European mainstream (Heidbreder and Schade 2020, 173). The results of the elections[38] reflected this new climate change focus on the part of some of the politicians and European society. With 22 additional mandates, the Greens were the big winners (Weyers et al. 2020, 296), although their respective country-specific results mirror the fragmented situation regarding the "economic affluence and environmental issue salience" of the people (Pearson and Rüdig 2020, 340). Together with the Liberals' strong election results, Johannes Hillje (2020, 438)

---

[38] The constituent session of the new parliament was made up as follows: EPP 182 seats/ 24.3% of all seats, S&D 154/20.51%, Renew Europe (former ALDE) 108/14.38%, Greens 74/9.85%, ID (Identity and Democracy) 73/9.72%, ECR (European Conservatives and Reformists Group) 62/8.26%, GUE/NGL (Confederal Group of the European United Left—Nordic Green Left) 41/5.46% and non-attached members 57/7.59% (European Parliament 2019b).

attests an overall voter desire for more climate protection and greater commitment to fundamental rights, which might be able to shape the future EU climate discourse. However, the increase of Eurosceptic and right-wing populist parties may on the contrary change the EU's climate change discourse and weaken its pace.[39]

However, due to the simultaneous losses by the EPP and S&D, the Greens had the chance to play a more important role than after the 2014 elections. Since the EP was unable to use the momentum in 2019 after the elections to push one of the *Spitzenkandidaten* through as the future Commission President, EUCO defeated the seemingly divided EP with a negotiated personnel tableau led by the German Christian Democrat Ursula von der Leyen—who did not run as *Spitzenkandidatin*—as candidate for the Commission Presidency (Müller Gómez and Thieme 2020). Circumventing the EP's weight in this regard, the results of the elections were more clearly reflected only in the Commission's future shape and agenda, as von der Leyen had to "reach out for support across political parties and climate was the principal unifying theme" (Bongardt and Torres 2022, 173).

It is therefore hardly surprising that out of her six priorities presented in mid-July 2019, von der Leyen first named the European Green Deal, the necessity of which she explained stating that "the message from Europe's voters—and those too young to vote—is loud and clear: they want real action on climate change and they want Europe to lead the way" (Von der Leyen 2019a, 5). Combining economic incentives and climate protection measures in the EGD's outline on the one hand and promoting it not least with the picture of a leading EU on the other hand, von der Leyen tried to secure the support of the EP while giving a glimpse into her probable future hegemonic project.

Although the calculation did not work out completely, since the Greens did not support von der Leyen—even though they can be described as originating the idea of an EGD in their 2014 manifesto—, she was able to secure the votes of the EPP, large parts of the S&D and of Renew Europe.[40] Von der Leyen was thus elected as Commission President. Against the background of this result, the effect of von

[39] Since far-right and right-wing populist parties deal with the climate change issue in different ways (Forchtner 2019), not least due to different national contexts, it is impossible to predict how the EP's policy will change. Denial, but also delaying strategies are conceivable (Forchtner 2020), although the connection between right-wing populist parties and climate policy has been little researched so far (Forchtner 2019, 7–8) The gendered dimension of climate denial among populists has already been explored, although not exclusively for EU actors (Agius, Rosamond, and Kinnvall 2020).

[40] Renew Europe (Renew) is the successor of the former ALDE group, now extended by the French party Renaissance, founded by French President Emmanuel Macron.

der Leyen's programmatic orientation and the announced focus on EU climate policy should not be underestimated as an important performative contribution to the discourse. The fact that the future Vice-President of the Commission Frans Timmermans as EGD Commissioner even bears the EGD in his nomenclature underlines the performative rearticulation of the various demands of the EGD which will later be analysed. Presenting her outline of the new college, von der Leyen also reorganised the subject area of climate and sustainability compared to the setup of the previous Commission (cf. Sect. 4.1.3).

After lengthy hearings in the EP and the replacement of three nominated Commissioners, the members of the future Commission were confirmed by the EP at the end of November 2019. The EP immediately increased the pressure for net-zero GHG emissions by 2050 and therewith needed ambitious climate policy proposals from the Commission, adopting the first EU-wide resolution declaring the state of a climate and environmental emergency (European Parliament 2019d) just one day after the confirmation of von der Leyen's college. Though this is a rather symbolic contribution to the discourse by the EP, it is an important one, lending weight to the demands rearticulated by the EP to "ensure a fair and equitable transition (…) before it is too late" (European Parliament 2019d, 2).[41]

Within its first 100 days in office, still in 2019 and before the hammer fell in Madrid at COP25, the Commission tabled its concept for the EGD.[42] This leads us to the question of whether a hegemonic strategy for creating an EU identity can be reconstructed in the EGD, which undoubtedly represents the central pillar of EU climate policy for implementing the PA. The communication of the EGD published by the Commission is to be understood as a roadmap outlining the path for an EGD uniting different future legislative projects and programmes (European Commission 2019b, 2)—which admittedly requires a long and multi-partite process. According to the communication, the EGD should address various policy areas interconnected with climate policy such as energy, housing, agriculture, the economy and also the financial sector, reforming them holistically to realise

[41] "2019 was the year of climate emergency declarations", as Justine Calma (2019) put it, pointing to 1288 declarations by municipalities, countries and states worldwide in that year. In May 2023, the worldwide emergencies declared covered more than one billion people (Climate Emergency Declaration 2023). Hence, it is still a popular movement to formally declare a climate emergency with the aim to increase both the political pressure and the public awareness for the fight against the climate catastrophe.

[42] In addition to the political pressure and the desire to achieve visible success within the first 100 days, the rapid pace is also due to the course set by Juncker's presidency (see instead of many Rietig and Dupont 2021, 30).

a just transition so that the EU's PA targets can be achieved and increased in the long term (European Commission 2019b, 3).

Interestingly, the general introduction of the EGD is used by the Commission to construct a robust chain of equivalence touching all facets of the to-be-outlined policy fields and work packages. Accordingly, the ecofeminist hegemony analysis focuses on the introductory chapter of the seminal document. First, the Commission labels the tackling of "climate and environmental-related challenges" as "this generation's defining task" (European Commission 2019b, 2), raising furthermore the visibility of the close link between diversity loss and the climate catastrophe. Using strategeme I, the Commission constructs a chain of equivalence which is conspicuous in its diversity. The demand *a more ambitious climate policy* with *increased NDCs* (European Commission 2019b, 4) is equivalised with economy-related demands and the demand *general well-being of the society*, to be created and strengthened by the EGD (European Commission 2019b, 2, 3). Although the Commission is increasingly focusing on feminised demands such as *well-being, just and inclusive transition* and the *"do no harm" principle* (European Commission 2019b, 2), masculinised patterns like *innovation* and *technical solutions* (European Commission 2019b, 9) as well as economy-related demands still prevail when concrete answers to the climate crisis are formulated. Though the EGD breaks "with previous more technocratic governance strategies" (Gengnagel and Zimmermann 2022, 272), it is regarded as still relying "explicitly on the further stimulation of growth" and meeting the related eco-social challenges with "green capitalism" (Gengnagel and Zimmermann 2022, 294). Another centrally formulated demand, which has already been articulated in the past by various actors, is the demand *circular economy* (European Commission 2019b, 3, 7), which is also woven into the chain of equivalence. Building on its recently calculated scenarios, the Commission announces a new circular economy action plan as one future project (European Commission 2019b, 7). Its proposal will be analysed in more detail later (cf. Sect. 4.2.9).

Another repeatedly rearticulated demand is to position or recognise the *EU as a leader in global climate policy* (European Commission 2019b, 2, 3, 20), supposed to unite all the demands outlined above. Explicitly following strategeme III, the Commission uses the entirety of Chapter 3 to represent "The EU as a Global Leader" (European Commission 2019b, 20). Within this chapter, three pages outline various diplomatic, economic and financial strategies which should secure this leadership and also demonstrate it to various actors such as the immediate neighbouring states, the G20 and partners on the African continent (European Commission 2019b, 20–23). Another very prominent example of the Commission's reliance on the *leader* as its hegemonic project can be seen in today's widely used graphic of the EGD outlining its architecture which suggests that the EU as a global leader is the circle encompassing all aspects of the EGD (cf. Fig. 4.3).

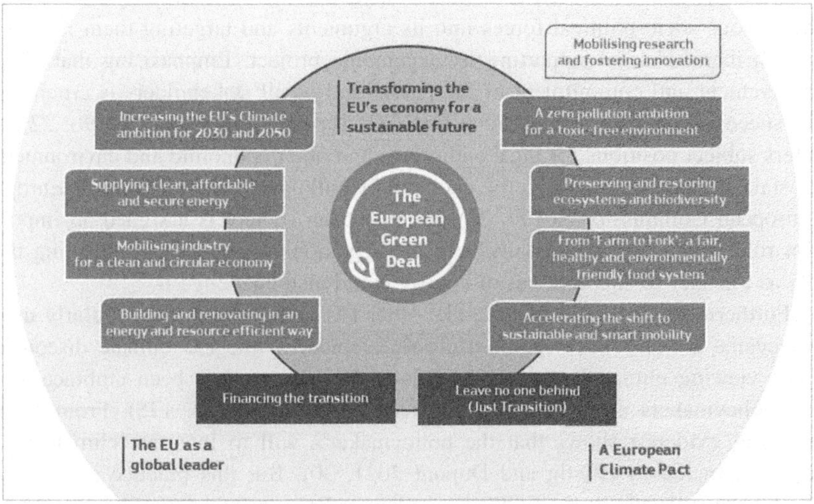

Fig. 4.3 The European Green Deal as illustrated in the Commission's Communication. (European Commission 2019b, 3)

While strategeme III can also be traced in the introductory section when the EGD stresses that the EU "can build on its strengths as a global leader on climate and environmental measures" (European Commission 2019b, 2) strategeme II is used rather discreetly. Emphasising the EU's pioneering role in being able to "mobilise its neighbours and partners to join it on a sustainable path" (European Commission 2019b, 2) brands them automatically as less innovative or successful in climate protection, so the Commission delimits the EU from these actors. This type of demarcation in the sense of strategeme II is used to establish the Commission's common point of reference, the empty signifier *leader*.

Extending the outreach and generating more support for its climate policy and the hegemonic project, the Commission uses the additional strategemes X, V and VI besides the already reconstructed core strategemes. Strategeme V, enlarging the leeway of the empty signifier, is used in the EGD when the Commission shows that the demand *leader* must be understood as standing both for the transformation of the economy and for society reaching a sustainable future (European

Commission 2019b, 2). Hence, by filling the empty signifier with a load of different demands, it becomes more and more emptied of its concrete meaning. Stressing economic and social aspects of a sustainable future reached by leading the global climate change policy, the Commission furthermore uses both masculinised and feminised demands. Strategeme VI is followed by incorporating various socio-political forces into its arguments and targeting them in order to win them over to supporting its hegemonic project. Emphasising that "The involvement and commitment of the public and of all stakeholders is crucial to the success of the European Green Deal" (European Commission 2019b, 22), it offers subject positions for local authorities and other economic and environmental stakeholders. Referring to the upcoming Conference on the Future of Europe (European Commission 2019b, 22), the European society is assigned an important role which can additionally help the hegemonic project of establishing the EU as a leader to succeed, but of course, could also challenge it.[43]

Furthermore, the Commission, like other EU institutions, very regularly uses strategeme X. This has become a classic element of the EU climate discourse since viewing climate policy as a cross-cutting theme "has been embraced by EU policymakers as a desirable practice" (Allwood 2020b, 175). From 2018 onwards, evidence shows that the policymaker's will to integrate climate policy even increased (Rietig and Dupont 2021, 30). But this practice is actually reaching new levels in the EGD (Kreienkamp, Pegram, and Coen 2022, 736) as it "aspires to extend CPI [climate policy integration; AZ] to virtually all policy sectors" (Oberthür and Homeyer 2023, 458). This becomes evident when the Commission demands that climate should be widely mainstreamed (European Commission 2019b, 15) and should therefore always be considered and, if necessary, addressed.[44] In Sect. 2.2, the EGD enumerates the fields of public investments, taxes, research and innovation, education as well as future law-making to be climate mainstreamed (European Commission 2019b, 15–19).

---

[43] The Conference on the Future of Europe (CoFE), in particular the wide-ranging discussions of randomly selected citizens on diverse topics such as EU climate policy, as well as the presentation of its results in the form of various recommendations to the EU institutions, lie outside the period of the study. Following this project, it would be worthwhile to analyse the CoFE process and its recommendations in the framework of an ecofeminist hegemony analysis. For an overview of CoFE see Greubel 2023.

[44] The best example from EU practice is of course gender mainstreaming, which was already introduced EU-wide, with the Treaty of Amsterdam aiming to anchor the norm of gender equality in all EU policies (European Union 1999). Even though scholars attest that it is today more of a "box-ticking exercise" (Allwood 2020b, 178), losing its originally transformative or critical power (Vida 2021; Rao et al. 2016; Guerrina and Wright 2016), mainstreaming is a first step to integrate cross-cutting themes in different policy areas.

Concerning the technical use of strategeme X, the EGD case differs from the Indian case. There, cultural roots are integrated into the climate change discourse (cf. Sect. 3.2.8), opening the limits of the climate discourse rather briefly in order to integrate one specific demand. In the case of the EGD, strategeme X is used in reverse: the relatively stable limits of the EU climate discourse are opened in an effort to export the demand *climate protection* or, even more precisely, *net-zero by 2050* into a multitude of other discourses with the aim to mainstream climate perspectives widely.

Therefore, strategeme X needs to be further specified at this point, because the ecofeminist hegemony analysis of the EU shows that strategeme X works in both directions, both importing and exporting a specific demand. Consequently, and in extension of the definition introduced in Sect. 3.2.8, strategeme X must be called the "occasional but deliberate breach of a specific discourse's limits to integrate or export a specific demand". Of course, it is too early and also far too huge an undertaking to reconstruct whether the Commission is successful in exporting the demand of *net-zero by 2050* into many other discourses, but it is certain that the strategeme is pursued in trying to stabilise the EU as a leader, since all policies helping to reach net-zero advance the potential leadership role. Furthermore, mainstreaming climate policy demands in other discourses can increase the acceptance of the EU's climate policy commitments.

In summary, the EGD can be identified not only as a key document of the EU's climate discourse and a far-reaching vision of restructuring the EU's economy, which at the same time is depicted as a sustainable growth strategy (Fischer 2020)—but also as a discourse contribution that uses the three core strategemes and several other strategemes to help the hegemonic project of positioning the EU as a leader to succeed. To this end, the Commission "opted for language of business, institutional diplomacy, economy, and technology" (Eckert and Kovalevska 2021, 3), and hence it continues to use predominantly masculinised patterns. But feminised aspects are gaining importance.

We may trace a change of direction in this regard in the reconstruction of the remaining two years of the research period. Some critical scholars argue that the EGD is merely a green growth project which does not question the economic system, and that therefore hardly any serious transformation is to be expected (Khoury 2023; Eckert and Kovalevska 2021). In any case, the EGD is a central component of the EU's future identity formation as it operates with an "us-versus-them approach" (Gengnagel and Zimmermann 2022, 281), which distinguishes the forward-going EU from other entities.

Those familiar with the EGD will have noticed that not all future thematic areas identified in the EGD have been reconstructed here, such as the outline of the farm to fork strategy, the idea of the carbon border tax or future initiatives in the areas of housing and transport. Some, though not all, will be taken up when the specific ideas of the different stakeholders are discussed in the context of the upcoming years.

The publication of the EGD provoked a variety of reactions. The ENGOs were divided in their opinions. While the ideas of the EGD were generally evaluated favourably (WWF European Policy Office 2020), critical voices commented on, for instance, the EGD's pace of implementation. Jagoda Munić (2019), Director of Friends of the Earth Europe, stated: "Finally, the EU is waking up to rising public concern about the planetary emergency. However, the promises are too small, too few, and too far off". The non-parliamentary left-wing movement Diem25 called the EGD "a colossal exercise in greenwashing" (Varoufakis and Adler 2020). On the contrary, Eurochambres welcomed the EGD, positively highlighting the planned mechanisms of the carbon border tax and the announced fight against carbon leakage (Eurochambres 2019b, 1).

The EGD was also recognised as an important step forward on the global stage when the UNFCCC parties met in Madrid at the end of 2019. Although COP25 did not have a major impact on the EU climate discourse, it was perceived positively that the EU had produced momentum to build international pressure for concrete climate measures to be taken by other countries (WWF European Policy Office 2020, 18). Just a few days after the launch of the EGD proposal, EUCO met the first time with its new president Charles Michel and agreed on the historic 2050 net-zero goal. However, Poland reserved the right to opt out, which is why the time until mid-2020, when the target was to be officially adopted, had to be used to convince Poland to agree (Rankin 2019).

These are interesting omens for the years 2020 and 2021, analysed below, which were supposed to unfold in the light of the concrete elaboration of the EGD and the increase of the NDCs, but were then quickly overshadowed by the global Corona pandemic.

4.2.8 The Corona Crisis Meets the Climate Crisis—and a Way to Build Back Better?

According to its work programme, the Commission wanted to quickly move forward with a total of approximately 50 initiatives and measures realising the EGD at the beginning of 2020 and kick off the processes with the respective proposals. Although the Commission repeatedly emphasised that the EGD would be maintained, the EGD work plan was adjusted (Fischer and Hoppe 2020) to tackle the pandemic-induced political consequences first (Nae and Panie 2021, 9). Apart from a handful of initiatives considered to be particularly important, the tabling of many projects had been postponed, either to the third and fourth quarters of 2020 or to subsequent years (European Commission 2020a).

For this reason, and also because most of the planned measures, programmes and laws have not been finally adopted by the end of the period under review in 2021 (Paleari 2022, 199), the following analysis of the EU climate discourse will focus on the one hand on most of those initiatives that have been largely completed in 2021; these are the Just Transition Fund (JTF), the European Climate Law (ECL), the European Climate Pact (ECP) and the Adaptation Strategy. Individual strands from other debates, such as the Circular Economy Action Plan (CEAP) or the Carbon Border Adjustment Mechanism (CBAM) will only be considered eclectically if they provide important information on the hegemonic project or aspects of the puzzle of gendered discourse patterns to be unravelled here.

On the other hand, the EU's climate discourse of that time is of course influenced by the Corona pandemic so that the question of what significance it has for the future EU climate policy and its goals must be tackled as well in the following analysis. Two partly antagonistic opinions are conspicuous here: The conviction that the EGD serves as an instrument to overcome the problems arising from the Corona pandemic and the view that the EGD and thus the overarching goal of net-zero by 2050 must take second place to solving the Corona-related economic crisis (Kurze 2020, 370, 371). The analysis will show that this new Corona-related antagonism is additionally suitable for structuring the EU climate discourse.

The pandemic not yet being foreseen, the Commission started 2020 with rapid steps; on January 14, 2020, it launched the proposals of the European Green Deal Investment Plan and the JTM, and by March 11, the proposals for the ECL, the European industrial strategy and the CEAP were presented, before the pace of the EGD-related initiatives was massively slowed down (European Commission 2023). But in an effort to overcome the challenges of the Covid crisis, the Commission returned to intensive work on the implementation of the EGD relatively quickly (Dupont, Oberthür, and Homeyer 2020, 1103). The following further analysis looks at whether this concern was successful or whether the fight against the Corona crisis changed the climate discourse of the EU in the sense of inactivity.

In the Commission's proposal for the future ECL, published on March 4, 2020, the continuation of the Commission's hegemonic strategy from the EGD can be reconstructed. Using the three core strategemes, the Commission produces the image of the EU as a leader in the global fight against climate change. The chain of equivalence discursively constructed with the help of the first strategeme consists inter alia of *climate neutrality by 2050* (European Commission 2020e, 14), *adaptation measures* (European Commission 2020e, 15), *regular future climate measure assessments* for both the Union level and the member state level (European Commission 2020e, 16–17), *public participation* (European Commission 2020e, 17) and an outline of different aspects fulfilling the future mitigation obligations (European Commission 2020e, 15). This enumeration can be understood as guard rails for designing future mitigation measures, including *cost-efficiency*, *competitiveness*, the *best available technology* and *a just and socially fair transition* or *fairness and solidarity between the member states* (European Commission 2020e, 15). These demands are also a part of the chain of equivalence reflecting the trend started with the EGD, still using particularly masculinised patterns focusing on technical and economic solutions but keeping a wider eye on feminised aspects such as *fair transition* and *equity* as central characteristics of the future climate framework. Using strategeme II, the Commission stresses that the other major emitters should follow the EU's example in setting and reaching ambitious climate goals (European Commission 2020e, 3), which are important elements of the ECL. When the EU is called a leader in this process, the Commission also uses strategeme III, as it does in additional parts of the document (European Commission 2020e, 1, 3, 11). Furthermore, strategeme VI, the (Re-) Articulation of subject positions for politico-societal forces, is also used, demanding the participation of the public (European Commission 2020e, 17) in order to win them for the Commission's hegemonic project. This is certainly to be understood as an important framework for the Climate Pact, which is intended to more closely involve the citizens, but it can also be addressed independently of this process.

Surprisingly, this strategeme succeeds when Greta Thunberg gives a well-received speech at the EP the day it discussed the Commission's proposal. Of course, Thunberg constructed a way more progressive chain of equivalence than that of the Commission. Demanding *higher mitigation goals* reflecting the limited *CO_2 budget* left before reaching the 1.5-degree threshold, she introduces a reference frame which drastically elucidates the urgency of climate action.[45] Hence, Thunberg attests the Commission's proposal that it is not enough to meet

[45] The German Advisory Council on the Environment describes the concept of CO_2 budgets computing the necessary climate goals as more appropriate than climate goals solely referencing preceding GHG emissions (Sachverständigenrat für Umweltfragen 2020, 40). They argue that mitigation efforts become more transparent and traceable if they draw on the calculated CO_2 budget left, also pointing to the IPCC's 1.5-degree report. Therefore, the Advisory

its Paris-obligations, and that this is why the EU cannot yet be recognised as a leader (Thunberg 2020, 4:50). After further criticism and reference to the EP's declaration of climate emergency, she nevertheless concludes that the ECL must now be improved as it is the only instrument currently available. Out of this logic, she continues that "the EU must lead the way. You have the moral obligation to do so. And you have the unique economic and political opportunity to become a real climate leader" (Thunberg 2020, 6:50), using strategeme III and V. On the one hand, she represents the EU as a future leader (strategeme III), on the other hand she puts more expectations into the empty signifier *leader*, insinuating that only an EU with ambitious mitigation goals, which are as high as needed in order to be consistent with the lasting CO_2 budget in a 1.5-degree framework, can be recognised as a global leader. As Thunberg also follows strategeme II, when she distances the EU from other actors, she uses all necessary strategemes of the hegemonic strategy in trying to establish the EU as a global climate leader—admittedly in order to reach the FFF's goal: a secure future in accordance with the lasting CO_2 budget. This is one of the evident cases that a common hegemonic project can nevertheless have different meanings for the actors.

Thunberg is backed in her reasoning by 33 prominent members of the FFF movement from different European countries, who published an open letter criticising the ECL proposal just one day before Thunberg—who also signed the letter—gave her speech. Although no hegemonic strategy can be proven within the FFF letter, it is clear that various demands such as *higher EU mitigation goals*, *climate justice*, *equity* and the *serious consideration of scientific findings* (Fridays for Future 2020) are overlapping in both discourse contributions. Stressing that "distant net-zero emission targets will mean absolutely nothing if we just continue to ignore the carbon dioxide budget—which applies for today, not a faraway future", (Fridays for Future 2020) FFF demands no less than a new calculation base determining the future EU's mitigation goals, as Thunberg also did. The letter furthermore states that the CO_2-storing technologies are not yet available, calling for *immediate degrowth measures* in order not to exceed the remaining emissions budget. FFF completely challenges the well-known EU dictum to fight the climate crisis with technology and economic mechanisms. According to the open letter, the student climate movement reinforces feminised patterns such as *equity*, *degrowth* and *climate justice*, while challenging well-established masculinised demands.

Council insists that the EU's targets must clearly refer to the CO_2 budget in order to immediately being able to review their scope (Sachverständigenrat für Umweltfragen 2020, 43). For these and other reasons, some European ENGOs see the CO_2 budget as a better alternative for measuring and naming the EU's GHG reduction frame.

On the contrary, Eurochambres insists on *economic success,* stressing that the future ECL must be linked to "an economically viable and socially acceptable way" (Eurochambres 2020, 2) demanding the explicit *inclusion of carbon sinks, lower requirements and larger intervals of future assessments* and *technology neutrality* (Eurochambres 2020, 3) which basically means not to phase-out fossil-based or ineffective technologies. However, no hegemonic project or strategy can be reconstructed within the contribution of Eurochambres, as in a lot of former cases.

ETUC firmly calls for linking climate action and social issues, in the sense of the demand to *leave no one behind*, which ETUC puts in a very broad context that would mean not only lifestyle changes but also the reorganisation of societies in a political, social and economic manner (European Trade Union Confederation 2020, 4). Following strategeme I, ETUC equivalises these demands with the demand "interim target of −55% GHG emissions by 2030" (European Trade Union Confederation 2020, 1). I can also reconstruct the use of strategeme II, even though it is used implicitly when ETUC points to third countries competing with the EU's economy through subsidies. Strategeme III is used explicitly when ETUC represents the EU as a leader. However, it is not the same understanding of the leader as in the case of the Commission. Instead, ETUC uses strategeme V generating greater leeway to the empty signifier *leader EU* in the light of its own political priorities stressing that *the fulfilment of the EU's commitments* and *significant financial and technical assistance to developing countries* are conditions of being recognised as a leader (European Trade Union Confederation 2020, 4). Hence, ETUC does use more feminised discourse patterns in constructing the EU as a leader, in comparison with FFF, who are nevertheless more progressive regarding the demanded extent of the mitigation goals.

The WWF also goes in this direction when it calls for net-zero by 2040 anchored in the future ECL (WWF European Policy Office 2021, 23). Within its annual report, the WWF again follows a hegemonic strategy constructing the EU as a leader in global climate change policy. Interestingly, the ENGO links this role with the need to act as a good example demanding that the EU pledge its updated NDCs in advance, embedded in the upcoming ECL (WWF European Policy Office 2021, 7)—which was indeed not the case. However, together with Greenpeace Europe, the WWF had already made key demands on the ECL in 2019, such *as deadlines to phase out fossil fuels, meaningful participation by the European society, an at least 65% goal by 2030* and *guaranteeing a fair transition* (Greenpeace Europe and WWF European Policy Office 2019). It is therefore not surprising that Greenpeace Europe was disappointed by the Commission's proposal in March 2020 (Greenpeace Europe 2020). Remarkably, the ENGO refers to the European Environmental Agency (Greenpeace Europe 2020), which states in its annual assessment of EU climate and environmental policy that

> *"Europe will not achieve its sustainability vision of 'living well within the limits of the planet' by continuing to promote economic growth and seeking to manage the environmental and social impacts" (European Environment Agency* 2019*).*

Agreeing with EEA's statement means that Greenpeace Europe, like FFF, is calling for a less growth-oriented new economic order. Hence, within the EU discourse, a variety of voices can be heard calling for feminised discourse patterns and wishing to expand the future ECL to include them. Whether this is the case in the subsequently adopted ECL will be taken up again later (cf. Sect. 4.3.2). First, however, another EGD proposal of the Commission will be analysed: the Circular Economy Action Plan (CEAP), which was adopted on March 11, 2020.

Building on the Commission's 2015 Action Plan for the Circular Economy (European Commission 2015c),[46] it adopted a much more ambitious plan outlining different future legislation projects contributing to the EU's transition towards net-zero in 2050. Interestingly, the Commission concretely formulates that the economic system has to change to install a real CE, because otherwise the EU's goals are not reachable (European Commission 2020b, 2). Overall, analysing the CEAP, a hegemonic strategy can be reconstructed—although it is followed relatively implicitly. When the Commission addresses the need for CE at the global level, it calls on the EU to play a leading role globally, both in terms of the EU's competitiveness and in particular to protect the climate through circularity (European Commission 2020b, 3). Besides following the three core strategemes shaping the EU's identity as a leader, the Commission entrenches a specific understanding of CE—which is particularly interesting when analysed with the ecofeminist lens.

Regarding the second dimension of gendered discourse patterns, I can attest that the CEAP focuses highly on technical innovation as the solution for tackling climate change, although CE conceptually consists of four "Rs": reduce, reuse, recycle (Kirchherr, Reike, and Hekkert 2017, 226) and recover (Kirchherr et al. 2023, 5). These are all mentioned within the CEAP (European Commission 2020b). Hence, we can therefore assume that CEAP is increasingly focusing

[46] The 2015 Action Plan has not yet been taken up, since the EGD firmly labelled the CE as part of the EU's climate policy for the first time. Of course, the conceptualisation of the CE has contributed to the EU's climate discourse from today's point of view; however, the Action Plan from 2015 does not provide additional insights regarding the EU's hegemonic project back then.

on feminised demands, because reuse, recycle or recover can be assigned to the sphere to which femininity is also assigned, in the sense of nurturing the environment. However, the Commission also points to the need for a "regenerative growth model" which can be accomplished by introducing an encompassing CE system (European Commission 2020b, 2). Although growth is to be regenerative, the focus at CEAP is still on economic success, seeing the reduction of resources as a welcomed co-benefit instead of a potential for strong sustainability, degrowing the economy and changing the structures for the sake of the climate and the ecological balance.

Besides addressing the industry and announcing future initiatives such as the review of the Industrial Emissions Directive (European Commission 2020b, 6), the CEAP first and foremost targets consumers regarding the role they play in the circular economy when it comes to recycling or reducing the need for resources (European Commission 2020b, 5). Hence, the CEAP runs the risk of diminishing the responsibility of the industry and in parallel transferring it into the consumer's private sphere. Due to "the feminization of the private sphere, and of the responsibility for household management in most cultures" (MacGregor 2014, 625), it is very likely that the responsibility is allocated to the sphere to which women and femininity are also assigned, not least as care work is located in this sphere. In the EU, women are still doing the bulk of a household's care work (European Commission 2017a, 8; European Institute for Gender Equality 2021, 15) and thus make the majority of consumer decisions in the family. Consequently, there is an implicit individualisation of responsibility to the detriment of women. Alongside the socially produced injustice, the individualisation of responsibility also obscures the view of necessary structural changes (Rask 2022, 1295) suggesting that all people can contribute equally to the CE, without taking into account the fact that individuals are not able to transform a whole economic system which builds primarily on fossil fuels.

This analysis therefore shows two kinds of gendered structures in the EU's understanding of CE: On the one hand, feminised patterns are used to shift responsibility to the private sphere. On the other hand, masculinised demands are used when primarily technological answers in the sense of CE are formulated in response to the climate crisis. Hence, we can understand the CEAP as a continuation of already prevailing masculinised patterns structuring the current economic system "based on technocratic and productivist approaches"—as Andrea Genovese and Mario Pansera (2021, 109) attest regarding the EU's overall CE discourse—but supplemented by feminised aspects, insofar as these alleviate the necessary concessions of the economy.

Just one day before CEAP was tabled, the New Industrial Strategy was presented by the Commission. Although the Communication does not provide any additional insights into the Commission's general hegemonic strategy, it does interpellate women as participants in the labour market. The strategy presents the contribution of industrial transformation to ecological transformation and aims to attract more women to study "Science, Technology, Engineering and Mathematics" and to pursue careers in these fields, "thus improving the gender balance in creating and leading businesses" (European Commission 2020f, 11). Similar arguments are put forward in the proposal of the Renovation Wave published on October 14, 2020, and the Sustainable and Smart Mobility Strategy tabled on December 9, 2020. These plans also address the empowerment of women, winning them for the labour market (European Commission 2020h, 15, 2020i, 21). Climate protection is thus also made an issue for women, but—and this would have to be clarified from a feminist point of view—primarily when labour is likely to be needed in the respective area due to demographic change (Zürn 2023). Nevertheless, the EGD is accompanied by a diverse interpellation of women (Carroll 2022). In the later discussion, it must be clarified as to whether or not this interpellation has an impact on the second dimension of gendered discourses, that is: whether this shaping of women's subject position influences perceptions of and responses to climate change.

Parallel to the discussion of the CEAP and the ECL, the revised NDCs still to be negotiated were an additional discourse-shaping topic. The Commission did not integrate a respective proposal into the ECL, as it wanted to wait for the assessment in autumn. In its amendments discussing the ECL proposal, the EP nevertheless called for the anchoring of a 60% mitigation target by 2030 compared with 1990 (European Parliament 2020, 28) as the centrepiece of the updated NDCs. Since they had to be reported to the UNFCCC in 2020, an interesting exchange of open letters developed between a progressive group of several member states' environmental and climate ministers around the Austrian Federal Minister Leonore Gewessler and the Visegrád states, extended to Bulgaria and Romania in this case. The former demanded ambitious NDCs with a mitigation goal of 50–55% compared to 1990 as the answer needed to counter the consequences of the Covid pandemic (Gewessler et al. 2020). The Visegrád Group stressed that the Corona effects must be addressed first and that the impact assessment, which should later serve as the basis for calculating the updated NDCs, must therefore be interpreted in this light (Visegrád Group, Republic of Bulgaria, and Romania 2020). The open letter remained very vague, but the reasoning echoes the world-wide growing "discourse of delay"[47] (King, Janulewicz, and Arcostanzo 2022) as it seems that delaying possible measures and goals is the foremost interest. Czech Prime Minister Andrej Babiš was more specific, saying on March 16 that the EU should "forget about" the EGD and focus on the virus instead (Babiš in Oroschakoff and Mathiesen 2020).

Comparing both open letters, it is notable that the Visegrád Group does not follow any hegemonic project in its letter, in contrast to the Gewessler Group, which presents the EU as a leader using the three core strategemes and pointing to the "need to send a strong political signal to the world and our citizens that the EU will lead by example even in difficult times like the present" (Gewessler et al. 2020).

It is not only Gewessler's group that links the fight against the Corona consequences with ambitious EU climate policy; in early April 2020, more than 160 European ENGOs around the Green 10 appealed for a green recovery, trying to initiate an even more ambitious EGD than originally planned as the answer to the Corona crisis. The group constructs—with the supporting signatures of more than 1.3 million citizens (WWF European Policy Office 2021, 13)—the EU as a potential leader, asking for its "leadership and foresight by directing all public stimulus investments at accelerating the transition to a just, resilient and sustainable economy, boosting the European Green Deal to deliver even more" (Green 10 et al. 2020). The short positioning aims in particular for the rapid implementation of the investment-related measures of the EGD, such as the taxonomy but also at the introduction of Eurobonds, with the overarching goal of a fair transition that should particularly work in times of multiple crises (Green 10 et al. 2020). It seems that the group expects a chance of succeeding by pursuing the hegemonic project of a *leader* EU, as they use the three core strategemes to shape the empty signifier according to their own objectives. The letter ends with pointing to different business actors, some EU member states' governments, the Club of Rome and further actors also demanding the EU's green recovery from the Covid crisis. Hence, the group around Green 10 also uses strategeme VI, inscribing these actors as potential supporters of their understanding of the hegemonic project of *the EU as a leader*. Interestingly, this open letter shows that even

[47] An increasing trend can be observed that climate-sceptic actors, for example populist-governed states, but also societal forces are turning away from the strategy of climate change denial and instead attempting to delay concrete measures, debates on more ambitious climate protection policies and thresholds (King, Janulewicz, and Arcostanzo 2022; Lamb et al. 2020). Aron Buzogány and Christoph Mohamad-Klotzbach (2022) provide an overview of further explanations of the restrained climate policy of conservative and (right-wing) populist parties, such as the rejection of climate policy as a critique of supposed elites.

though it focuses on financial aspects of the crisis, a lot of feminised demands such as a *just transition* and the *people's well-being* are pushed as part of the answer to combatting climate change.

Of course, there were also non-governmental voices opposing this, albeit less clearly. BusinessEurope (2020), for example, called on Executive Vice-President Timmermans to extend the various consultation processes for the EGD projects. What at first sounds like an understandable request—BusinessEurope also emphasises several times that it will continue to stand by the EGD—is, however, highly problematic. Putting the timetable of the mammoth project EGD behind schedule could lead to significant weakening of projects due to changing political circumstances or social acceptance.

How have these different views on recovery been reflected in the wider EU climate discourse? The Commission positioned itself early when Timmermans stressed in a consultation with the EP at the end of April 2020:

> *"We can help after the crisis, for example by installing more solar panels, helping families buy new and clean cars or by ensuring farmers work in a more environmentally friendly way. But for that we need the discipline of a climate law." (Timmermans in Schulz* 2020*)*

Ursula von der Leyen also argued this way when the first measures to deal with the Corona-related economic crisis were presented by the Commission President in the EP on April 16, 2020. In her speech, she emphasised on the one hand that the European budget must be "the mothership" (Von der Leyen 2020a, 13:40) of the recovery efforts and on the other hand that the EGD is the means of choice to invest these funds in a resilient, sustainable and more digital future EU (Von der Leyen 2020a, 15:10). Likewise, this was a central point of reference in her State of the Union speech in mid-September, also referring to the later Fit for 55 package and the NextGeneration EU with its tremendous budget (Von der Leyen 2020b, 10). Here it becomes clear how *green recovery* is entering the climate discourse as a new demand. Indeed, it is not only von der Leyen and Commissioner Timmermans who use strategeme X, taking this specific demand out of the economic or Covid crisis discourse and adopting it into the EU climate discourse. As the previous examples of more than 160 ENGOs and environment ministers show, they also followed this strategeme to support their hegemonic projects. Overall, the EU succeeded in using the EGD as a tool to address the consequences of the Covid crisis (Khoury 2023, 2).

Returning to the ECL-related debate of the updated NDCs to be pledged in 2020, the Commission evaluated the assessment it received in autumn 2020 of the targets submitted for Paris in 2015. Based on this, the Commission formulated the recommendation to save "at least 55%" by 2030, reformulating the NDCs (European Commission 2020g, 2), as von der Leyen had already announced during her State of the Union address (Von der Leyen 2020b, 9). The Commission's stance was thus closest in nature to the proposal of the group around the Austrian Environment Minister Gewessler. While the defensiveness of the Visegrád states and their allies has already been reconstructed, other actors were disappointed by the Commission's recommendations, regarding them as too tame. These were for instance different ENGOs such as Friends of the Earth Europe (FotEE). They stressed that the recommended 55% goal "would fall well short of our fair share of global emission reductions, and the global leadership the EU likes to proclaim" (Friends of the Earth Europe 2020b). FotEE follows the same strategy the ENGO used in discussing the EGD, constructing the EU as a future leader, if it fulfils its fair share (Friends of the Earth Europe 2020a, 5). Accordingly, FotEE wants to shape the understanding of what actually makes the EU a leader, to be able to call for these activities.

The same strategy, albeit with different interests, is being pursued by the loose association of 170 CEOs from various EU business sectors who are calling for the "at least 55%" target (Mazzoncini et al. 2020, 1). Following strategeme I, a chain of equivalence is produced that focuses on *technological innovation, competitiveness* and *green recovery* and *to act as a leader* (Mazzoncini et al. 2020, 1, 2). Using strategeme II implicitly, the CEOs demand that the EU not only act as a leader but also to generate followership of other states with elevated NDCs (Mazzoncini et al. 2020, 1). Interestingly, the letter uses strategeme III several times by representing the EU as a current, past and future leader (Mazzoncini et al. 2020, 1, 2), not only linking leadership to the *at least 55% target*, but rather focusing on *competitiveness* and *high green investments* in the wake of the Corona crisis. Thus, we can conclude at this point that many, sometimes very diverse actors construct the EU as a leader in order to charge this empty signifier with their own concerns.

Later in the year it was EUCO and the Council under the German Presidency negotiating the final "at least 55%" goal of mitigation by 2030 compared to the 1990 level. The new NDCs also address the EU's 32.5% targets regarding energy efficiency and the share of renewable energy of the EU's overall energy mix (Germany and the European Commission 2020, 4), both of which had already been adopted in 2018 (cf. Sect. 4.2.6). Regarding the content of the NDCs, it should be noted that the EU continues to keep them very concise, as was the case in 2015. Remarkably new is the inclusion of gender aspects in the NDCs (Germany and the European Commission 2020, 2). Although these do not lead to the interpellation of subject positions, it is clear that the EU is increasingly turning towards social dimensions of climate change, although the solutions formulated in the renewed NDCs continue to be particularly characterised by technological and market-based aspects.

Pledging the updated NDCs to the UNFCCC, the EU was among the first 20 entities following the ambition mechanism of the PA (Deutsche Präsidentschaft im Rat der Europäischen Union 2020) and the first actor of the industrialised countries and entities to do so. Hence, we can attest the performative rearticulation in an effort to be recognised as a leader. However, this is not necessarily a genuine EU success. The revised NDCs were originally to be submitted in February 2020, which was still shortly before the start of the Corona crisis in Europe. This is why it cannot serve as a credible justification for the delay. But since almost all member states of the Paris Agreement were late, the EU was able to claim leadership through its pledging date. Svenja Schulze, then German Council President of ENV, stressed that "this sends an important message internationally, as we can now credibly urge our international partners to follow us on this path" (Schulze in Deutsche Präsidentschaft im Rat der Europäischen Union 2020) and thereby supported the idea of the EU's identity as a leader. Even though COP26 in Glasgow had been postponed due to the Corona pandemic, the EU has underlined its international role with its elevated targets, which indeed do not meet the EU's fair share, but which are ambitious compared to other major economies and their respective meanwhile revised NDCs. Another success of the EU threatened to be overlooked at the end of 2020, namely the achievement of the 2020 EU climate targets—perhaps justifiably because the energy efficiency target in particular could only be achieved due to the economic impact of the Corona crisis, as has been attested by the European Court of Auditors (2023, 16). Nevertheless, the EU has achieved the targets it set for 2020 and can also refer to this when it discursively constructs itself as a leader.

4.2.9 2021: Bringing the European Green Deal to Life

In light of the Corona crisis, and against the background that many actors see the fight against its economic consequences as a contribution to combatting the climate crisis, the climate discourse in 2021 was dominated by the implementation of the Green Deal and the Fit for 55 package, which was designed to implement the now ambitious climate goal of at least 55% mitigation by 2030. Fit for 55 includes, in particular, revisions of already existing climate policy measures like the ETS or the complex of land use and land use change and forestry (LULUCF), as well as some new initiatives such as a carbon offset system and a social facility for climate protection. Not all of the measures of the package and the discourse that arose around them are reconstructed here. As in the case of the EGD, the focus is on identifying discourse patterns that either influence the overall focus of EU climate policy or reproduce or change the gendered identity and are therefore also likely to modify the responses that the EU formulates to the climate crisis.

Referred to as an integral part of the EGD (European Commission 2019b, 5), the Commission adopted the new Adaptation Strategy in February 2021. The Commission invited the Parliament and the Council to support it, which both actors did during the year, although the strategy is not a formal legislative act. The new strategy builds on the 2013 Adaptation Strategy and its impact assessment as well as a public consultation phase, so it already includes a wide range of feedback from different actors (Zürn 2023, 239–41). In the strategy, the central demands are to make the EU adaptation actions *smarter*, *more systemic*, *swifter*, and *more extensive* regarding the EU's international engagement (European Commission 2021d, 4). We can furthermore attest a particular focus on better digital connectivity and exchange between member states and municipalities which should lead to a more resilient society (European Commission 2021d, 9). Although I cannot reconstruct the use of a hegemonic strategy, as the second strategeme is not traceable, the Adaptation Strategy provides insights into how the Commission wants to shape the discourse. Stressing that "Our climate change adaptation ambition must match our global leadership in climate change mitigation" (European Commission 2021d, 17), the Commission describes the EU as a leader (European Commission 2021d, 4, 17)—even if (so far) only in the field of mitigation. This finding is in line with the general analysis of EU climate policy, which traditionally concentrates more on mitigation aspects and less on adaptation action (Lenaerts, Tagliapietra, and Wolff 2022, 2; Biesbroek and Swart 2019, 194–96)—though through the EGD and the ECL the EU now provides a greater emphasis on adaptation (Rayner et al. 2023, 9).

In addition, the strategy addresses the issue of the vulnerability of different groups and how this is reinforced through varying dimensions of discrimination such as age, gender, displacement or disabilities (European Commission 2021d, 9). Here, the Adaptation Strategy breaks the dualistic structure of the discourse and chooses an intersectional approach insisting on the mutually reinforcing forms of marginalisation and that these "pre-existing inequalities and vulnerabilities" are amplified by the impacts of climate change (European Commission 2021d, 9). These are perhaps some of the first officially anchored discursive steps towards overcoming the dichotomous production of climate policy by the

Commission. But since the Commission does not demonstrably pursue any hegemonic strategy within the document, this can only be an indication of changing gendered discourse patterns of the first dimension, leaving the need for further research beyond the year 2021. Having in mind that the aspect of adaptation as the discursively produced antagonism of mitigation can be assigned to the cultural sphere, in which femininity is also placed (Djoudi et al. 2016, 257), the Adaptation Strategy itself—understood as a performative articulation of the demand to adapt—and its new importance given by the ECL and the EGD contribute to the understanding of the empty signifier *leader*, which seems to be constructed as giving adaptation a growing importance.

Against the backdrop of the discussed need to economically recover from the Corona crisis (Dupont, Oberthür, and Homeyer 2020) and in the sense of the EGD's aim "to leave no one behind" (European Commission 2019b, 16), the Just Transition Fund (JTF) adopted at the end of June 2021 is an important mosaic stone of the EU's new climate change policy architecture. The JTF was already envisaged within the framework of the EGD to shape the upcoming ecological and economic transformation in a socially acceptable and just manner (European Commission 2019b, 16). Swiftly after its inauguration, the Commission tabled the first proposal to establish the JTF in January 2020 (European Commission 2020d). However, the Commission was not the mother of the JTF idea. Rather, such a fund had been demanded for several years by different EU institutions and individuals. For instance, the Polish EPP MEP and former Prime Minister Jerzy Buzek initiated the 2018 EP resolution for the creation of a Just Energy Transition Fund (European Parliament 2018d, 7, 11; Leppänen and Liefferink 2022, 59). S&D parliamentarians also called for a JTF back in 2016 (European Parliament 2016b, 105–6) and onwards (Kirton-Darling 2015; van Brempt 2018) and were greatly flanked by labour unions such as ETUC, as frequently reconstructed in this analysis. Apart from these main actors, different ENGOs articulated the demand *just transition* (Spring Alliance 2014; Coalition for Higher Ambition 2019; WWF European Policy Office 2019) as did the Greens in their election manifesto in 2019 (European Green Party 2019, 2, 5, 10) and of course different MEPs from S&D, the Greens and EPP in several EP debates, for instance in the run up to the yearly COPs.

This alliance fighting for a just transition mechanism appears at first glance to be surprisingly politically diverse. But it can be explained, as all those different actors used two different framings: Buzek and other EPP MEPs as well as business advocates demanded a JTF that would particularly serve the industry and its foreseeable losses due to the phase-out of coal mining (Kyriazi and Miró 2023, 122)—not least in his highly affected home country Poland; the labour unions and S&D parliamentarians, on the contrary, followed a social framing in which a JTF would especially cushion the consequences of the structural change for the workers and the people living in the affected areas (Leppänen and Liefferink 2022). Which framing finally prevails gives an insight into the gendered identity of the EU, since the industrial reading can be assigned to the masculinised sphere, while the social reading is located in the feminised sphere.

Without going through the entire development process of the JTF,[48] it should be pointed out that the drafting of the Commission's proposal was in the hands of DG for Regional and Urban Policy (DG REGIO) and thus of S&D Commissioner Elisa Ferreira and her DG's cohesion policy expertise. Therefore, the chances were good for a proposal favouring the social framing of the JTF and addressing more the people affected than the industry (Leppänen and Liefferink 2022)—thus following a feminised connotation. After lengthy discussions, the final version of the JTF was adopted by the Council and the EP in June 2021 regulating the provision of a relatively large budget of 17.5 billion Euro (European Union 2021a, 13).[49]

Article 8 of the JTF provides an overview of the measures to be supported by the Fund (European Union 2021a, 8, 9). Articulating different demands and equivalising them, Article 8 generates parts of the chain of equivalence. Interestingly, the first ten supported activities listed are dedicated to the needs of industry and the economy. The legislative act calls for the *promotion of enterprises, jobs, investments in research and innovation* or *renewable energy* with the aim of stabilising the economy and achieving growth (European Union 2021a, 9). Only the last five supported activities listed concentrate on social aspects; for example, *upskilling of workers* is demanded and equivalised with the demand *job-search assistance* as well as "other activities in the areas of education and social inclusion" (European Union 2021a, 9). Accordingly, the focus of investments under the JTF can be said to be primarily assigned to the goals of growth and economic success, and the social dimension is only given secondary importance. This impression is also shared by Simo Sarkki et al. (2022, 780) who conclude in their analysis that the JTF in its final form, due to limited financial resources, risks becoming an instrument of "retraining" of workers alone, rather than a broad social policy instrument that leads to the green transition and putting

[48] See for instance Kyriazi and Miró 2023; Leppänen and Liefferink 2022.

[49] The original budget foreseen by the Commission was only about half of this total. Later, the budget was increased through additional resources from the NextGenerationEU recovery programme.

people's needs at the centre.[50] Indeed, some aspects of the JTF coincide with the demands of business advocates such as Eurochambres, particularly demanding *retraining measures* for the employees and *support for the businesses* (Eurochambres 2021b). On the contrary, ETUC called—before and after the JTF was finally adopted—for greater incorporation of the social dimension in working towards a just transition for the people (European Trade Union Confederation 2020, 7–9, 2021a, 2–5). While the understanding of a just transition differs between the different actors, a surprisingly broad alliance of ENGOs, educational organisations and the European parties and political groups of the Greens, Social Democrats and Left built the "European Alliance for a Just Transition" in September 2021. They called for the JTF to *reduce social injustice*, *improve people's wellbeing*, and generally demanded a far more *social face of the JTF* and the overall Just Transition Mechanism (European Alliance for a Just Transition 2021). Nonetheless—just as with the other actors discussing the future shape of the EU's just transition—in their Joint Declaration, the Alliance does not follow any specific hegemonic project.

As the JTF also lacks the second and third core strategemes and therefore any traceable hegemonic strategy—probably due to its technical character regulating the exact conditions and distribution mechanisms—, my findings analysing the JTF and the reactions on it are only indications regarding the construction of the EU's identity, albeit with a palpable tendency towards masculinised demands. Nevertheless, the mere adoption of the JTF shows that in the successive implementation of the EGD, the importance of demands in the greater social context is reinforced (Oberthür and Homeyer 2023, 458). Overall, however, it is clear from the reconstruction of the JTF and its discussion that the actors do not share a common understanding of what exactly *just transition* means. Here we can observe very well how a single demand is contested in its meaning. Still, we must note at this point that the adoption of the JTF sends a twofold signal: on the one hand the demand *just transition* can be understood as a demand located in the feminised sphere, whereas on the other hand, the critical reconstruction of the meaning different actors put into this nodal point shows that the masculinised aspects are dominating the feminised ones. Hence, the JTF contributes to a more

[50] Karen Moesker and Udo Pesch (2022) also conclude that the JTF is suitable for implementing just policy, but that there is a danger in some places in the concrete architecture of the JTF that it will not meet this goal due to insufficient funding and differing levels of stakeholder participation in the formulating process.

feminised EU identity, but only to limited extent, as the JTF is actually dominated by the ideas of securing growth and economic success.

Another interesting aspect of the JTF is Article 7. It stipulates that states will only receive their full allocation from the JTF if they accept the EU's net-zero target, otherwise the country's budget will be reduced by 50% (European Union 2021a, 8). Without naming the states concerned, it is clear that with the adoption of the JTF in its final form, Poland's acceptance of the European Climate Law (ECL) legally fixing the net-zero target and parts of the EGD was secured. Just a few weeks later, on July 9, 2021, it was possible to pass the ECL, designed to write the EGD into law, thereby enshrining the net-zero 2050 goal and the binding 2030 goal with at least 55% GHG removal compared to 1990 (European Union 2021b, 8–10). Within the ECL, the institutions involved in the legislative process rearticulated the hitherto well-known demand to *consider scientific evidence* by establishing the first independent European Scientific Advisory Board on Climate Change providing the EU expertise and advice regarding its climate targets and their implementation (European Union 2021b, 9).[51] The chain of equivalence built in the ECL furthermore consists of demands for *ambitious mitigation goals* and for *assessment mechanisms* regarding both the EU and the member states (European Union 2021b, 11–13). Further equivalised demands following strategeme I are: more *adaptation activities of the EU* pointing to the Adaptation Strategy (European Union 2021b, 11), *public participation* integrating many different actors such as citizens and the social partners (European Union 2021b, 13), and *future sectoral roadmaps* outlining different programmes and strategies for the manifold sectors contributing to the EU's climate policy. Without reproducing all demands and arguments already pointed out in the previous chapter analysing the ECL proposal, it can be concluded from the ECL that the now

[51] The introduction of the Advisory Board builds on the former Scientific Committee of the European Environmental Agency. This Committee was not independent; it rather "assisted" the Management Board and the Executive Director on their request (European Union 2009, 4). Thus, the Board is now upgraded and able to close the gap of a well-known group of European scientists being recognised for shaping the EU's climate discourse. Indeed, before the new Advisory Board was established, scientists also took part in the discourse, however they were hardly publicly recognised and intervened more at the member state level within their advisory boards or as experts in cooperation or within ENGOs. As the Advisory Board was only established in 2022, it is not a part of this analysis, although activities from 2023 (Süddeutsche Zeitung 2023) suggest that the Board rapidly gained an important subject position in the EU climate discourse.

legal anchoring of the ambitious EU NDCs is the performative articulation of the demand to *lead*.

The adoption of the ECL was a great success and was followed only a few days later by the Commission's next major project, the Fit for 55 package, which was proposed on July 14, 2021. Although the individual components of the package were not adopted within the period under review, a brief look at Fit for 55 and its contribution to the EU identity will be taken.

At the beginning of the proposal, the Commission states in bold letters that it conceives the EU as a *leader* (European Commission 2021a, 2). Using all three core strategemes throughout the document, the Commission pursues this identity construction as its hegemonic strategy. Strategeme I is used when the Commission equivalises the various demands; one of the new ones is the Carbon Border Adjustment Mechanism (CBAM) (European Commission 2021a, 6, 12), which represents the concrete implementation of the well-established demand to *prevent carbon leakage*. Within the proposal for the CBAM which was tabled the same day, the Commission again constructed the EU as a leader (European Commission 2021b, 1). Within the Fit for 55 proposal, a long chain of equivalence is built, equivalising inter alia the *circular economy* (European Commission 2021a, 1, 12), *innovation* (European Commission 2021a, 11, 12), *strengthening EU climate diplomacy* (European Commission 2021a, 13) and considering the *social dimension* of the transition via the new future Social Climate Fund (European Commission 2021a, 4, 5).

Remarkably, the Commission's proposal also addresses the diverse facets and its understanding of the EGD, underlining its character as the EU's growth strategy but also stressing that it is "contributing to transform the global narrative, affecting political and business markets, and offering an example to follow" (European Commission 2021a, 12). Hence, the EGD and all related aspects thus serve to construct the EU's identity as a leader—not only in climate change policy but particularly regarding its growth-related side effects. Overall, the Commission continues to focus on discourse patterns with masculinised connotations, such as innovation and investment. But it does also continue to follow the trend initiated by the EGD of relying on the strategy to integrate more demands and programmes, which is connoted with femininity via the new Social Climate Fund, prominently placed in the package.

Which reactions did the package generate? ETUC took its position regarding the upcoming Fit for 55 package already in spring 2021 in favour of mainstreaming the *just transition* approach in the document. Furthermore, it asked for wide-ranging social mechanisms managing the needed transition (European Trade Union Confederation 2021b). However, ETUC saw its own demands as not being sufficiently taken into account by the Commission. In contrast, Eurochambres reacted relatively positively to the package. Moreover, it uses the three core strategemes to construct the EU as a *leader* in international climate policy (Eurochambres 2021a, 1). A closer analysis of the position, however, reveals that Eurochambres wants to stick to old mechanisms, equivalising different demands such as the *free allowances in the ETS* (Eurochambres 2021a, 1), *technical innovation*, and *technological neutrality* (Eurochambres 2021a, 2). In fact, demanding technological neutrality can mean that technologies based on fossil fuels should not be banned, whereby an important regulatory aspect of the EU could be lost. Keeping that aspect in mind, it is not surprising that Eurochambres' position produces the impression that they are primarily focusing on economic success through innovation but also traditional mechanisms, insofar as they appear to be more profitable. Whereas ETUC draws on feminised as well as masculinised discourse patterns, Eurochambres relies entirely on demands located in the sphere perceived as masculinised.

The Coalition of Higher Ambition bringing together European cities and regions, business associations and civil society organisations formulated in May 2021 its demands for the Fit for 55 package. It is worth noting that CAN Europe (2023), one of the prominent members, describes the Coalition as "demonstrating the breadth and depth of the call for climate leadership" of the EU. Without pursuing a concrete hegemonic project, some demands were articulated in the coalition's open letter, such as *leaving no one behind* and *creating resilient communities and jobs* (Coalition for Higher Ambition 2021, 1). In addition, a better *energy efficiency* and a *higher share of renewable energy* was demanded and the *best and highest possible climate action* in line with a *just transition* (Coalition for Higher Ambition 2021, 2). In the end, the impressively large alliance also refers to *green recovery* in the wake of the ongoing Covid pandemic and thus sets some priorities in the social sphere. Although no hegemonic project is being pushed, the alliance shows that it demands the inclusion of the social dimension of a comprehensive transition in the Fit for 55 package within the wider picture of wanting the EU to lead.

In its Annual Report 2021, the European Climate Foundation (ECF) critically reviewed the Commission's proposal and demanded to focus more on the *social dimension* in the package (European Climate Foundation 2022, 12). Furthermore, ECF with its prominent chairperson Laurence Tubiana uses the three core strategemes to construct the EU as a leader in international climate policy (European Climate Foundation 2022, 3, 5). Pointing to the tactical benefit for NGOs to frame the EU as a leader, ECF gives insights into its strategic concern to reproduce the EU as such (European Climate Foundation 2022, 8). Since ECF also provides knowledge to ENGOs and other actors and furthermore tries to influence the climate discourse by advising journalists (European Climate Foundation 2022, 12, 13, 15), this aspect is particularly interesting, also for the later reflection on the hegemony in the EU climate discourse.

One year late due to the Covid pandemic, COP26 took place in Glasgow in November 2021. We will focus here on two aspects of the EU's—or its member states'—performance: First on fossil fuel subsidies and the demand from some actors to phase them out, and second on the debate to set an end-date for coal powering. Both themes' discourse contributions are able to shape the EU's identity.

Only a few days after the start of the summit, eleven EU member states participated in the pledge entitled "end to finance for overseas fossil fuel projects", initiated by the host country UK.[52] The declaration, signed jointly with some developing countries and development banks, declares that from the end of 2022 "we will end new direct public support for the international unabated fossil fuel energy sector" (United Kingdom 2021). Through this globally noticed pledge, the eleven EU member states are performatively setting an example and, according to the declaration, also want to motivate other states to work towards an end to support fossil energies within the framework of COP26. Subsequently, it must be assumed that the EU member states are divided into two groups. This impression is also confirmed by the fact that the Council in its pre-Glasgow position did not comment in any way on ending fossil fuel subsidies or projects (Council of the European Union 2021), while the EP (European Parliament 2021b, 9, 21) and some ENGOs (WWF European Policy Office 2022, 10) demanded exactly this. Interestingly, the European Investment Bank has also joined this pledge (United Kingdom 2021), so it is not only an intergovernmental dispute—but also an interinstitutional one.

[52] The eleven member states part of this pledge are: Belgium, Denmark, Finland, France, Germany, Ireland, Italy, Portugal, Slovenia, Spain and Sweden (United Kingdom 2021). Among the 20 signatory states were also the US, Canada, New Zealand and the UK (Kottasová et al. 2021) as well as some states hitherto receiving this financing, such as Costa Rica.

Knowing that the different actors support the demand to end financing overseas fossil fuel projects, what role does this play in building the EU's identity? The EP (2021b, 9), for example, even links the necessary pioneering role of the G7 countries with the end of the support for fossil fuel projects, so this demand can partly be seen as stabilising the EU's claim to leadership. But what is its significance at the level of the climate negotiations in Glasgow?

Within the proposal for the final summit document, the chief negotiators wanted to anchor the *phase-out* of coal power. As discussed in detail in Sect. 3.2.7, besides China, it was India who opposed this formulation. In the end, it was stated in the mitigation chapter that future measures need to include accelerated "efforts towards the phasedown of unabated coal power and phase-out of inefficient fossil fuel subsidies" (United Nations Framework Convention on Climate Change 2021a, 3). Although no concrete target year is named in the Glasgow Climate Pact, it was a novelty to include these demands—which are decades old and regularly articulated by different actors of the EU climate change discourse.

While the Council neither advocated for the phase-out of coal nor the phase-down of fossil fuel subsidies in its position for Glasgow, the EP has regularly raised the issue (European Parliament 2012a, 2014a, 2016a, 2017, 2018a) and indeed called in its positioning for Glasgow for an end of fossil fuel subsidies in the EU by 2025 at the latest, pointing out that the subsidies amount "to some EUR 50 billion" a year in the EU (European Parliament 2021b, 21). Besides addressing the EU member states, the EP demands the same withdrawal world-wide (European Parliament 2021b, 21). And as the previous chapters analysing EP debates showed, except for the Christian conservatives, most of the democratic parties made the end of fossil fuel subsidies part of their portfolio regarding the EU's future climate change policy.

Hence, it is not surprising that the outcome of COP26 concerning the kick-off thematising the end of fossil fuel subsidies in the Glasgow Climate Pact was welcomed by the EPP (Pereira 2021), while other parties—such as the Greens (Eickhout 2021; Bloss 2021), the Left (Kokkalis 2021; Modig 2021) and S&D (López 2021)—demanded even more action in this regard. The Council Presidency and Frans Timmermans, both acting as chief negotiators at COP26, also welcomed the achieved fossil fuel subsidies phase-out (European Parliament 2021a). Remarkably, all the former Climate Commissioners had articulated this demand already during their terms in office (Cañete 2015a; Hedegaard 2012b). The debate on the subsidies of fossil fuels renews classic findings of EU climate policy: On the one hand, the role of the Commission as a climate entrepreneur (Bürgin 2023) and the EP as a particularly progressive force (Petri, Zapletalová, and Biedenkopf 2023, 63)—albeit recently with limitations (Burns 2021)—and, on the other hand, the role of some member states that are currently standing on the brakes (Wurzel, Di Lullo, and Liefferink 2023, 46–48).

Of course, the various NGOs, such as CAN Europe (Clima Action Network Europe 2013a, 35, 2015a, 13), WWF (WWF European Policy Office 2014, 18, 2018, 3, 2019, 14, 2021, 26), and ENGO alliances (Coalition for Higher Ambition 2020, 1; WWF European Policy Office et al. 2020, 2) have also been calling for an end to the promotion of fossil energies for some time. This aspect is interesting because most of the actors, like the EP, but also the NGOs, charge the empty signifier of the *leader* with this demand, insinuating that both demands go hand-in-hand. Nevertheless, the demand to end fossil fuel subsidies abroad is fragile within the chain of equivalence, as it is contested by a not insignificant number of MS. The position of EU citizens, on the other side, is clearer. The special Eurobarometer 2021 found that "slightly more than eight in ten Europeans believe that more public financial support should be given to the transition to clean energies even if it means fossil fuel subsidies should be reduced" (European Commission 2021c, 60). Indeed, this question was aimed at the EU area and not at fossil fuel subsidies abroad, but it is an indication that the demand to end fossil fuel subsidies has a positive resonance and definite support.

Although it cannot yet be attested that the fossil fuel (subsidies) phase-out is a strongly anchored demand in the EU discourse, we see how it is increasingly reproduced—and that the discourse in increasingly receptive to it. This is particularly interesting in light of the fact that the move away from fossil fuels is also the move away from the classical economic system based primarily on fossil energies and therefore from discourse patterns with masculinised connotations. Of course, this is not yet the complete questioning of the EU economic system, but it is certainly the breaking open and questioning of certainties—and thus the start of a real transformation. In 2021, the discourse is still characterised by the idea that the EU is acting as a *leader* in global climate policy, and various masculinised and feminised discourse patterns are cited—which support their own reading and shaping of the *leader*. Table 4.2 shows the strategemes reconstructed over the 12-year study period.

How the different identities have developed is shown in the following chapter and discussed using the ecofeminist foil.

Table 4.2 Strategemes used by the various actors involved in the respective hegemonic project or hegemony of the EU

Years 2010–2021		10	11	12	13	14	15	16	17	18	19	20	21
I	Articulation of equivalences between different demands made with regard to the universal	+	+	+	+	+	+	+	+	+	+	+	+
II	Antagonistic division of the discursive space	+	+	+	+	+	+	+	+	+	+	+	+
III	Representation	+	+	+	+	+	+	+	+	+	+	+	+
IV	Super-differential demarcation of a specific discourse's limits	–	–	–	–	–	–	–	–	–	–	–	–
V	Gradual constitution of greater leeway for interpreting the symbolic equivalent of the universal	+	–	+	+	+	+	+	+	+	+	+	+
VI	(Re-)Articulation of subject positions for politico-societal forces	–	+	+	–	+	–	–	+	+	+	+	–
VII	Occasional but deliberate breach of the antagonistic line of division	–	–	–	–	–	+	–	–	–	–	–	–
VIII	Strategeme of true championship	–	–	–	–	–	–	–	–	–	–	–	–

(continued)

Table 4.2 (continued)

Years 2010–2021		10	11	12	13	14	15	16	17	18	19	20	21
IV	Strategeme of true meaning	–	–	–	–	–	–	–	–	–	–	–	–
X	Occasional but deliberate breach of a specific discourse's limits to integrate or export a specific demand	+	+	+	+	+	+	–	–	–	+	+	+

4.3 The Gendered EU Climate Policy Identity

The ecofeminist hegemony analysis of the EU climate discourse covers the years 2010–2021. In the following, I will discuss the genesis of the gendered climate policy identity of the EU with a decidedly ecofeminist foil—which indeed incorporates normative elements (MacGregor 2014, 627; Cuomo 2002, 9). The ecofeminist approach allows a critical analysis of the "hierarchical power relations, and value systems that [...] cause, and are standing in the way of addressing, the current predicament" (MacGregor 2014, 627) of the climate crisis, and also shape the solutions the EU formulates to fight climate change. Consciously or unconsciously, gender structures are also contributing to the justification of one's climate policy (Kangasluoma 2020). The gendered demands and argumentation patterns already reconstructed in the discourse analysis will be recapitulated and put into a close context of the discursively-built identity justifying the EU's answers to climate change. Due to the discourse's nature, which is more formal and more concentrated on legislative details than the Indian discourse, the focus is on the second dimension of gendered discourse patterns, since the first dimension is only present in isolated cases regarding the EU's discourse. This means that the discourse patterns that influence the perception of and reactions to climate change and determine the EU's climate policy are foregrounded, while the patterns that shape the subject positions are less frequently addressed.

The genesis of the EU's climate identity can be divided into five phases:

(1) 2010–2011: First, the EU was constructed as a *leader* as well as a *leader by example*, which is why we are not able to trace a hegemony after the shock of Copenhagen (cf. Sect. 4.2.1 above).

(2) 2011–2012: Under the impact of Rio + 20, a short phase of stronger struggles can be identified, when some of the EU institutions and also NGOs pursued the new hegemonic project of the *green economy* as the EU's identity, which of course could mean new foci of the EU's climate change policy (cf. Sect. 4.2.2 above).
(3) 2013–2015: The years that followed, however, showed that especially in preparation for Paris this hegemonic project was again replaced by the *leader*. Since it was particularly stable, we must speak of a discursively created hegemony of the *leader* in and around Paris, as the reconstruction of the strategemes used and also the broad support of the empty signifier *leader* by the EU institutions, NGOs as well as business actors and others disclosed (cf. Sect. 4.2.4 above).
(4) 2016–2018: After Paris, the attention was on the implementation of the commitments, and one could observe—at least partially—a reorientation towards *leading by example*. But this short phase was replaced by the need to act more strongly as a leader again due to global events such as the US withdrawal from Paris and the Brexit voting (cf. Sect. 4.2.5).
(5) 2019–2021: With the EGD, it could be assumed that the hegemony of the EU as a *leader* that had been established in the meantime could be broken up while the institutions and other actors would make the EGD their new empty signifier. However, it is rather the case that a well-implemented EGD is regarded as being part of the leadership role.

Accordingly, at the end of the period under study, it can be stated that the EU is constructed as a *leader* of global climate policy by very different actors of the EU climate change discourse. And if there were some critical voices, for instance demanding other priorities, they have been overshadowed by the common project painting the EU as a leader. How this identity and also the other hegemonic projects that have been pursued in the meantime are gendered and what this means for EU climate policy will be discussed in the following.

This will be done in five chronological steps, pursuing the five phases of different hegemonic projects. Some of them will overlap as a discourse normally develops rather slowly, while some of them are characterised by clear cuts and obvious hegemonies. Table 4.3 below gives a concise overview of the different hegemonic projects within these five phases and of their supporters.

Table 4.3 Overview of different hegemonic projects, (re-)producing actors and hegemonies of the EU-discourse

Hegemonic project(s)	**(Re-)producing actors** (selected)	**Hegemony** Intersubjectively produced and temporary stable identity
(1) 2010–2011		
Leader	• European Parliament • EPP, S&D, Greens, Left, ALDE • WWF • Spanish Council Presidency	Leader / Leader by example: No definite decision possible
Leader by example	• Commissioner for Climate Action Hedegaard • Commission	
(2) 2011–2012		
Green Economy	• Commission • WWF	Neither leader nor green economy: Green economy was just of short duration, inspired by the Rio + 20 conference
Leader	• Occasionally Commission (at COP) • EP and its democratic factions • EIGE • Commissioner for Climate Action Hedegaard	
(3) 2013–2015		
Leader	• President of the Commission Barroso (in 2014) • WWF • S&D • Council Presicency • Commission (in 2015) • Council (in 2015) • President of the Commission Juncker (sporadically) • ETUC • EP • Commissioner for Energy and Climate Action Cañete	Leader

(continued)

Table 4.3 (continued)

Hegemonic project(s)	**(Re-)producing actors** (selected)	**Hegemony** Intersubjectively produced and temporary stable identity
Leader by example	• Commission (until 2014) • President of the Commission Barroso (until 2014) • EUCO	
(4) 2016–2018		
Leader by example	• Commissioner for Energy and Climate Action Cañete (in 206) • Commission (in 2016) • DG Climate (in 2016) • President of the Commission Juncker (in 2016) • WWF (in 2016) • EP (in 2016); some MEPs • Press before Marrakech • GGG (in 2016)	Leader by example was a hegemonic project of short duration. The leader became the hegemony quickly after the withrawal of the USA
Leader	• Commissioner for Energy and Climate Action Cañete • MEPs (after election of Trump) • ECF • Greens • EP (from 2017 onwards) • Council (from US withdrawal onwards) • ETUC • Commission (from 2018 at the latest)	

(continued)

Table 4.3 (continued)

Hegemonic project(s)	**(Re-)producing actors** (selected)	**Hegemony** Intersubjectively produced and temporary stable identity
(5) 2019–2021		
Leader	• Commission • President of the Commission von der Leyen • Executive Vice-President Timmermans • Council • Council Presidency (in 2020) • EP • Most MEPs of the democratic EP groups • ETUC • ECF • WWF • Gewessler Group • Green 10 et al. (around 160 ENGOs) • FotEE (from 2020 onwards) • Greta Thunberg • Group of 170 CEOs (cited as Mazzoncini et al. 2021)	Leader

Each period will be introduced with its respective chain of equivalence depicted in the form of the spider's web. The nucleus of the web is the empty signifier, or in cases where multiple hegemonic projects are being pushed, the nucleus is split representing the ambiguity. In the spider's web, only the equivalences between the different nodal points will be delineated, but not the potential differences between the demands. These are over-formed by the reference to the empty signifier and therefore only of interest regarding the preliminary work reconstructing the discourse (see for the theoretical background Sect. 2.2.3). The subsequent discussion of each discursively produced identity with the help of the ecofeminist lens and further concepts follows each spider's web. For a better overview, the chapters will be divided in a period before and after the preparations for the Paris Summit.

4.3.1 The EU Identity until Paris: Old Masculinised Demands in New Bottles?

The chain of equivalence for the first phase (2010–2011) reconstructed in Sect. 4.2.1 is shown in the form of a spider's web (cf. Fig. 4.4). Depicted in particular are the frequently rearticulated and broadly asserted demands, while only individually expressed demands that have not yet been able to establish themselves as temporary stable nodal points of the chain of equivalence remain unsketched. These were, for example, in 2010 the feminised demands *climate justice* or *a good life for citizens* articulated by Rovana Plumb (S&D). As has been reconstructed, the actors such as the EP and all its democratic factions (EPP, S&D, the Greens, the Left and ALDE) agreed on *the leader* as the identity of the EU at that time. They were supported by, for instance, the WWF and the Spanish Council Presidency, though neither actor could be proven to have a hegemonic project. Conversely, the unity in the EP does not match the direction of the Commission. The Commission and its Climate Action advocat Connie Hedegaard relied on their hegemonic project *leading by example*; in the end, however, neither hegemonic project prevailed. Instead, the graph shows that the EU as *leader* or as a *leader by example* has not clearly been constructed as its identity. Moreover, it remains open at this point which—if either—of the two hegemonic projects will find the necessary broad support in the reordering discourse after Copenhagen, while both nodal points fulfil the otherwise necessary aspects being established as empty signifier (cf. Sect. 4.2.1).

Although no hegemony was traceably formed by the end of 2010, the gendered structure of the discourse is quite revealing. The actors all primarily articulate masculinised demands, trying to anchor them in their version of the empty signifier and justifying their favoured response to climate change through the EU's identity. These demands are particularly technical, scientific and economic in nature—aspects that can clearly be located in the masculinised sphere (MacGregor 2010). Demands such as adaptation, sustainable lifestyles or the precautionary principle, which can be located in the feminised sphere (Djoudi et al. 2016, 257; Nelson 2012; Johnsson-Latham 2007, 6), were hardly articulated at all and therefore could not claim a firm place in the chain of equivalence (cf. Sect. 4.2.1). How can these findings be interpreted through an ecofeminist lens?

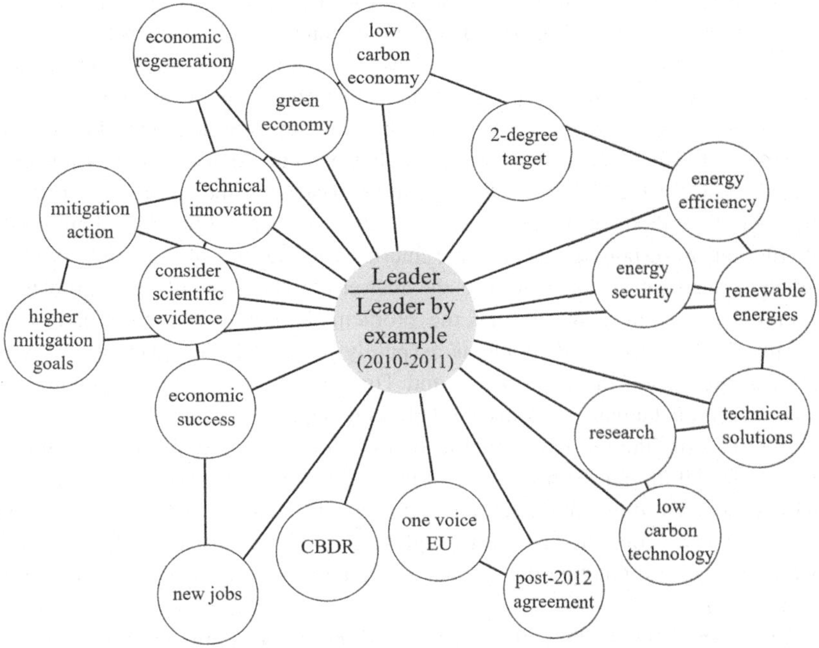

Fig. 4.4 EU chain of equivalence: spider's web of phase 1, 2010–2011

The triad of technological advancement, for which sciences are an important factor, market-based mechanisms improving the economy and cooperative governments supporting these 'win-win scenarios' is subsumed under the concept of ecological modernisation (Bäckstrand and Lövbrand 2006, 52–53; Buttel 2000; Hajer 1995). These ecological modernisation elements, which can also be understood as an outflow of an "advanced liberal government" (Oels 2005, 185), are the basis of the EU's climate change discourse of that time (Machin 2019). Accordingly, the masculinised demands are not randomly articulated, but rather serve a specific thrust heading towards weak sustainability, which does not operate within the planetary boundaries. As the necessary technical solutions following this particular concept of dealing with climate change are advocated for via masculinised discourse patterns pointing to the chance of mitigation via innovation, we can trace the "ecomodern masculinity" (Hultman and Anshelm 2017, 32) constructed and reflected through the EU discourse.

Importantly, if the discourse is narrowed in this way, this in turn means that only certain ways of reading and answering climate change are perpetuated. Furthermore, pushing technical innovation as the main answer to climate change suggests that it is defined as a scientific rather than a political problem. This "scientific framing of climate change has been achieved in large part through the discourse of ecological modernisation that has become dominant in (…) Europe" (MacGregor 2010, 230). Therefore, we can likewise attest the discourse's underlying idea of solutionism. Evgeny Morozov (2013) elaborated his critical approach of solutionism responding to the large internet corporations of the Silicon Valley, which exclusively see technology as the answer to today's multifaceted problems. Solutionism therefore means that the problem definition is already narrowed, framing multidimensional problems as purely technically decipherable. This is how, according to Oliver Nachtwey and Timo Seidl (2017, 21), the promise of solving central human problems through technological means arises. As shown in the analysis, this specific framing of the climate change problem is implemented by falling back on the masculinised discourse patterns that are socially acknowledged as dominating the sphere perceived as feminised. Science, the mind or reason and rationality is constructed here as the antithesis of emotion and nature, the first dominating the second, as Plumwood had already worked out (cf. Sect. 2.3.3).

In our case, this dichotomisation is based on the prevailing economic system within the EU, which continuously strives for growth in the sense of increasing the GDP. The resulting narrowing of the problem, as variously criticised by ecofeminists (Huyer et al. 2020, 575; Dankelman and Naidu 2020, 452), thus leads to the marginalisation of the social dimension of climate change consequences but also of specific solutions, which are connotated with femininity (Alston 2014, 287). For instance, sufficiency approaches or justice aspects are marginalised since technological responses are discursively constructed and socially recognised as part of the master identity (Plumwood 1993, 23), which can be traced back to the gendered dichotomous structure of the world. Thus, the struggle for the EU identity either as *leader by example* or as *leader* hides a significant number of the possible responses to the climate crisis due to masculinised hegemony embedded within society.

Has the marginalisation of feminised solutions combatting the climate crisis shifted in the second phase (2011–2012) with the discussion of the green economy? Undisputedly, the chain of equivalence expanded the debate on the alternative economic concept of a green economy in 2011 and 2012 to include social and ecological demands. In addition, a gender perspective was also included in the discourse. Nevertheless, the analysis in Sect. 4.2.2 showed that no hegemony within the EU climate discourse could be clearly established during this period. Rather, these years, particularly against the background of Rio + 20, were marked by a debate about the two rival hegemonic projects of EU identity construction either as a *green economy* or as a *leader*, as the chain of equivalence, again depicted in the spider's web model, reveals (cf. Fig. 4.5).

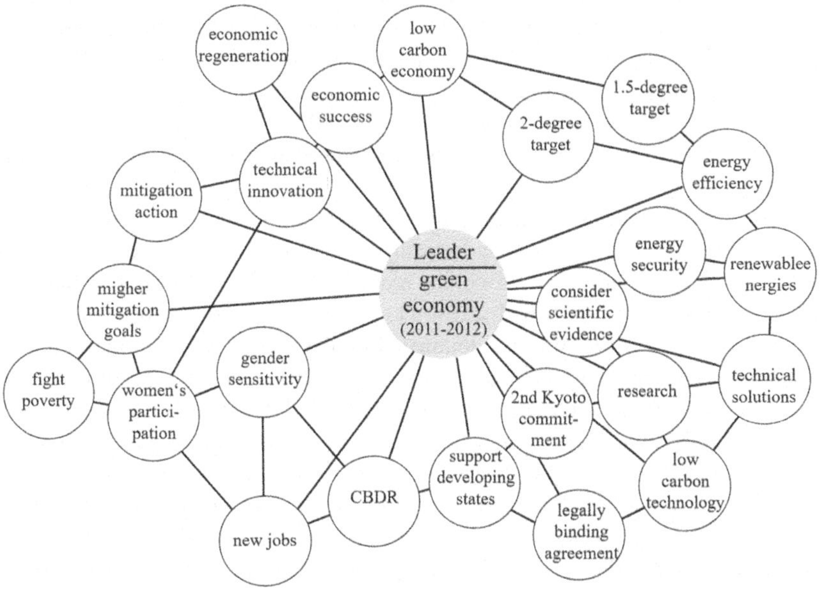

Fig. 4.5 EU chain of equivalence: spider's web of phase 2, 2011–2012

The central elements of ecological modernisation and solutionism can again be identified here, although they find themselves now in a larger field of different demands, as more feminised demands have entered the chain of equivalence. The hegemonic project of painting the EU as a leader is once again followed by the EP and on some occasions by the Climate Action Commissioner. At the same time, the hegemonic project of the green economy is being pursued by the Commission and also by NGOs such as the WWF. Newly or hitherto insufficiently anchored demands located in the feminised sphere of domination, such as gender sensitivity or the fight against poverty, have found their way into the discourse.

Though the masculinised demands regarding energy needs, technological answers and economic success are still strongly represented in the discourse, the discourse is now more multifaceted, as these demands are equivalised with feminised aspects, regardless of whether the actor is trying to shape the EU's identity as a *leader* or as a *green economy*. In the case of leadership, it should come as little surprise that the demand for economic success as well as those demands supporting economic means are still well-anchored, since most understandings of leadership in global climate change policy encompass economic prosperity as a vital part (Liefferink and Wurzel 2017, 957). In contrast, the green economy concept claims—to a greater or lesser extent, depending on definition—to engage with the environmental and social dimensions of sustainability (D'Amato et al. 2017, 725), incorporating the potential to decouple the seemingly antagonistic poles of growth and ecology (Ferguson 2015, 22).[53] For these reasons, it is important to examine the different definitions of the green economy in the analysis.

While the EP addresses the triad of ecological, social and economic progress, also within its hegemonic project painting the EU as a leader, the Commission clearly focuses on economic recovery following the global financial crisis (cf. Sect. 4.2.2; Allwood 2014, 11). Despite these different nuances, the EU's discourse was nevertheless dominated by the well-established masculinised demands. Granting that there are some other actors pushing the transformative potential of the green economy concept, the EU institutions nevertheless agree with UNEP, which does not fundamentally question the primacy of growth (Tomaselli et al. 2017, 144). Since the mainstream GE discourse of that time does not unleash a transformative moment towards a sustainable society asking questions of power (Wanner 2015, 35, 36), the EU actors, who have been part of the mainstream discourse, additionally stick to the hegemonic masculinised discourse patterns of the EU climate change discourse of earlier years. Accordingly, the discourse is still to be classified as characterised by an ecomodern masculinity. Hence, economy-centred perspectives on ecological or climate issues obscure the need for greater transformation (Coffey 2016, 219), as advocated in the sense of an ecofeminist understanding of the green economy and as demanded by activists for instance in the course of Rio + 20 (Goodman and Salleh 2013). The economy-centred perspective is also reflected in the fact that work is almost exclusively discussed as paid labour, leaving out care work, which is culturally located in the feminised sphere (Bauhardt 2019).

[53] Some scientists completely reject the possibility of decoupling these poles altogether as a "myth", pointing to the necessary condition of exploiting the environment in order to achieve growth (Wanner 2015, 31).

Against this background, the dichotomous and gendered order of the analysed discourse is evident, as is the marginalisation of climate protection—if it is understood as caring for the environment. Or in Beate Littig's (2017, 324) words criticising the green economy: It "only caters to abstract value creation and not the satisfaction of needs". Intriguingly, this finding did not change with the raised focus on the gender dimension in EU climate policy, initiated by the EIGE Report. Nevertheless, the interpellation of women as participants in decision-making as well as in technology-related professions (cf. 4.2.2) leaves open the possibility that these gendered discourses of the first dimension will in the future change those of the second—for example by introducing new perspectives into policy-making and technological development (for an in-depth discussion see the following chapter).

What is the preliminary résumé of the EU identity construction up to the end of 2012? In summary, my conclusions are corroborated by other findings (Benson et al. 2021, 18; Wilson and Chu 2020, 1098; Bina 2013, 1026) that the understanding of both hegemonic projects—the *green economy* and the *leader*—as they have been pursued by the majority of the analysed actors must be seen as old wine in new bottles: Relying on masculinised discourse patterns to overshadow the rare feminised ones to promote economic growth that is not confronted with planetary boundaries and is merely painted green.

In this section, the ecofeminist foil was particularly placed on the hegemonic project of the green economy and further concepts shaping the discourse, also in the framework of the figure of *leading* climate change policy. The following years bore the imprint of the hegemonic projects of *leadership* and *leading by example*—and not only regarding Paris. I will therefore place the focus on these hegemonic projects determining the EU climate discourse, discussing in the next chapter the spider's webs of 2013 to 2021 from an ecofeminist point of view.

4.3.2 Leadership through Ecomodern Masculinity with a Pinch of Feminised Demands

To recap the analysis of the years from 2013 to 2015 (cf. Sects. 4.2.3 and 4.2.4), the development and struggle of EU identity-building in the third phase is evident. Some of the important actors, such as the Commission, its President and some MEPs, initially worked towards establishing the EU identity as a *leader by example*. However, the election campaign already shows that some of the EP factions such as the Greens and S&D wanted to establish the identity of the EU as a *leader*. When the preparations for Paris were in full swing, a few more stakeholders followed this hegemonic project. Then, the Commission, particularly in the person of Cañete and Juncker, but also the Council and EUCO pursued the hegemonic project of anchoring the EU's identity as a *leader*, aligning the other articulated demands with it. As Sect. 4.2.4 showed, many different actors supported this project, for example ETUC and ENGOs, who also saw their respective demands represented by the establishment of the identity of the EU as a *leader* in global climate change policy. Of course, this surprising unanimity and, in the cases of EUCO and the Council, unambiguity, must be seen in the context of the Paris Summit and the absolute will to make it a success. Moreover, climate change can be considered one of the few foreign policy areas in which the EU has the potential to position itself as a leader (Müller-Brandeck-Bocquet, Pietzko, and Zürn 2018) and thus as an important actor in the concert of international politics. The following spider's web shows the demands used, which in their gendered nature establish and also stabilise the hegemony of the leader (cf. Fig. 4.6).

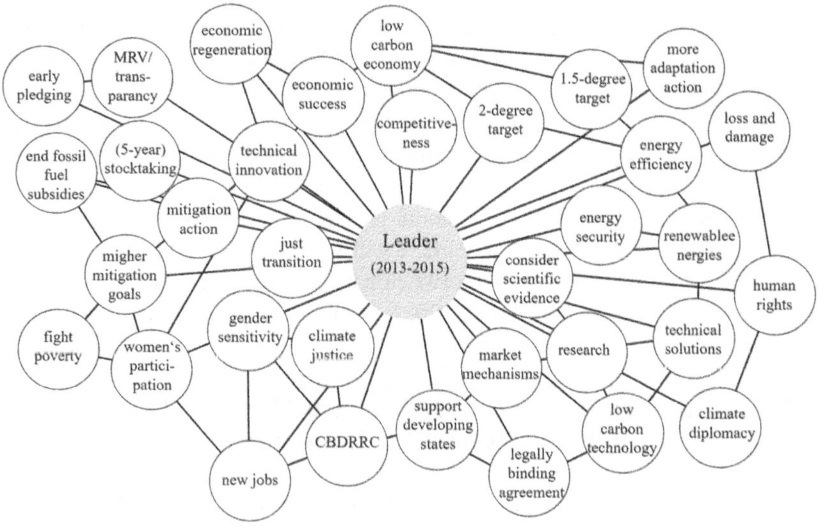

Fig. 4.6 EU chain of equivalence: spider's web of phase 3, 2013–2015

As in previous years, the reconstructed chain of equivalence continues to be shaped by masculinised demands such as growth, renewable energy,[54] technological innovation, and mitigation. Nevertheless, Paris and the preparations for it with partners such as the High Ambition Coalition brought more feminised demands into the discourse. Due to the rearticulation by various actors, these were also relatively consolidated. Rearticulating more feminised demands means in this specific case to call for more adaptation action, stronger engagement of the EU in loss and damage, and a just transition. Hence, while the actors shaping the EU's identity almost unanimously strove for the *leader* at that time, it did not rely exclusively on masculinised demands, as it had done in former years. Instead, the specific gendered nature of the EU's understanding of leadership is visible in two cases: When it still comes to demands assigned to the masculinised sphere such as competitiveness or the strong will (Bruff and Wöhl 2016, 96), and also in the insistence on loss and damage or stronger adaptation—both feminised demands—which are also part of the leadership construction. Does this outcome contradict the former finding that the figure of the leader in climate policy has mostly masculinised connotations, particularly due to its economic focus?

First and foremost, it must be stated that the type of leader, as it is drawn in the EU discourse, continues to be based on and simultaneously reproduces ecomodern masculinity (cf. Sect. 4.3.1), invoking the central pillars of ecological modernisation in its masculinised nature. In this respect, the EU-specific understanding of the leader with all its masculinised connotations persists, and the focus continues to be narrowed to technical and economic solutions. The more frequent feminised patterns are to be understood as an additional justification, a positive add-on that is intended to increase the acceptance of the EU's traditional answers countering climate change.

Therefore, the more feminised demands do not automatically lead to a more diverse identity with more diverse response options. Rather, a new feminised network of nodal points has been created to support the masculinised identity. Although it is certainly not the will of supporters of demands such as a just transition or more adaptation action, the feminised patterns must be understood as solely supporting the still-dominant masculinised strategic thrust of technical solutions, economic growth and innovation. The discourse analysis showed that

---

[54] For a detailed explanation of why renewable energies as part of the technological fixes is a masculinised idea to combat climate change, see MacGregor 2010, 231.

the masculinised ideas overlay the feminised ones, for example with more financial resources or with concrete measures. Hence, the discourse is changing, but its basic character remains even in times of the Paris Agreement, and it continues to force a certain direction for the different possibilities with which to meet the climate crisis. This mechanism can be traced back to the structure of Western societies, in which values with a masculine connotation such as innovation, technology and the spirit clearly outweigh the supposedly softer topics such as adaptation and the precautionary principle (Tickner 1993; Hekman 1990, 5). This finding is underlined by the fact that the identity of the leader EU cites various aspects such as the success of the agreement and the strengthening of international cooperation as components of leadership—which are still rearticulated, pushing for economic success and securing European prosperity.

In sum, the EU has developed from the struggle of two hegemonic projects to constructing itself now clearly as a *leader* in the sense of a hegemony and has also been recognised as such by many actors, as the decisive moments in Paris showed. Hence, the EU sticks to its basic idea of its identity in climate change policy: Building on economic success and therefore competitiveness. But the EU is also encompassing more social aspects in order to assert itself overall as a leader in global climate policy, justifying its own climate policy decisions at a time when different actors are calling for more climate justice and social safeguarding of climate policy.

How did this identity evolve in the fourth phase (2016–2018), when the first priority was to fill the Paris Agreement with life? As shown in the analysis (cf. Sect. 4.2.5), some actors such as the Commission, the WWF and the EP seemed to agree to modify the EU's identity after Paris. Departing from the well-anchored *leader* in 2015 they discursively moved in the direction of a *leader by example,* on the one hand creating the internal motivation to realise the EU's Paris goals and on the other hand to actively generate global followership in implementing the Paris Agreement. At the same time, Cañete initially stuck to his hegemonic project, the *leader*. While he joined the Commission's line just a few months later, first it was a matter of stabilising the Paris successes and thus presumably also his own position.

Although many actors shared the hegemonic project of installing the EU's identity as a *leader by example,* this project lasted only a relatively short time of one-and-a-half years, until mid-2017 (cf. Sect. 4.2.5). After this period, most actors shaping the EU's identity quickly turned to the hegemonic project inscribing the identity as a *leader* not only due to Brexit, but particularly induced by the interruptive discursive impulse of Trump's announcement that the US will withdraw from the PA. The new hegemonic project was largely unanimously followed and supported by most of the actors. Hence, it rapidly grew into the hegemony of the EU climate discourse. What the discourse in the associated spider's web looked like is shown in Fig. 4.7 and the question of how these developments can be estimated from an ecofeminist perspective will be discussed.

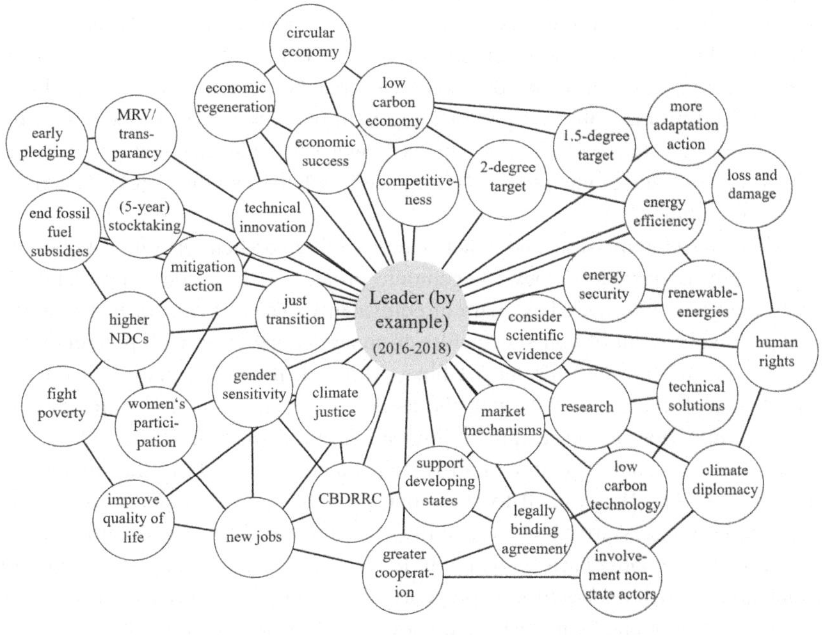

Fig. 4.7 EU chain of equivalence: spider's web of phase 4, 2016–2018

Considering the chain of equivalence, the hegemonic project of leading by example increasingly relies on feminised discourse patterns, for example because cooperation or adaptation are becoming even more important after Paris in order to win other actors as followers. While sharing more attention with the demands for cooperation, adaptation and the circular economy as a supposedly more sustainable economic model, the picture the discourse paints is the opposite. The hegemonic project of *leading by example* is still based in particular on the well-known approach of solving the climate crisis in the sense of ecomodern masculinity, which Hultman (2017, 97) attests. Nevertheless, the feminised patterns have contributed to a new connotation of identity as a *leader by example*. This form of leader favoured here was not 'bluntly' aimed at a strong role, but at cooperation and collaboration, values that can certainly be understood as feminised. While the feminised demands therefore do not significantly shape the responses formulated by the EU to the climate crisis, they are again able to support their implementation or acceptance, as they were used during Paris. Indeed, regarding the EU discourse in 2016 and the first half of 2017, the triad of economic success, innovation and technological greenhouse gas reduction continues to be shaped by masculinised patterns and is to be exported as a model for others; at the same time, it represents the harbinger for the subsequent phase of the EU climate discourse when the US announced its Paris exit and Brexit became a near-future reality.

While the Commission and other actors calibrated their priorities regarding the abruptly reintroduced 'new' hegemonic project of the leader, new demands found their way into the EU's chain of equivalence. These were, for instance, the stronger involvement of non-state actors or greater cooperation with partners other than the USA—first and foremost China. Although feminised patterns were not cancelled with the US withdrawal, the actors still focused on economic success through mitigation action, technical innovation and further masculinised aspects of the climate change discourse in order to set a counterweight to the supposedly failing USA. The EU wanted to show that growth, innovation and climate protection do not necessarily have to contradict each other. To this end, although the adherence to the noble Paris goals might lead one to believe otherwise, the idea of sustainable growth anchored in ecomodern masculinity, which does not pose any system-shattering questions, was again strongly emphasised. Nevertheless, it was not exactly the same identity that—due to the broad support—relatively quickly became a hegemony of the *leader*, as that that had followed Paris. Rather, the increased emphasis on cooperation and also the element of setting a good example left its mark on the empty signifier *leader*, which was appropriately charged using strategeme V. Thus, this identity continues to be strongly masculinised, which in turn pushes and justifies the EU's 'classic' responses to climate change, albeit with a new feminised focus on cooperation in concrete implementation and support for developing countries. The feminised discourse patterns are thus still placed in service to secure the necessary support; the discourse and also the identity have been expanded but not significantly changed when it comes to formulating responses to the climate crisis, but their justification relies more on feminised discourse patterns.

How has the identity of the EU and its gendered structure changed with the introduction of the EGD and the impact of the Corona crisis during the fifth and final phase of identity formation (2019–2021)? During the course of the year 2019, with the EP being newly elected, the general direction of the discourse developed further when citizens called for stronger climate action, the EP declared the climate emergency and at the end of the year the Commission announced "Europe's 'man on the Moon' moment" (Von der Leyen 2019b), namely the European Green Deal, which should now reach a new level of climate policy integration. As described in Sect. 4.2.7, the new Commission wanted to begin its work with a 'big bang' in 2020, quickly starting to implement some of EGD's manifold components. The discourse and the identity changed not only in this respect, but also in the light of the Corona crisis. As the spider's web (cf. Fig. 4.8) shows, the chain of equivalence has been slightly adapted due to the pandemic, for instance regarding the incorporation of the demand to leave no one behind, not only but particularly when it comes to the economic regeneration. Furthermore, the examination of the discourse from 2019–2021 also disclosed changes in its gendered structure not least because of the EGD and its partly newly created, described as socially just underpinnings of EU climate policy. What did not change was the EU's identity as a *leader*, which became an even stronger hegemony—perpetuated and recognised by a critical mass. The leader EU solidified and held on to its masculinised discourse patterns while strengthening feminised discourse patterns. The first dimension of gendered discourse elements led to a multidimensional interpellation of women as political and working participants in the climate discourse.

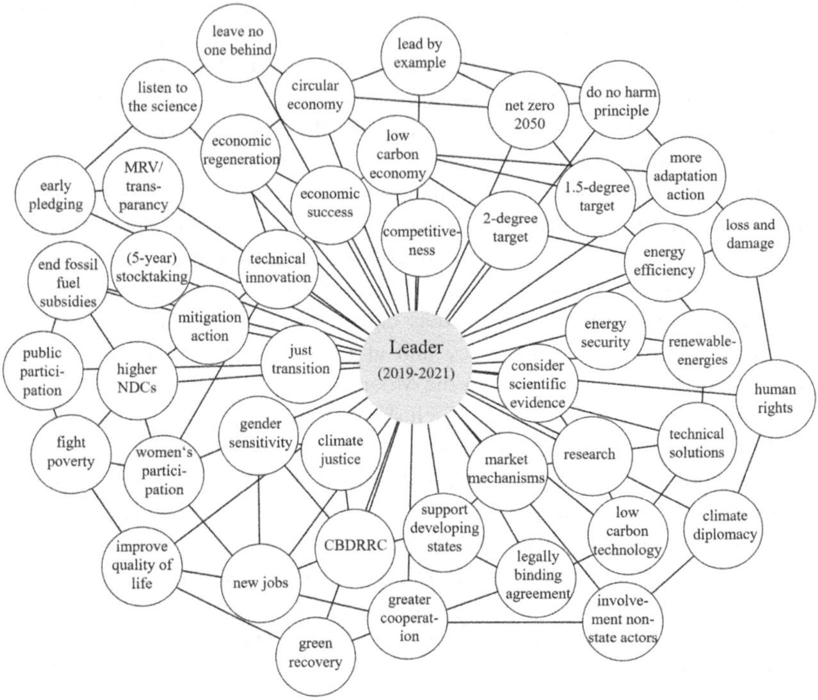

Fig. 4.8 EU chain of equivalence: spider's web of phase 5, 2019–2021

Those climate protection aspects aiming for the social dimension to be fulfilled become more prominent with the EGD. These are the just transition, institutionalised with the eponymous Fund, or greater participation of women in different fields, significantly strengthened in the chain of equivalence and attributed to the feminised sphere. This also applies to the demand for adaptation (Djoudi et al. 2016, 257), which has gained greater importance in the EU climate discourse as well. However, not all demands intuitively perceived as feminised actually belong to the feminised and therefore dominated sphere. As discussed above, circular economy (CE) is one such example. Its concrete implementation, which in the case of the EU is primarily aimed at perpetuating ecological modernisation (Leipold 2021; Fitch-Roy, Benson, and Monciardini 2020), is to be assigned to the masculinised sphere of domination, because the feminised aspects such as recycling and reduction of resource use are solely pushed either for the aim of economic success—constructed as dominating—or as responsibilities of the consumers. In the first case, feminised aspects are reinterpreted as masculinised elements because savings are used, for example, for additional growth—which has a masculine connotation—, and not to protect the climate thanks to the savings. In the second case, the CE is transferred from the public to the private sphere, transferring the responsibility to act into the private sphere of individual households, and therefore feminising it (cf. Sect. 2.3.3). The masculinised goal of growth thus relieves the public and the economy of its responsibility, whereby CE can continue to be understood as a masculinised demand. This example makes it particularly clear how the feminisation of responsibility leads to the reproduction of the dichotomy of private and public and thus shifts responsibility from the masculinised, politicised public sphere into the private, feminised sphere. This also creates additional work for the people entrusted with reproductive work such as caring for the environment (MacGregor 2010, 233; Prügl 2021, 302).

This is where the fundamental (eco-)feminist critique of privatising and therefore depoliticising (Fraser 2016; MacGregor 2010, 2014; Peterson 2000, 14) the fight against climate change comes into play. In the case of the CE this means that if recycling is primarily imposed on private households, the economy can continue to operate and act relatively unhindered to the detriment of the environment, as its protection is discursively depoliticised. Hence, to load the bulk of protecting the environment onto women's shoulders means that CE can almost completely contribute to the accumulation of capital and must therefore be understood to have masculine connotations.

Because the EGD is concerned with anchoring climate change policy even more strongly as a cross-cutting issue than in former times, further 'new' topics entered the discourse and influenced its gendered nature. But as Sect. 4.2.9 showed, it was again for reasons of supporting the well-known economy-related goals of the EU in the new form of the EGD that the actors integrated more feminised discourse patterns. Thus, Vincent Gengnagel and Katharina Zimmermann (2022) note that the EGD representing its different programmes and initiatives is primarily a capitalist project that neither identifies the limits of growth nor wants to comply with them. The EGD—like the previous EU climate policy measures—must be understood as an ecological modernisation project pushing for technical solutions while following the idea of weak sustainability,[55] but now by integrating climate policy even more strongly, for instance by responding nimbly to social dynamics (Gengnagel and Zimmermann 2022, 272).

[55] Regarding the distinction between weak and strong sustainability, see Sect. 1.1.

Hence, it is not surprising, but important to note, that most actors who contribute to the stabilisation of the hegemonic project and thus also to the hegemony of *leadership* continue to rely on masculinised patterns while using the increasingly articulated feminised patterns to support the triad of economic success, technical innovation and research instead of formulating new or additional solutions for fighting climate change. The empty signifier *leadership* can be classified as having additional masculine connotations. The understanding of leadership in Western cultures is characterised by values such as competitiveness, (Bruff and Wöhl 2016, 100) "control, assertiveness, and skills of advocacy and domination" (Fletcher 2004, 650)—which are culturally connoted as masculine and part of the EU leadership, albeit not in an aggressively formulated way. Following Plumwood's approach, such a leader embodies the values ascribed to the "master identity" (Plumwood 1993, 23).

After sketching the nuances of the second dimension of gendered discourses, we must also discuss the gendered aspects of the first dimension of the EU climate change discourse, that is: discourse patterns that interpellate women in different climate-related roles (cf. Sect. 2.3.3). The three proposals of the EGD strategies—the Industrial Strategy, Renovation Wave and the Sustainable and Smart Mobility Strategy—exhibit gendered discourses of the first dimension (cf. Sect. 4.2.8). In all three proposals, women are called upon to study and later work in the fields of engineering and mathematics and are thus interpellated as participants in the climate-related economy and furthermore as politicians. Following this approach, the EU creates an opportunity to promote more diversity (Carroll 2022, 13), countering the masculinised "techno-scientific solutions, from which there is money to be made" (MacGregor 2014, 626), articulated in particular by male researchers from the Global North. Of course, we could assume that with an increased proportion of women in these fields the responses to climate change, then more often formulated by them, would change (Carroll 2022, 13; Michael et al. 2020), since they no longer particularly originate from an androcentristic stance. This could also gradually break down the dichotomous separation of private and public, from which policy formulation would benefit in terms of a new diversity (Prügl 2021, 302).

However, as a study of gendered decision-making in the mobility sector in various Scandinavian countries showed, near-parity does not automatically lead to the breaking down of encrusted structures oriented towards the masculine-connotated ideal of economic success (Magnusdottir and Kronsell 2015). Rather, the researchers were able to show that internalised gendered social norms play an important role in deciding how to shape mobility and other climate-relevant policies. In this respect, we can only state here that a participation level of approximately 30 percent women in decision-making bodies reliably represents their point of view (Ergas and York 2012, 968; Burk and Hartmann 2017, 143), but possibly only that of the privileged women of a society. Hence, the question of whether the first dimension of gendered discourse influences the second dimension must be examined again at a later point in time with the help of qualitative instruments—when the outlined programmes are activated.[56] Conversely, we can already note two important points from an ecofeminist perspective. First, it is not enough to solely integrate more women; rather, norms, values, and structures—all part of our understanding of the discourse—of climate policy must be deconstructed, otherwise the same one-dimensional responses to climate change will continue to be formulated in terms of weak sustainability. Or as Simon Irving and Jenny Helin (2018, 275) put it: "the economic empowerment of women denies their independence through a process of incorporation into the masculine centre". Second, there is nevertheless a potential to gradually break through these structures by increasing the representation of women in engineering and politics, if they have powerful positions with decision-making abilities. Only in this way can the ecofeminist idea of questioning the current economic model with its masculinised and androcentristic ideals develop an effect for the sake of a more sustainable and socially just fight against climate change. Therefore, the interpellation of women as participants that has now started must be advanced intersectionally considering more social dimensions than just gender and increasing the chances of leading to a more sustainable economic system. However, if this is not the case, it must be assumed that the EU wants to recruit more women as engineers in order to compensate for the demographically-induced shortage of skilled workers interested in accumulating further wealth—ultimately at the expense of the environment and climate (Zürn 2023).

Of course, this also applies to the identity of *leader* in general. Since the EU links prosperity very closely with economic growth in its current form, the ecofeminist demand for reconciling the environment and people must also include questioning here once again the current Western economic model. The major project of the EGD, which symbolises the understanding of the EU as a technological pioneer, underlines its leadership aspirations (Weyers et al. 2020, 297) in the sense of a "race for climate neutrality" (Gengnagel and Zimmermann 2022, 281) and represents the main characteristic of economic interests of the EU identity as a *leader*. Striving for a sustainable economy by overcoming masculinised patterns is not only complicated because the power structures are gendered and it

[56] Paula Carroll (2022, 13) identifies the same research gap for the energy sector of the EGD.

is culturally determined which goals are considered as desirable, but also because the EU climate discourse is deemed to be depoliticised due to its economic dominance (Machin 2019). This has also been demonstrated through the few attempts within the discourse to challenge the primacy of growth and the prevailing idea of economic success and therewith closely related idea of leadership: Neither the green economy nor the circular economy have been able to develop their potential to change the economic model in the sense of pushing, for instance, the precautionary principle or degrowth attempts—demanded inter alia by the Left, ENGOs and grassroot initiatives such as FFF as well as some parts of the society—which are both feminised in their character. Instead, the feminised aspects within the green economy and circular economy initiatives were mostly reshaped by masculinised aspects within the respective concept and incorporated into the logic of economic growth.

My assessment is supported by others (Lorek and Fuchs 2013) who also find that green and circular economy are not able to maintain strong sustainability—because "unlimited growth cannot be reconciled with strong sustainability" (D'Amato et al. 2017, 725). This failure must also be attributed to the EGD, first and foremost as it "still relies explicitly on the further stimulation of growth and conceptualizes the eco-social challenge as a challenge to be met with green capitalism" (Gengnagel and Zimmermann 2022, 294), but also as it "has failed to address how social and gender justice intersects with climate justice" (Khoury 2023, 11); an ecofeminist EGD would "aim to restructure social reproduction, degender and collectivise care work, and radically transform production and consumption" (Khoury 2023, 13). To achieve a more sustainable economy, the culturally firmly-anchored masculinised beliefs must be questioned and the dichotomous thinking reproducing inherent power structures must be deconstructed, as will be shown below in the final reflexion of the cases of India and the EU.

References

acatech. 2018. "CCU und CCS – Bausteine für den Klimaschutz in der Industrie." acatech POSITION. Accessed May 02, 2023. https://www.wwf.de/fileadmin/fm-wwf/Publikationen-PDF/CCU_und_CCS_Bausteine_fuer_den_Klimaschutz_in_der_Industrie.pdf.

Affronte, Marco. 2016. "Speech." In European Parliament (EP) 2016.

Agius, Christine, Annika B. Rosamond, and Catarina Kinnvall. 2020. "Populism, Ontological Insecurity and Gendered Nationalism: Masculinity, Climate Denial and Covid-19." *Politics, Religion & Ideology* 21 (4): 432–50. https://doi.org/10.1080/21567689.2020.1851871.

Agrawala, Shardul. 1998. "Context and Early Origins of the Intergovernmental Panel on Climate Change." *Climatic Change* 39 (4): 605–20. https://doi.org/10.1023/A:1005315532386.

Alliance of Liberals and Democrats for Europe. 2013. "A Europe that works: European Liberals are the first to adopt Election Manifesto." Accessed March 21, 2021. https://assets.nationbuilder.com/aldeparty/pages/1959/attachments/original/1596050894/2014_manifesto_-_a_europe_that_works.pdf?1596050894.

Alliance of Liberals and Democrats for Europe. 2018. "Freedom, opportunity, prosperity: the Liberal vision for the future of Europe." Accessed March 21, 2021. https://assets.nationbuilder.com/aldeparty/pages/1589/attachments/original/1594138326/2019_freedom_opportunity_prosperity_the_liberal_vision_for_the_future_of_europe_0.pdf?1594138326.

Allwood, Gill. 2014. "Gender Mainstreaming and EU Climate Change Policy." *European Integration online Papers*, 1–26. https://doi.org/10.1695/2014006.

Allwood, Gill. 2020a. "Gender Equality in European Union Development Policy in Times of Crisis." *Political Studies Review* 18 (3): 329–45. https://doi.org/10.1177/1478929919863224.

Allwood, Gill. 2020b. "Mainstreaming Gender and Climate Change to Achieve a Just Transition to a Climate-Neutral Europe." *JCMS: Journal of Common Market Studies* 58 (S1): 173–86. https://doi.org/10.1111/jcms.13082.

Allwood, Gill. 2022. "The EU's Transition to Climate Justice & Gender Equality. How Just and How Equal?" Policy Brief. Accessed December 23, 2023. https://feps-europe.eu/wp-content/uploads/2022/06/FINAL_DIGITAL_21.6.22_How_just_and_how_equal.pdf.

Alston, Margaret. 2014. "Gender mainstreaming and climate change." *Women's Studies International Forum* 47: 287–94. https://doi.org/10.1016/j.wsif.2013.01.016.

Ammann, Odile, and Audrey Boussat. 2023. "The Participation of Civil Society in European Union Environmental Law-Making Processes: A Critical Assessment of the European Commission's Consultations in Connection with the European Climate Law." *Eur. j. risk regul.* 14 (2): 235–52. https://doi.org/10.1017/err.2022.39.

Arora-Jonsson, Seema. 2011. "Virtue and vulnerability: Discourses on women, gender and climate change." *Global Environmental Change* 21 (2): 744–51. https://doi.org/10.1016/j.gloenvcha.2011.01.005.

Ayuzo, Pilar. 2015. "Speech." In European Parliament (EP) 2015.

Bäckstrand, Karin, and Ole Elgström. 2013. "The EU's role in climate change negotiations: from leader to 'leadiator'." *Journal of European Public Policy* 20 (10): 1369–86. https://doi.org/10.1080/13501763.2013.781781.

Bäckstrand, Karin, and Eva Lövbrand. 2006. "Planting Trees to Mitigate Climate Change: Contested Discourses of Ecological Modernization, Green Governmentality and Civic Environmentalism." *Global Environmental Politics* 6 (1): 50–75. https://doi.org/10.1162/glep.2006.6.1.50.

Baiardi, Donatella, and Claudio Morana. 2021. "Climate change awareness: Empirical evidence for the European Union." *Energy Economics* 96: 1–16. https://doi.org/10.1016/j.eneco.2021.105163.

Barroso, Manuel. 2012a. "Speech at the opening session of the Rio+20 conference." Accessed January 19, 2023. https://ec.europa.eu/commission/presscorner/detail/en/SPEECH_12_478.

Barroso, Manuel. 2012b. "State of the Union 2012 Address." Accessed October 22, 2022. https://ec.europa.eu/commission/presscorner/api/files/document/print/en/speech_12_596/SPEECH_12_596_EN.pdf.
Barroso, Manuel. 2013a. "State of the Union address." Accessed April 03, 2021. https://ec.europa.eu/commission/presscorner/detail/en/SPEECH_13_684.
Barroso, Manuel. 2013b. "Speech." In European Parliament (EP) 2013.
Barroso, Manuel. 2014. "Speech by President Barroso at the Climate Summit 2014." Accessed March 12, 2021. https://ec.europa.eu/commission/presscorner/detail/en/SPEECH_14_616.
Bauhardt, Christine. 2019. "Nature, Care and Gender: Feminist Dilemmas." In *Feminist Political Ecology and the Economics of Care: In Search of Economic Alternatives*, edited by Christine Bauhardt and Wendy Harcourt, 16–35. Routledge Studies in Ecological Economics Ser. Milton: Routledge.
BBC. 2015. "COP21: Rallies call for Paris climate change action." *BBC*, November 29. Accessed February 22, 2023. https://www.bbc.com/news/world-34956825.
BBC. 2016. "Donald Trump would 'cancel' Paris climate deal." May 27. Accessed April 16, 2021. https://www.bbc.com/news/election-us-2016-36401174.
Becker, Markus. 2016. "Europa kann auch schnell: Ratifizierung des Pariser Klimavertrags." *Spiegel online*, September 30. Accessed March 12, 2021. https://www.spiegel.de/wissenschaft/natur/eu-ratifiziert-klimaschutz-abkommen-europa-kann-auch-schnell-a-1114712.html.
Beisheim, Marianne, Birgit Lode, and Nils Simon. 2012. "Rio+20-Realpolitik und die Folgen für 'Die Zukunft, die wir wollen'." SWP-Aktuell 39. Accessed January 24, 2023. https://www.swp-berlin.org/publications/products/aktuell/2012A39_bsh_lod_sin.pdf.
Belet, Ivo. 2015. "Speech." In European Parliament (EP) 2015.
Benson, David, Jenny M. Fairbrass, Irene Lorenzoni, Tim O'Riordan, and Duncan Russel. 2021. "The Green Economy." In *Emerging Governance of a Green Economy*, edited by Jenny M. Fairbrass and Nicholas Vasilakos, 14–33: Cambridge University Press.
Biedenkopf, Katja. 2015. "The European Parliament in EU External Climate Governance." In *The European Parliament and Its International Relations*, edited by Stelios Stavridis, 92–108. Routledge Advances in European Politics. Hoboken: Taylor and Francis.
Biedenkopf, Katja, and Hayley Walker. 2016. *Playing to one's strengths: The implicit division of labor in U.S. and EU climate diplomacy.* AICGS policy report 64. Washington, D.C. American Institute for Contemporary German Studies, Johns Hopkins University.
Biermann, Frank. 2013. "Curtain down and Nothing Settled: Global Sustainability Governance after the 'Rio+20' Earth Summit." *Environ Plann C Gov Policy* 31 (6): 1099–1114. https://doi.org/10.1068/c12298j.
Biesbroek, Robbert, and Rob Swart. 2019. "Adaptation policy at supranational level? Evidence from the European Union." In *Research Handbook on Climate Change Adaptation Policy*, edited by E. C. Keskitalo and B. L. Preston, 194–211: Edward Elgar Publishing.
Bina, Olivia. 2013. "The Green Economy and Sustainable Development: An Uneasy Balance?" *Environment and Planning C: Government and Policy* 31 (6): 1023–47. https://doi.org/10.1068/c1310j.
Bloss, Michael. 2021. "Speech." In European Parliament (EP) 2021.

Bocquillon, Pierre, and Tomas Maltby. 2017. "The more the merrier? Assessing the impact of enlargement on EU performance in energy and climate change policies." *East European Politics* 33 (1): 88–105. https://doi.org/10.1080/21599165.2017.1279605.

Bocse, Alexandra-Maria. 2020. "The UK's decision to leave the European Union (Brexit) and its impact on the EU as a climate change actor." *Climate Policy* 20 (2): 265–74. https://doi.org/10.1080/14693062.2019.1701402.

Bonafè, Simona. 2016. "Speech." In European Parliament (EP) 2016.

Bongardt, Annette, and Francisco Torres. 2022. "The European Green Deal: More than an Exit Strategy to the Pandemic Crisis, a Building Block of a Sustainable European Economic Model." *JCMS: Journal of Common Market Studies* 60 (1): 170–85. https://doi.org/10.1111/jcms.13264.

Boyle, Jessica. 2013. "Gathering in the Gloom of Warsaw: Assessing the outcomes of COP19: A Climate Insights Product." Accessed February 11, 2023. https://www.iisd.org/system/files/publications/gathering_in_the_gloom.pdf.

Bruff, Ian, and Stefanie Wöhl. 2016. "Constitutionalizing Austerity, Disciplining the Household: Masculine Norms of Competitiveness and the Crisis of Social Reproduction in the Eurozone." In *Scandalous Economics: Gender and the Politics of Financial Crises*, edited by Aida A. Hozic and Jacqui True, 92–108. Oxford studies in gender and international relations. New York, NY: Oxford University Press.

Brun, Aslak. 2016. "Conference Diplomacy: The Making of the Paris Agreement." *PaG* 4 (3): 115–23. https://doi.org/10.17645/pag.v4i3.649.

Brún, Bairbre de. 2010. "Speech." In European Parliament (EP) 2010.

Bürgin, Alexander. 2015. "National binding renewable energy targets for 2020, but not for 2030 anymore: why the European Commission developed from a supporter to a brakeman." *Journal of European Public Policy* 22 (5): 690–707. https://doi.org/10.1080/13501763.2014.984747.

Bürgin, Alexander. 2023. "The European Commission: a climate policy entrepreneur." In Rayner, Szulecki, Jordan, and Oberthür 2023, 23–37.

Burk, Martha, and Heidi Hartmann. 2017. "Corporate Boards and International Policies: Gender Parity on Corporate Boards: A Path to Women's Equality?" In *Gender in the twenty-first century: the stalled revolution and the road to equality*, edited by Davis Shannon, Sarah Winslow, and David J. Maume, 127-146. Oakland: University of California Press.

Burns, Charlotte. 2017. "The European Parliament and Climate Change: A Constrained Leader?" In Wurzel, Connelly, and Liefferink 2017, 52–65.

Burns, Charlotte. 2021. "The European Parliament." In *Environmental Policy in the EU*, edited by Andrew Jordan and Viviane Gravey, 128–46. New York: Routledge.

BusinessEurope. 2015. "On the Road to Paris: A Global Deal is our Business." Accessed June 12, 2021. https://www.businesseurope.eu/sites/buseur/files/media/imported/2015-00450-E.pdf.

BusinessEurope. 2020. "Letter to Frans Timmermans." Accessed June 29, 2030. https://www.businesseurope.eu/sites/buseur/files/media/public_letters/iaco/2020-04-10_businesseurope_letter_environment_and_climate_consultations_and_regulations_-_executive_vp_timmermans.pdf.

Buttel, F. H. 2000. "Ecological modernization as social theory." *Geoforum* 31 (1): 57–65. https://doi.org/10.1016/S0016-7185(99)00044-5.

Buzogány, Aron, and Christoph Mohamad-Klotzbach. 2022. "Environmental Populism." In *The Palgrave Handbook of Populism*, edited by Michael Oswald, 321–40. Cham: Springer International Publishing.

Calliari, Elisa. 2018. "Loss and damage: a critical discourse analysis of Parties' positions in climate change negotiations." *Journal of Risk Research* 21 (6): 725–47. https://doi.org/10.1080/13669877.2016.1240706.

Calma, Justine. 2019. "2019 was the year of 'climate emergency' declarations." *The Verge*, December 27. Accessed June 03, 2023. https://www.theverge.com/2019/12/27/21038949/climate-change-2019-emergency-declaration.

Campins Eritja, Mar. 2007. "Reviewing the Challenging Task Faced by Member States in Implementing the Emissions Trading Directive: Issues of Member State Liablility." In Peeters and Deketelaere 2007, 69–82.

Cañete, Miguel A. 2014. "Statement at the opening of the high-level segment of COP20." Accessed March 01, 2023. https://unfccc.int/files/meetings/lima_dec_2014/statements/application/pdf/cop20_hls_eu.pdf.

Cañete, Miguel A. 2015a. "Speech." In European Parliament (EP) 2015.

Cañete, Miguel A. 2015b. "Historic climate deal in Paris: speech by Commissioner Miguel Arias Cañete at the press conference on the results of COP21 climate conference in Paris." Accessed September 04, 2022. https://ec.europa.eu/commission/presscorner/detail/el/SPEECH_15_6320.

Cañete, Miguel A. 2016a. "Speech." In European Parliament (EP) 2016.

Cañete, Miguel A. 2016b. "How will the Paris agreement impact EU climate and energy policies? Speech at the Bruegel event." Accessed September 04, 2022. https://ec.europa.eu/commission/presscorner/detail/de/SPEECH_16_264.

Cañete, Miguel A. 2017. "Statement by Climate Action and Energy Commissioner Miguel Arias Cañete on the US announcement to withdraw from the Paris Agreement." Accessed February 12, 2021. https://ec.europa.eu/commission/presscorner/detail/en/STATEMENT_17_1513.

Cañete, Miguel A. 2019. "Speech." In European Parliament 2019.

Carroll, Paula. 2022. "Gender Mainstreaming the European Union Energy Transition." *Energies* 15 (21). https://doi.org/10.3390/en15218087.

Caspari, Stefan. 1995. *Die Umweltpolitik der Europäischen Gemeinschaft.* Baden-Baden: Nomos.

Çelik, Feyyaz B. 2022. "The EU's different faces in climate diplomacy: leadership, interests, and responsibilities." *Journal of European Integration* 44 (8): 1019–39. https://doi.org/10.1080/07036337.2022.2068538.

Christoforou, Lefteris. 2016. "Speech." In European Parliament (EP) 2016.

Ciot, Melania G. 2019. "Speech." In European Parliament 2019.

Clima Action Network Europe. 2012. "CAN Europe's position on: Post-2020 EU climate and energy targets." Accessed March 12, 2021. https://caneurope.org/content/uploads/2012/12/CAN-E-position-on-post-2020-targets.pdf.

Clima Action Network Europe. 2013a. "Adequacy and feasibility of the 1.5°C long-term global limit." Accessed March 12, 2021. https://caneurope.org/content/uploads/2013/07/Adequacy-feasibility-of-1.5c-long-term-global-limit-July-2013-v2.pdf.

Clima Action Network Europe. 2013b. "CAN Europe's Response to the European Commission's Green paper Consultation on a 2030 Climate and Energy Framework." Accessed

March 12, 2021. https://caneurope.org/content/uploads/2013/06/130626-CAN-Europe-Response-to-Green-paper-Final.pdf.

Clima Action Network Europe. 2015a. "Annual Report 2014." Accessed March 12, 2021. https://caneurope.org/content/uploads/2020/09/CanE_Annual_report_2014_FIN.pdf.

Clima Action Network Europe. 2015b. "Briefing to EU Environment Ministers on the European Commission's Road to Paris Communication and its draft EU INDC proposal." Accessed February 23, 2023. https://web.archive.org/web/20151117033201/http://www.caneurope.org/docman/position-papers-and-research/un-climate-negotiations-2/2542-can-europe-briefing-to-ministers-on-the-ec-road-to-paris-and-its-indc/file?force-download=1.

Clima Action Network Europe. 2023. "Coalition for Higher Ambition." Accessed July 18, 2023. https://caneurope.org/coalition-for-higher-ambition/.

Climate Action Network, ed. 2017. *Civil society strongly condemns Trump's decision to withdraw US from the Paris Agreement*. Accessed March 12, 2021. https://climatenetwork.org/wp-content/uploads/2020/11/media_release_can_reactive_trump_exit_01_06_2017.pdf.

Climate Action Tracker. 2015. "USA. Rating." Accessed April 06, 2021. https://climateactiontracker.org/media/documents/2018/4/CAT_2015-03-31_CountryAssessment_USA.pdf.

Climate Action Tracker. 2016. "EU. Rating." Accessed June 12, 2021. https://climateactiontracker.org/media/documents/2018/4/CAT_2015-06-07_CountryAssessment_EU.pdf.

Climate Action Tracker. 2022. "Fair Share." Accessed May 08, 2022. https://climateactiontracker.org/methodology/cat-rating-methodology/fair-share/#:~:text=The%20Climate%20Action%20Tracker's%20%E2%80%9Cfair,analysis%20to%20close%20data%20gaps.

Climate Emergency Declaration. 2023. "Climate Emergency Declaration (CED) data shee." Accessed June 03, 2023. https://climateemergencydeclaration.org/climate-emergency-declarations-cover-15-million-citizens/.

Coalition for Higher Ambition. 2018. "Unique 'Coalition for Higher Ambition' calls upon the EU to step up at COP24." Accessed September 05, 2022. https://caneurope.org/unique-coalition-for-higher-ambition-calls-upon-the-eu-to-step-up-at-cop24/.

Coalition for Higher Ambition. 2019. "Open letter from the Coalition for Higher Ambition." Accessed September 05, 2022. https://caneurope.org/content/uploads/2019/03/Coalition-March-EUCO-Open-Letter_FINAL.pdf.

Coalition for Higher Ambition. 2020. "Europe needs and wants an ambitious, effective and fossil free cohesion policy delivering on climate neutrality." Accessed September 05, 2022. https://caneurope.org/content/uploads/2020/01/190127-Open-Letter_Cohesion-Policy-for-Climate-Neutrality.pdf.

Coalition for Higher Ambition. 2021. "Letter from the Coalition for Higher Ambition on the 'Fit for 55' Package." Accessed September 05, 2022. https://caneurope.org/content/uploads/2021/05/Coalition-Letter_May-2021_final.pdf.

Coffey, Brian. 2016. "Unpacking the politics of natural capital and economic metaphors in environmental policy discourse." *Environmental Politics* 25 (2): 203–22. https://doi.org/10.1080/09644016.2015.1090370.

Cornwall, Andrea, and Jenny Edwards. 2014. "Introdution: Negotiating Empowerment." In *Feminisms, Empowerment and Development: Changing Women's Lives*, edited by Andrea Cornwall and Jenny Edwards, 1–31. Feminisms and development. London, England: Zed Books; Bloomsbury Publishing.

Council of the European Union. 2010. "Council Conclusions: Preparation for the 16th Conference of the Parties to the UN Framework Convention on Climate Change Cancún, 29 November to 10 December 2010." https://www.consilium.europa.eu/uedocs/cms_data/docs/pressdata/en/envir/117096.pdf.

Council of the European Union. 2011a. "Council Conclusions: Preparations for the 17th session of the Conference of the Parties (COP 17) to the United Nations Framework Convention on Climate Change (UNFCCC)." Accessed October 22, 2022. https://www.consilium.europa.eu/uedocs/cms_data/docs/pressdata/en/envir/125026.pdf.

Council of the European Union. 2011b. "Presidency Conclusions: A Roadmap for moving to a competitive low-carbon economy in 2050." Accessed January 12, 2023. https://www.consilium.europa.eu/uedocs/cms_data/docs/pressdata/en/envir/122956.pdf.

Council of the European Union. 2012. "Gender equality and the environment: enhanced decision-making, qualifications and competitiveness in the field of climate change mitigation policy in the EU: Council Conclusions." Accessed January 22, 2023. https://data.consilium.europa.eu/doc/document/ST%2011638%202012%20INIT/EN/pdf.

Council of the European Union. 2013. "Council Conclusions: Preparations for the 19th session of the Conference of the Parties (COP19) to the United Nations Framework Convention on Climate Change and the 9th session of the Meeting of the Parties to the Kyoto Protocol." Accessed October 10, 2022. https://www.consilium.europa.eu/uedocs/cms_data/docs/pressdata/en/envir/139002.pdf.

Council of the European Union. 2014. "Council Conclusions: Preparations for the 20th session of the Conference of the Parties to the United Nations Framework Convention on Climate Change and the 10th session of the Meeting of the Parties to the Kyoto Protocol." Accessed October 10, 2022. https://www.consilium.europa.eu/uedocs/cms_data/docs/pressdata/en/envir/145508.pdf.

Council of the European Union. 2015a. "Council Conclusions: EU position for the UN climate change conference in Paris." Accessed October 26, 2022. https://www.consilium.europa.eu/en/press/press-releases/2015/09/18/counclusions-un-climate-change-conference-paris-2015/.

Council of the European Union. 2015b. "Outcome of the Council Meeting Environment." Accessed February 23, 2023. https://data.consilium.europa.eu/doc/document/ST-6932-2015-INIT/en/pdf.

Council of the European Union. 2016a. "European climate diplomacy after COP21: Elements for continued climate diplomacy in 2016. Annex." Accessed February 19, 2023. https://data.consilium.europa.eu/doc/document/ST-5853-2016-INIT/en/pdf.

Council of the European Union. 2016b. "Council Decision: on the conclusion, on behalf of the European Union, of the Paris Agreement adopted under the United Nations Framework Convention on Climate Change." https://eur-lex.europa.eu/legal-content/EN/TXT/PDF/?uri=CELEX:32016D1841&from=EN.

Council of the European Union. 2017. "Council Conclusions: The Paris Agreement and preparations for the UNFCCC meetings (Bonn, 6–17 November 2017)." Accessed April 03, 2021. https://data.consilium.europa.eu/doc/document/ST-13198-2017-INIT/en/pdf.

Council of the European Union. 2018. "Council Conclusions: Preparations for the UNFCCC meetings in Katowice (2 - 14 December 2018)." Accessed April 03, 2021. https://data.consilium.europa.eu/doc/document/ST-12901-2018-INIT/en/pdf.

Council of the European Union. 2019. "Outcome of the Council Meeting Environment." Accessed May 09, 2023. https://www.consilium.europa.eu/media/38980/st07171-en19.pdf.

Council of the European Union. 2021. "Council Conclusions: Preparations for the United Nations Framework on Climate Change (UNFCCC) meetings." Accessed October 10, 2022. https://data.consilium.europa.eu/doc/document/ST-12594-2021-REV-1/en/pdf.

Creighton, Lucinda. 2013. "Speech." In European Parliament (EP) 2013.

Creutzig, Felix, Marcus Hedahl, James Rydge, and Kacper Szulecki. 2014. "Challenging the European Climate Debate: Can Universal Climate Justice and Economics be Reconciled with Particularistic Politics?" *Global Policy* 5: 6–14. https://doi.org/10.1111/1758-5899.12156.

Cuomo, Chris. 2002. "On Ecofeminist Philosophy." *Ethics and the Environment* 7 (2): 1–11. https://www.jstor.org/stable/40339033.

D'Amato, D., N. Droste, B. Allen, M. Kettunen, K. Lähtinen, J. Korhonen, P. Leskinen, B. D. Matthies, and A. Toppinen. 2017. "Green, circular, bio economy: A comparative analysis of sustainability avenues." *Journal of Cleaner Production* 168: 716–34. https://doi.org/10.1016/j.jclepro.2017.09.053.

Dalli, Helena. 2017. "Speech." In European Parliament (EP) 2017.

Dankelman, Irene, and Kavita Naidu. 2020. "Introduction: Gender, development, and the climate crisis." *Gender & Development* 28 (3): 447–57. https://doi.org/10.1080/13552074.2020.1843830.

Darby, Megan. 2014. "Ambitious or weak? Reaction to the EU's 2030 climate package." October 24. https://www.climatechangenews.com/2014/10/24/ambitious-or-weak-reaction-to-the-eus-2030-climate-package/.

Debusscher, Petra. 2021. "Development Policy." In *The Routledge Handbook of Gender and EU Politics*, edited by Gabriele Abels, Andrea Krizsán, Heather MacRae, and Anna van der Vleuten, 290–301. Routledge international handbooks. London: Routledge Taylor & Francis Group.

Del Castillo Vera, Pilar. 2010. "Speech." In European Parliament (EP) 2010.

Della Porta, Donatella, and Martín Portos. 2023. "Rich kids of Europe? Social basis and strategic choices in the climate activism of Fridays for Future." *Italian Political Science Review/Rivista Italiana di Scienza Politica* 53 (1): 24–49. https://doi.org/10.1017/ipo.2021.54.

Delreux, Tom, and Charlotte Burns. 2019. "Parliamentarizing a Politicized Policy: Understanding the Involvement of the European Parliament in UN Climate Negotiations." *PaG* 7 (3): 339–49. https://doi.org/10.17645/pag.v7i3.2093.

Delreux, Tom, and Sander Happaerts. 2016. *Environmental Policy and Politics in the European Union.* The European Union series. London: Palgrave Macmillan education.

Deutsche Präsidentschaft im Rat der Europäischen Union. 2020. "Bundesumweltministerin Schulze: Klimaschutz wird in Europa unumkehrbar." Accessed July 10, 2023. https://www.eu2020.de/eu2020-de/aktuelles/pressemitteilungen/klimaschutz-klimagesetz-europa-treibhausgasneutralitaet/2428836.

DG Climate Action. 2017. "Annual Activity Report 2016." Accessed September 05, 2022. https://ec.europa.eu/info/sites/default/files/file_import/aar-clima-2016_en_0.pdf.

Diez, Thomas. 2005. "Constructing the Self and Changing Others: Reconsidering 'Normative Power Europe'." *Millennium* 33 (3): 613–36. https://doi.org/10.1177/03058298050330031701.

Diez, Thomas. 2013. "Normative power as hegemony." *Cooperation and Conflict* 48 (2): 194–210. https://doi.org/10.1177/0010836713485387.

Djoudi, Houria, Bruno Locatelli, Chloe Vaast, Kiran Asher, Maria Brockhaus, and Bimbika Basnett Sijapati. 2016. "Beyond Dichotomies: Gender and Intersecting Inequalities in Climate Change Studies." *Ambio* 45 (Suppl 3): 248–62. https://doi.org/10.1007/s13280-016-0825-2.

Douma, Wybe T. 2007. "The European Union, Russia and the Kyoto Protocol." In Peeters and Deketelaere 2007, 51–66.

Dröge, Susanne, Thomas Spencer, Alexandra Deprez, Gallagher Liz, Artur Gradziuk, Andrei Marcu, Sebastian Oberthür, Oliver Sartor, Henri Waisman, and Tomas Wyns. 2015. "The EU's INDC and its contribution to a successful deal in Paris 2015." Working Paper 8. Accessed June 12, 2021. https://www.swp-berlin.org/publications/products/arbeitspapiere/Droege_et_al2015-The_EU_s_INDC_and_its_contribution_to_a_successful_deal_in_Paris.pdf.

Dupont, Claire, and Sebastian Oberthür. 2017. "The Council and the European Council: Stuck on the Road to Transformational Leadership." In Wurzel, Connelly, and Liefferink 2017, 66–79.

Dupont, Claire, Sebastian Oberthür, and Ingmar von Homeyer. 2020. "The Covid-19 crisis: a critical juncture for EU climate policy development?" *Journal of European Integration* 42 (8): 1095–1110. https://doi.org/10.1080/07036337.2020.1853117.

Earsom, Joseph. 2023. "It's not as simple as copy/paste: the EU's remobilisation of the High Ambition Coalition in international climate governance." *Int Environ Agreements* 23 (1): 27–42. https://doi.org/10.1007/s10784-023-09592-z.

Eckert, Eva, and Oleksandra Kovalevska. 2021. "Sustainability in the European Union: Analyzing the Discourse of the European Green Deal." *JRFM* 14 (2): 1–22. https://doi.org/10.3390/jrfm14020080.

Eickhout, Bas. 2010. "Speech." In European Parliament (EP) 2010.

Eickhout, Bas. 2021. "Speech." In European Parliament (EP) 2021.

Ellerman, A. D., Claudio Marcantonini, and Aleksandar Zaklan. 2016. "The European Union Emissions Trading System: Ten Years and Counting." *Review of Environmental Economics and Policy* 10 (1): 89–107. https://doi.org/10.1093/reep/rev014.

ENV. 1998. "2106th Council meeting." Accessed October 31, 2022. https://ec.europa.eu/commission/presscorner/api/files/document/print/en/pres_98_205/PRES_98_205_EN.pdf.

Ergas, Christina, and Richard York. 2012. "Women's Status and Carbon Dioxide Emissions: a Quantitative Cross-National Analysis." *Social science research* 41 (4): 965–76. https://doi.org/10.1016/j.ssresearch.2012.03.008.

Estrela, Edite. 2012. "Speech." In European Parliament (EP) 2012.

Eurochambres. 2013. "Towards a Business-freindly Approach an EU Climate Action." Accessed March 12, 2021. https://www.eurochambres.eu/wp-content/uploads/2020/02/ECH_PP_2030targets_final-2013-00861-01.pdf.

Eurochambres. 2014. "Eurochambres position on the Commission's proposal for a policy framework for climate and energy in the period from 2020 to 2030." Accessed March 12, 2021. https://www.eurochambres.eu/wp-content/uploads/2014/06/EUROCHAMBRES_

Position_Paper_on_Commission_proposal_2030_Policy_Framework_Climate_and_Energy-1.pdf.
Eurochambres. 2018. "Position on the Strategy for longterm EU greenhouse gas emissions reduction." Accessed March 12, 2021. https://www.eurochambres.eu/wp-content/uploads/2018/10/EUROCHAMBRES_Position_Strategy_Long-term__EU_GHG-emissions-reduction-2018-00230-01.pdf.
Eurochambres. 2019a. "Eurochambres signs 2019 elections partnership with the European Parliament." Accessed May 25, 2023. https://www.eurochambres.eu/wp-content/uploads/2019/01/ECH_PressRelease_EPMoU_30-01-2019-2019-00019-01-1.pdf.
Eurochambres. 2019b. "European Chamber network delivers recommendations for EU 2019–2024 term under the von der Leyen Commission." Accessed March 21, 2021. https://www.eurochambres.eu/wp-content/uploads/2019/11/ECH_Press_Release_European_Chamber_network_delivers_recommendations_for_EU_2019-2024_term_27-11-19-2019-00285-01.pdf.
Eurochambres. 2020. "Eurochambres' input to the public consultation on the 2030 Climate Target Plan." Accessed June 12, 2021. https://www.eurochambres.eu/wp-content/uploads/2020/06/200623_-_ECH_input_2030_Climate_Target_Plan_consultation-2020-00068-01.pdf.
Eurochambres. 2021a. "Position: Fit-for-55 Package." Accessed September 16, 2021. https://www.eurochambres.eu/wp-content/uploads/2021/07/Fit-for-55-Package_Paper_v5.pdf.
Eurochambres. 2021b. "The Just Transition must stimulate entrepreneurship and creation of new jobs." Accessed March 14, 2022. https://www.eurochambres.eu/wp-content/uploads/2021/11/211108-ECH-Position-on-the-social-impact-of-the-Green-Deal.pdf.
European Alliance for a Just Transition. 2021. "A Call to Deliver a Just Transition to a Social and Sustainable Europe." Accessed July 17, 2023. https://alliancejusttransition.eu/wp-content/uploads/2023/03/joint-declaration-EN-03-2023.pdf.
European Climate Foundation. 2017. "10 Years Promoting European Climate Leadership: 2017 Annual Report." Accessed March 12, 2021. https://europeanclimate.org/wp-content/uploads/2019/11/2017-ecf-annual-report.pdf.
European Climate Foundation. 2022. "Shaping a carbon-free society: 2021 Annual Report." Accessed April 25, 2023. https://europeanclimate.org/wp-content/uploads/2022/11/ecf-2021-annual-report-03-10-22.pdf.
European Commission. 1995. "Eurobarometer 43.1: Europeans and the Environment in 1995." Accessed April 21, 2021. https://europa.eu/eurobarometer/surveys/detail/143.
European Commission. 2008. "Communication from the Commission to the European Parliament, the Council, the European Economic and Social Committee and the Committee of the Regions." Accessed March 19, 2022. https://eur-lex.europa.eu/legal-content/EN/TXT/PDF/?uri=CELEX:52008DC0030&from=EN.
European Commission. 2009. "Special Eurobarometer: Europeans' attitudes towards climate change." Accessed January 06, 2024. https://climate.ec.europa.eu/system/files/2022-02/ebs_322_en.pdf.
European Commission. 2010. "International climate policy post-Copenhagen: Acting now to reinvigorate global action on climate change." Accessed September 05, 2022. https://eur-lex.europa.eu/legal-content/EN/TXT/PDF/?uri=CELEX:52010DC0086&from=EN.

European Commission. 2011a. "Rio+20: towards the green economy and better governance." Accessed March 04, 2021. https://eur-lex.europa.eu/LexUriServ/LexUriServ.do?uri=COM:2011:0363:FIN:EN:PDF.
European Commission. 2011b. "Special Eurobarometer 372: Climate change." Accessed January 06, 2024. https://climate.ec.europa.eu/system/files/2022-02/ebs_372_en.pdf.
European Commission. 2011c. "A Roadmap for moving to a competitive low carbon economy in 2050." Accessed January 11, 2023. https://eur-lex.europa.eu/LexUriServ/LexUriServ.do?uri=COM:2011:0112:FIN:en:PDF.
European Commission. 2013. "Green Paper: A 2030 framework for climate and energy policies." Accessed January 12, 2023. https://eur-lex.europa.eu/LexUriServ/LexUriServ.do?uri=COM:2013:0169:FIN:en:PDF.
European Commission. 2014a. "Special Eurobarometer 409: Climate Change." Accessed September 12, 2023. https://europa.eu/eurobarometer/api/deliverable/download/file?deliverableId=40929.
European Commission. 2014b. "A policy framework for climate and energy in the period from 2020 to 2030: Communication from the Commission." Accessed January 31, 2023. https://data.consilium.europa.eu/doc/document/ST%205644%202014%20REV%201/EN/pdf.
European Commission. 2015a. *Special Eurobarometer 435: Climate change*. Special Eurobarometer 435. Luxembourg: Publications Office. Accessed May 16, 2021. https://climate.ec.europa.eu/system/files/2016-11/report_2015_en.pdf.
European Commission. 2015b. "Communication: The Paris Protocol – A blueprint for tackling global climate change beyond 2020." https://eur-lex.europa.eu/resource.html?uri=cellar:e27fdb4d-bdce-11e4-bbe1-01aa75ed71a1.0003.03/DOC_1&format=PDF.
European Commission. 2015c. "Closing the loop - An EU action plan for the Circular Economy." Accessed June 03, 2023. https://eur-lex.europa.eu/resource.html?uri=cellar:8a8ef5e8-99a0-11e5-b3b7-01aa75ed71a1.0012.02/DOC_1&format=PDF.
European Commission. 2016a. "The Road from Paris: assessing the implications of the Paris Agreement and accompanying the proposal for a Council decision on the signing, on behalf of the European Union, of the Paris agreement adopted under the United Nations Framework Convention on Climate Change: Communication from the Commission to the European Parliament and the Council." Accessed February 14, 2023. https://eur-lex.europa.eu/legal-content/EN/TXT/PDF/?uri=CELEX:52016DC0110&from=EN.
European Commission. 2016b. "Clean Energy For All Europeans: Communication." https://eur-lex.europa.eu/resource.html?uri=cellar:fa6ea15b-b7b0-11e6-9e3c-01aa75ed71a1.0001.02/DOC_1&format=PDF.
European Commission. 2017a. "Gender Equality 2017: Gender Equality, Stereotypes, and Women in Politics." Accessed August 11, 2023. https://www.globalwps.org/data/HRV/files/Eurobarometer%20in%202017.pdf.
European Commission. 2017b. "Special Eurobarometer 459: Climate change." Accessed January 06, 2024. https://climate.ec.europa.eu/system/files/2017-09/report_2017_en.pdf.
European Commission. 2018a. "A Clean Planet for all: A European strategic long-term vision for a prosperous, modern, competitive and climate neutral economy." Accessed September 14, 2021. https://eur-lex.europa.eu/legal-content/EN/TXT/PDF/?uri=CELEX:52018DC0773.

European Commission. 2018b. "Commissioners Miguel Arias Cañete and Carlos Moedas welcome the UN climate change report on 1.5° C global warming limit." Accessed April 12, 2023. https://ec.europa.eu/commission/presscorner/detail/en/STATEMENT_18_6043.

European Commission. 2019a. "Special Eurobarometer 490: Climate Change." Accessed May 02, 2023. https://climate.ec.europa.eu/system/files/2019-09/report_summary_2019_en.pdf.

European Commission. 2019b. "The European Green Deal." Accessed March 15, 2021. https://eur-lex.europa.eu/resource.html?uri=cellar:b828d165-1c22-11ea-8c1f-01aa75ed71a1.0002.02/DOC_1&format=PDF.

European Commission. 2020a. "Annex 1: new initiatives." Accessed June 15, 2021. https://www.klimareporter.de/images/dokumente/2020/04/Legislativ_Proposal_EU_COM-1.PDF.

European Commission. 2020b. "Circular Economy Action Plan: For a cleaner and more competitive Europe." Accessed June 03, 2023. https://eur-lex.europa.eu/resource.html?uri=cellar:9903b325-6388-11ea-b735-01aa75ed71a1.0017.02/DOC_1&format=PDF.

European Commission. 2020c. "Special Eurobarometer: Attitudes of European citizens towards the Environment." Accessed September 01, 2022. https://europa.eu/eurobarometer/api/deliverable/download/file?deliverableId=72616.

European Commission. 2020d. "Proposal for a Regulation of the European Parliament and of the Council establishing the Just Transition Fund." Accessed July 25, 2023. https://eur-lex.europa.eu/resource.html?uri=cellar:b82780d8-3771-11ea-ba6e-01aa75ed71a1.0003.02/DOC_1&format=PDF.

European Commission. 2020e. "Proposal for a Regulation of the EP and the Council establishing the framework for achieving climate neutrality and amending Regulation." Accessed June 26, 2023. https://eur-lex.europa.eu/legal-content/EN/TXT/PDF/?uri=CELEX:52020PC0080.

European Commission. 2020f. "Communication. A New Industrial Strategy for Europe." Accessed August 12, 2023. https://eur-lex.europa.eu/legal-content/EN/TXT/PDF/?uri=CELEX:52020DC0102.

European Commission. 2020g. "Communication: Stepping up Europe's 2030 climate ambition. Investing in a climate-neutral future for the benefit of our people." Accessed July 10, 2023. https://eur-lex.europa.eu/legal-content/EN/TXT/PDF/?uri=CELEX:52020DC0562.

European Commission. 2020h. "Communication. A Renovation Wave for Europe - greening our buildings, creating jobs, improving lives." Accessed August 12, 2023. https://eur-lex.europa.eu/resource.html?uri=cellar:0638aa1d-0f02-11eb-bc07-01aa75ed71a1.0003.02/DOC_1&format=PDF.

European Commission. 2020i. "Communication. Sustainable and Smart Mobility Strategy – putting European transport on track for the future." Accessed August 12, 2023. https://eur-lex.europa.eu/resource.html?uri=cellar:5e601657-3b06-11eb-b27b-01aa75ed71a1.0001.02/DOC_1&format=PDF.

European Commission. 2021a. "'Fit for 55': delivering the EU's 2030 Climate Target on the way to climate neutrality." Accessed April 11, 2023. https://eur-lex.europa.eu/legal-content/EN/TXT/PDF/?uri=CELEX:52021DC0550&from=EN.

European Commission. 2021b. "Establishing a carbon border adjustment mechanism." Accessed April 11, 2023. https://eur-lex.europa.eu/resource.html?uri=cellar:a95a4441-e558-11eb-a1a5-01aa75ed71a1.0001.02/DOC_1&format=PDF.

European Commission. 2021c. "Special Eurobarometer 513. Climate Change." Accessed September 14, 2023. https://europa.eu/eurobarometer/api/deliverable/download/file?deliverableId=75838.

European Commission. 2021d. "Forging a climate-resilient Europe - the new EU Strategy on Adaptation to Climate Change." Accessed July 12, 2023. https://eur-lex.europa.eu/legal-content/EN/TXT/PDF/?uri=CELEX:52021DC0082.

European Commission. 2022. "Kyoto 1st commitment period (2008–12)." https://climate.ec.europa.eu/eu-action/climate-strategies-targets/progress-made-cutting-emissions/kyoto-1st-commitment-period-2008-12_en.

European Commission. 2023. "A European Green Deal: Striving to be the first climate-neutral continent." Accessed June 15, 2023. https://commission.europa.eu/strategy-and-policy/priorities-2019-2024/european-green-deal_en.

European Council. 2009. "Presidency Conclusions." Accessed February 16, 2023. https://www.consilium.europa.eu/uedocs/cms_data/docs/pressdata/en/ec/110889.pdf.

European Council. 2014a. "Conclusions." Accessed February 12, 2021. https://data.consilium.europa.eu/doc/document/ST-7-2014-REV-1/en/pdf#page=8.

European Council. 2014b. "Conclusions." Accessed February 21, 2021. https://data.consilium.europa.eu/doc/document/ST-169-2014-INIT/en/pdf.

European Council. 2015. "VN-Klimakonferenz in Paris, 30. November - 12. Dezember 2015." https://www.consilium.europa.eu/de/meetings/international-summit/2015/11/30-12/.

European Court of Auditors. 2023. "EU Climate and Energy Targets: 2020 targets achieved, but little indication that actions to reach the 2030 targets will be sufficient." Accessed December 10, 2023. https://www.eca.europa.eu/ECAPublications/SR-2023-18/SR-2023-18_EN.pdf.

European Environment Agency. 2014. "Annual European Union greenhouse gas inventory 1990–2012 and inventory report 2014: Submission to the UNFCCC Secretariat." Accessed June 24, 2021. https://www.eea.europa.eu//publications/european-union-greenhouse-gas-inventory-2014.

European Environment Agency. 2019. "Europe's state of the environment 2020." Accessed June 27, 2023. https://www.eea.europa.eu/highlights/soer2020-europes-environment-state-and-outlook-report.

European Environmental Bureau. 2017. "EEB calls for tough response to US withdrawal from Paris Agreement." https://eeb.org/eeb-calls-for-tough-response-to-us-withdrawal-from-paris-agreement/.

European Green Party. 2014. "Change Europe, Vote Green. Green Common Manifesto European Elections 2014." Accessed April 12, 2021. https://europeangreens.eu/sites/europeangreens.eu/files/2014_Manifesto.pdf.

European Green Party. 2019. "Time to renew the promise of Europe." Accessed March 12, 2021. https://europeangreens.eu/sites/europeangreens.eu/files/8.%20PROOFREAD%20Adopted%20%20EGP%20Manifesto%202019.pdf.

European Institute for Gender Equality, ed. 2012. *Review of the Implementation in the EU of area K of the Beijing Platform for Action: Women and the Environment Gender Equality and Climate Change – Report.* Luxembourg. Accessed January 21, 2023. https://eige.europa.eu/sites/default/files/documents/Gender-Equality-and-Climate-Change-Report.pdf.

European Institute for Gender Equality. 2021. "Gender inequalities in care and consequences for the labour market." Accessed August 22, 2023. https://eige.europa.eu/sites/default/files/documents/20204948_mh0220888enn_pdf.pdf.

European Left. 2018. "Manifesto." Accessed March 12, 2021. https://www.european-left.org/wp-content/uploads/2018/11/Manifesto-European-Left_ENG.pdf.

European Parliament, ed. 2010a. *Debate: Outcome of the Climate Change Conference in Cancún. 14.12.2010.* Accessed September 05, 2022. https://www.europarl.europa.eu/doceo/document/CRE-7-2010-12-14-ITM-016_EN.html.

European Parliament. 2010b. "European Parliament Resolution of 25 November 2010 on the Climate Change Conference in Cancun (COP16)." Accessed March 14, 2021. https://www.europarl.europa.eu/doceo/document/TA-7-2010-0442_EN.pdf.

European Parliament. 2010c. "Resolution of 10 February 2010 on the outcome of the Copenhagen Conference on Climate Change (COP 15)." Accessed September 05, 2022. https://eur-lex.europa.eu/legal-content/EN/TXT/PDF/?uri=CELEX:52010IP0019&from=EN.

European Parliament, ed. 2010. *Preparations for Cancún Climate Change Conference.* Accessed March 04, 2021. https://www.europarl.europa.eu/doceo/document/CRE-7-2010-11-24-ITM-014_EN.html.

European Parliament. 2011. "Climate change conference in Durban: European Parliament resolution of 16 November 2011 on the climate change conference in Durban (COP 17)." Accessed March 13, 2021. https://eur-lex.europa.eu/legal-content/EN/TXT/PDF/?uri=CELEX:52011IP0504&from=EN.

European Parliament. 2012a. "Climate change conference in Doha (COP 18): European Parliament resolution of 22 November 2012 on the Climate Change Conference in Doha, Qatar (COP 18)." Accessed March 13, 2021. https://www.europarl.europa.eu/doceo/document/TA-7-2012-0452_EN.pdf.

European Parliament, ed. 2012. *Debate: Climate change conference in Doha (COP 18). 21.11.2012.* Accessed September 05, 2022. https://www.europarl.europa.eu/doceo/document/CRE-7-2012-11-21-ITM-010_EN.html.

European Parliament. 2012b. "Women and climate change." Accessed January 12, 2021. https://www.europarl.europa.eu/doceo/document/TA-7-2012-0145_EN.pdf.

European Parliament. 2012c. "Role of women in the green economy." Accessed January 12, 2021. https://www.europarl.europa.eu/doceo/document/TA-7-2012-0321_EN.pdf.

European Parliament. 2013. "Climate Change Conference: European Parliament resolution of 23 October 2013 on the climate change conference in Warsaw, Poland (COP19)." Accessed March 13, 2021. https://eur-lex.europa.eu/legal-content/EN/TXT/PDF/?uri=CELEX:52013IP0443&from=EN.

European Parliament, ed. 2013. *Debate: UN Framework Convention on Climate Change.* Accessed September 05, 2022. https://www.europarl.europa.eu/doceo/document/CRE-7-2013-02-06-ITM-015_EN.html.

European Parliament, ed. 2013. *State of the Union (debate).* Accessed April 14, 2021. Source: https://www.europarl.europa.eu/doceo/document/CRE-7-2013-09-11-ITM-004_EN.html.

European Parliament. 2014a. "2014 UN Climate Change Conference — COP 20 in Lima, Peru (1–12 December 2014): European Parliament resolution of 26 November 2014 on the 2014 UN Climate Change Conference — COP 20 in Lima, Peru (1–12 December 2014) (." Accessed February 21, 2021. https://eur-lex.europa.eu/legal-content/EN/TXT/PDF/?uri=CELEX:52014IP0063&from=EN.

European Parliament. 2014b. "Election Results." https://www.europarl.europa.eu/elections 2014-results/en/election-results-2014.html.

European Parliament. 2014c. "A 2030 framework for climate and energy policies: European Parliament resolution on a 2030 framework for climate and energy policies." Accessed February 12, 2021. https://www.europarl.europa.eu/doceo/document/TA-7-2014-0094_EN.pdf.

European Parliament, ed. 2015. *Debate: Towards a new international climate agreement in Paris*. Accessed February 21, 2023. https://www.europarl.europa.eu/doceo/document/CRE-8-2015-10-14-ITM-014_EN.html.

European Parliament. 2015. "EU position for COP 21 climate change conference." Briefing. Accessed October 26, 2022. https://www.europarl.europa.eu/RegData/etudes/BRIE/2015/572787/EPRS_BRI(2015)572787_EN.pdf.

European Parliament. 2016a. "2016 UN Climate change Conference in Marrakesh, Morocco (COP22): European Parliament resolution of 6 October 2016 on the implementation of the Paris Agreement and the 2016 UN Climate Change Conference in Marrakesh, Morocco (COP22)." Accessed March 21, 2021. https://www.europarl.europa.eu/doceo/document/TA-8-2016-0383_EN.pdf.

European Parliament, ed. 2016. *Debate: Outcome of COP 22 in Marrakesh*. Accessed April 12, 2021. https://www.europarl.europa.eu/doceo/document/CRE-8-2016-11-23-ITM-008_EN.html.

European Parliament, ed. 2016. *Debate: UN Climate change Conference in Marrakesh, Morocco (COP22)*. https://www.europarl.europa.eu/doceo/document/CRE-8-2016-10-05-ITM-019_EN.html.

European Parliament. 2016b. "AMENDMENTS 537 - 775: on the proposal for a directive of the European Parliament and of the Council amending Directive 2003/87/EC to enhance cost-effective emission reductions and low-carbon investments." Accessed July 26, 2023. https://www.europarl.europa.eu/RegData/commissions/itre/projet_avis/2016/585448/amendements/ITRE_AM(2016)585448_EN.pdf.

European Parliament. 2017a. "2017 UN Climate Change Conference in Bonn, Germany (COP23): European Parliament resolution of 4 October 2017 on the 2017 UN Climate Change Conference in Bonn, Germany (COP23)." Accessed March 21, 2021. https://www.europarl.europa.eu/doceo/document/TA-8-2017-0380_EN.pdf.

European Parliament, ed. 2017. *Debate: 2017 UN Climate Change Conference in Bonn, Germany (COP23)*. Accessed April 18, 2021. https://www.europarl.europa.eu/doceo/document/CRE-8-2017-10-03-ITM-013_EN.html.

European Parliament, ed. 2017. *Debate: President Trump's decision to withdraw the US from the COP 21 Climate agreement*. https://www.europarl.europa.eu/doceo/document/CRE-8-2017-06-14-ITM-006_EN.html.

European Parliament. 2018a. "2018 UN Climate Change Conference in Katowice, Poland (COP24): European Parliament resolution of 25 October 2018 on the 2018 UN Climate Change Conference in Katowice, Poland (COP24)." Accessed March 21, 2021. https://www.europarl.europa.eu/doceo/document/TA-8-2018-0430_EN.pdf.

European Parliament. 2018b. "Women, gender equality and climate justice: European Parliament resolution of 16 January 2018 on women, gender equality and climate justice (2017/2086(INI))." Accessed April 12, 2021. https://www.europarl.europa.eu/doceo/document/TA-8-2018-0005_EN.pdf.

European Parliament. 2018c. "Climate diplomacy: European Parliament resolution." Accessed April 12, 2021. https://www.europarl.europa.eu/doceo/document/TA-8-2018-0280_EN.pdf.

European Parliament. 2018d. "Resolution on the Multiannual Financial Framework 2021–2027 – Parliament's position with a view to an agreement." Accessed July 26, 2023. https://www.europarl.europa.eu/doceo/document/TA-8-2018-0449_EN.pdf.

European Parliament. 2019a. "Closer to the Citizens, Closer to the Ballot." Accessed May 23, 2023. https://www.europarl.europa.eu/at-your-service/files/be-heard/eurobarometer/2019/closer-to-the-citizens-closer-to-the-ballot/report/en-eurobarometer-2019.pdf.

European Parliament, ed. 2019. *Debate: Climate Change*. Accessed May 03, 2023. https://www.europarl.europa.eu/doceo/document/CRE-8-2019-03-13-ITM-008_EN.html.

European Parliament. 2019b. "European Parliament: 2019–2024 Constitutive session." Accessed June 03, 2023. https://www.europarl.europa.eu/election-results-2019/en/european-results/2019-2024/.

European Parliament. 2019c. "A European strategic long-term vision for a prosperous, modern, competitive and climate neutral economy in accordance with the Paris Agreement: Resolution." Accessed May 23, 2023. https://www.europarl.europa.eu/doceo/document/TA-8-2019-0217_EN.pdf.

European Parliament. 2019d. "Climate and environmental emergency: European Parliament resolution of 28 November 2019 on the climate and environment emergency." Accessed May 27, 2023. https://eur-lex.europa.eu/legal-content/EN/TXT/PDF/?uri=CELEX:52019IP0078(01).

European Parliament. 2020. "European Climate Law: Amendments adopted by the European Parliament." Accessed June 28, 2023. https://www.europarl.europa.eu/RegData/seance_pleniere/textes_adoptes/definitif/2020/10-08/0253/P9_TA(2020)0253_EN.pdf.

European Parliament, ed. 2021a. *Outcome of the COP26 in Glasgow (debate)*. Accessed July 20, 2023. https://www.europarl.europa.eu/doceo/document/CRE-9-2021-11-24-ITM-003_EN.html.

European Parliament. 2021b. "UN Climate Change Conference in Glasgow, UK (COP26): European Parliament resolution of 21 October 2021 on the 2021 UN Climate Change Conference in Glasgow, UK (COP26)." Accessed October 26, 2022. https://www.europarl.europa.eu/doceo/document/TA-9-2021-0437_EN.pdf.

European Parliament, and European Commission. 2008. "Special Eurobarometer 300: Europeans' attitudes towards climate change." Accessed January 06, 2024. https://climate.ec.europa.eu/system/files/2022-02/ebs_300_en.pdf.

European People's Party. 2014. "Why Vote for the European People's Party?" Accessed March 21, 2021. https://www.epp.eu/files/uploads/2017/04/EPPMANIFESTO_UK.pdf.

European People's Party. 2019. "EPP Manifesto: Let's open the next Chapter for Europe together." Accessed March 21, 2021. https://www.epp.eu/files/uploads/2019/04/EPP-MANIFESTO-2019-002.pdf.

European Trade Union Confederation. 2015. "ETUC key demands for the Climate COP21." Accessed March 12, 2021. https://etuc.org/sites/default/files/document/files/00_-_en_-_cop21_-_position_def.pdf.
European Trade Union Confederation. 2018. "ETUC Resolution on Strategy for long-term EU greenhouse gas emissions reduction: ETUC views." Accessed March 12, 2021. https://etuc.org/sites/default/files/document/file/2018-07/ADOPTED%20-%20Strategy%20for%20long-term%20EU%20greenhouse%20gas%20emissions%20reduction%20ETUC%20views_0.pdf.
European Trade Union Confederation. 2019. "ETUC Position Key demands to build a Just Transition and boost climate action after the EU elections." Accessed May 25, 2023. https://www.etuc.org/system/files/document/file2019-04/Adopted%20-%20Key%20demands%20to%20build%20a%20Just%20Transition%20and%20boost%20climate%20action_3.pdf.
European Trade Union Confederation. 2020. "ETUC resolution on European Green Deal initiatives: Climate Law, EU Sustainable Investment Plan, Just Transition Fund Regulation and new EU Industrial Strategy." Accessed June 20, 2023. https://www.etuc.org/system/files/document/file2020-06/Adopted%20-%20ETUC%20Resolution%20on%20EGD%20initiatives%20-%20EN_0.pdf.
European Trade Union Confederation. 2021a. "ETUC position: A Just Transition Legal Framework to complement the Fit for 55 package." Accessed March 12, 2022. https://etuc.org/sites/default/files/document/file/2021-12/ETUC%20position%20for%20a%20Just%20Transition%20Legal%20Framework%20to%20Complement%20the%20Fit%20for%2055%20Package_1.pdf.
European Trade Union Confederation. 2021b. "ETUC resolution on 'Fit for 55' package." Accessed March 12, 2022. https://etuc.org/sites/default/files/document/file/2021-03/ETUC%20resolution%20on%20Fit%20for%2055%20Package.pdf.
European Union. 1999. "Treaty of Amsterdam." Accessed June 06, 2023. https://eur-lex.europa.eu/legal-content/EN/TXT/PDF/?uri=CELEX:11997D/TXT.
European Union. 2007. "Treaty of Lisbon: Amending the Treaty on European Union and the Treaty Establishing the European Community." http://publications.europa.eu/resource/cellar/688a7a98-3110-4ffe-a6b3-8972d8445325.0007.01/DOC_19.
European Union. 2009. "Regulation (EC) No 401/2009 on the European Environment Agency and the European Environment Information and Observation Network." https://eur-lex.europa.eu/legal-content/EN/TXT/PDF/?uri=CELEX:32009R0401.
European Union. 2012. "Consolidated Version of the Treaty on the Functioning of the European Union." Accessed October 26, 2022. https://eur-lex.europa.eu/LexUriServ/LexUriServ.do?uri=CELEX:12012E/TXT:en:PDF.
European Union. 2016. "Shared Vision, Common Action: A Stronger Europe: A Global Strategy for the European Union's Foreign And Security Policy." Accessed February 12, 2022. https://eeas.europa.eu/archives/docs/top_stories/pdf/eugs_review_web.pdf.
European Union. 2021a. "Regulation (EU) 2021/1056 Establishing the Just Transition Fund." Accessed June 20, 2023. https://eur-lex.europa.eu/legal-content/EN/TXT/PDF/?uri=CELEX:32021R1056&qid=1687158942016.
European Union. 2021b. "Regulations. Establishing the framework for achieving climate neutrality and amending Regulations." Accessed June 20, 2023. https://eur-lex.europa.eu/legal-content/EN/TXT/PDF/?uri=CELEX:32021R1119.

Falkner, Robert. 2016. "The Paris Agreement and the new logic of international climate politics." *International Affairs* 92 (5): 1107–25. https://doi.org/10.1111/1468-2346.12708.
Ferguson, Peter. 2015. "The green economy agenda: business as usual or transformational discourse?" *Environmental Politics* 24 (1): 17–37. https://doi.org/10.1080/09644016.2014.919748.
Fernandes, José M. 2010. "Speech." In European Parliament (EP) 2010.
Fischer, Christine. 2020. "Der Green Deal – Europas 'Mann-auf-dem-Mond-Moment'?" https://maineuropa.de/european-green-deal-von-der-leyen-cop-25-cop25-eu-kommission-klima-klimaschutz-klimapolitik.
Fischer, Eva, and Till Hoppe. 2020. "EU-Kommission will Teile des Green Deals verschieben." *Handelsblatt*, April 16. Accessed June 15, 2023. https://www.handelsblatt.com/politik/international/coronakrise-eu-kommission-will-teile-des-green-deals-verschieben/25748146.html.
Fischer, Severin, and Oliver Geden. 2016. "Die Energie- und Klimapolitik der EU nach dem Brexit-Referendum." SWP-Aktuell 50. Accessed February 13, 2021. https://www.swp-berlin.org/publications/products/aktuell/2016A50_fischer_gdn.pdf.
Fitch-Roy, Oscar, David Benson, and David Monciardini. 2020. "Going around in circles? Conceptual recycling, patching and policy layering in the EU circular economy package." *Environmental Politics* 29 (6): 983–1003. https://doi.org/10.1080/09644016.2019.1673996.
Fletcher, Joyce K. 2004. "The paradox of postheroic leadership: An essay on gender, power, and transformational change." *The Leadership Quarterly* 15 (5): 647–61. https://doi.org/10.1016/j.leaqua.2004.07.004.
Florenz, Karl-Heinz. 2015. "Speech." In European Parliament (EP) 2015.
Forchtner, Bernhard. 2019. "Climate change and the far right." *WIREs Climate Change* 10 (5). https://doi.org/10.1002/wcc.604.
Forchtner, Bernhard, ed. 2020. *The Far Right and the Environment: Politics, Discourse and Communication.* Facism and the far right. London: Routledge Taylor & Francis Group.
Foster, Emma A. 2017. "Gender, Environmental Governmentality, and the Discourses of Sustainable Development." In MacGregor 2017, 216–28.
France, Germany, Italy. 2017. "Joint statement from Germany, Italy and France on Pres. Trump's climate decision." https://id.ambafrance.org/Joint-statement-from-Germany-Italy-and-France-on-Pres-Trump-s-climate-decision.
France24. 2015. "Rallies worldwide call for action at Paris climate summit." November 29. Accessed February 22, 2023. https://www.france24.com/en/20151129-rallies-worldwide-call-action-paris-climate-summit.
Fraser, Nancy. 2016. "Contradictions of capital and care." *New Left Review* (100): 99–117.
Fridays for Future. 2020. "Open letter to EU leaders on why their new climate law is 'surrender'." Accessed June 26, 2023. https://www.carbonbrief.org/climate-strikers-open-letter-to-eu-leaders-on-why-their-new-climate-law-is-surrender/.
Friends of the Earth Europe. 2012. "The pursuit of a 'Green economy' in Rio+20 debate." Accessed January 12, 2021. https://friendsoftheearth.eu/news/the-pursuit-of-a-green-economy-in-rio20-debate/.
Friends of the Earth Europe. 2013. "Annual Review 2012." Accessed January 12, 2022. https://friendsoftheearth.eu/wp-content/uploads/2013/06/foee_annual_review_network_2012.pdf.

Friends of the Earth Europe. 2019. "#FridaysForFuture – a new generation of People Power." It is likely that the pressure from the street with more than 1.5 million students took on the street on March 15th has further backed the EP's resolution.
Friends of the Earth Europe. 2020a. "Principle for Transformation: How the European Green Deal Can Achieve Systemic Change." Accessed June 14, 2023. https://friendsoftheearth.eu/wp-content/uploads/2020/11/PRINT-VERSION-European-Green-Deal-Principles-for-Transformation.pdf.
Friends of the Earth Europe. 2020b. "We're about to learn if Europe's Green Deal is the real deal." News release. September 9. Accessed July 10, 2023. https://friendsoftheearth.eu/news/were-about-to-learn-if-europes-green-deal-is-the-real-deal/.
Fuchs, Doris, and Berenike Feldhoff. 2016. "Passing the Scepter, not the Buck Long Arms in EU Climate Politics." *JSD* 9 (6): 58–74. https://doi.org/10.5539/jsd.v9n6p58.
Fuhr, Lili, Liane Schalatek, Katarzyna Urgryn, and Wanun Permpibul. 2013. "COP19: Warsaw." Accessed March 05, 2021. https://www.boell.de/en/2013/12/03/after-cop-19-warsaw-checkmate-international-climate-politics.
Gambús, Francesc. 2017. "Speech." In European Parliament (EP) 2017.
Gardini, Elisabetta. 2010. "Speech." In European Parliament (EP) 2010.
Geden, Oliver, and Severin Fischer. 2014. "Moving Targets. Die Verhandlungen über die Energie- und Klimapolitik-Ziele der EU nach 2020." SWP-Studie. Accessed May 14, 2021. https://www.swp-berlin.org/publications/products/studien/2014_S01_fis_gdn.pdf.
Gengnagel, Vincent, and Katharina Zimmermann. 2022. "The European Green Deal as a Moonshot – Caring for a Climate-Neutral yet Prospering Continent?" Historical Social Research Vol. 47, No. 4 (2022), Volumes per year: 1. https://doi.org/10.12759/hsr.47.2022.47.
Genovese, Andrea, and Mario Pansera. 2021. "The Circular Economy at a Crossroads: Technocratic Eco-Modernism or Convivial Technology for Social Revolution?" *Capitalism Nature Socialism* 32 (2): 95–113. https://doi.org/10.1080/10455752.2020.1763414.
Gentes, Timo, Lina Löning, and Alena Trapp. 2020. "Nähe und Distanz: Das Verhältnis zwischen Umwelt-NGOs und Fridays for Future." In *Fridays for Future – die Jugend gegen den Klimawandel: Konturen der weltweiten Protestbewegung*, edited by Sebastian Haunss and Moritz Sommer, 163–80. X-Texte zu Kultur und Gesellschaft. Bielefeld: transcript; Walter de Gruyter.
Gerbrandy, Gerben-Jan. 2013. "Speech." In European Parliament (EP) 2013.
Gerbrandy, Gerben-Jan. 2016. "Speech." In European Parliament (EP) 2016.
Germany and the European Commission. 2020. "The update of the nationally determined contribution of the European Union and its Member States." Accessed July 11, 2023. https://unfccc.int/sites/default/files/NDC/2022-06/EU_NDC_Submission_December%202020.pdf.
Gewessler, Leonore, Dan Jørgensen, Krista Mikkonen, and Sergio Costa. 2020. "Comment: We must not lose sight of the persisting climate and ecological crisis when working out how to spur the economy after the coronavirus pandemic." Accessed February 18, 2023. https://www.climatechangenews.com/2020/04/09/european-green-deal-must-central-resilient-recovery-covid-19/.
Gieg, Philipp, Timo Lowinger, Manuel Pietzko, Carolin Rüger, Manuela Scheuermann, and Anja Zürn, eds. 2023. *Jenseits der Krisen: Potenziale der europäischen Integration im 21. Jahrhundert.* Wiesbaden: Springer VS.

Goodman, James, and Ariel Salleh. 2013. "The 'Green Economy': Class Hegemony and Counter-Hegemony." *Globalizations* 10 (3): 411–24. https://doi.org/10.1080/14747731.2013.787770.
Grant, Wyn. 2017. "Business: Greening at the Edges." In Wurzel, Connelly, and Liefferink 2017, 207–20.
Green 10. 2014. "Open Letter: New Commission sidelining environment." http://awsassets.panda.org/downloads/14_089_g10_open__letter_to_president_elect_juncker_september_2014.pdf.
Green 10. 2019. "Green 10 European Parliament Election Manifesto." Accessed January 24, 2023. https://green10.org/wp-content/uploads/2017/06/GREEN-10-MANIFESTO-IN-ENGLISH-WITH-CONTACT-DATA.pdf.
Green 10, WWF European Policy Office, Transport &Environment, Bird Life Europe, Greenpeace European Unit, Friends of the Earth Europe, Clima Action Network Europe, and European Environmental Bureau. 2020. "Time to invest in a healthy future: Time for a green & just recovery." Accessed June 28, 2023. https://drive.google.com/file/d/1Vns5c-jhS65wINQuhFm5lwFNmn4Ygevw/view.
Green Growth Group. 2014. "Green Growth Group Ministers' statement on climate and energy framework for 2030." Accessed February 20, 2021. https://www.gov.uk/government/news/green-growth-group-ministers-statement-on-climate-and-energy-framework-for-2030.
Green Growth Group. 2018. "Common statement on the long-term strategy and the climate ambition of the EU." Accessed April 25, 2023. https://www.ecologie.gouv.fr/sites/default/files/2018.06.25_statement_ggg_climat.pdf.
Greenpeace Europe. 2018. "Media briefing: EU's last chance to tackle climate change." Accessed April 12, 2023. https://www.greenpeace.org/eu-unit/issues/climate-energy/1718/media-briefing-eus-last-chance-to-tackle-climate-change/.
Greenpeace Europe. 2020. "Draft climate law points to lost decade for EU action." Accessed June 27, 2023. https://www.greenpeace.org/eu-unit/issues/climate-energy/2640/draft-climate-law-points-to-lost-decade-for-eu-action/.
Greenpeace Europe, and WWF European Policy Office. 2019. "A European Climate Law." Accessed June 27, 2023. https://www.greenpeace.org/static/planet4-eu-unit-stateless/2020/01/20191120-GP-WWF-12-points-EU-climate-law.pdf.
Greubel, Johannes. 2023. "Neue Impulse für eine bürgernähere Europäische Union – Die Potenziale und Grenzen der Konferenz zur Zukunft Europas." In Gieg, Lowinger, Pietzko, Rüger, Scheuermann, and Zürn 2023, 365–87.
Groen, Lisanne, and Arne Niemann. 2013. "The European Union at the Copenhagen climate negotiations: A case of contested EU actorness and effectiveness." *International Relations* 27 (3): 308–24. https://doi.org/10.1177/0047117813497302.
Groen, Lisanne, Arne Niemann, and Sebastian Oberthür. 2012. "The EU as a Global Leader? The Copenhagen and Cancun UN Climate Change Negotiations." *Journal of Contenporary European Research* 8 (2): 173–91.
Groote, Matthias. 2013. "Speech." In European Parliament (EP) 2013.
Grubb, Michael, and Joyeeta Gupta. 2000. "Climate Change, Leadership and the EU." In *Climate Change and European Leadership: A Sustainable Role for Europe?* edited by Joyeeta Gupta and Michael Grubb, 3–14 27. Dordrecht: Springer.

Guerrina, Roberta, and Katharine A. M. Wright. 2016. "Gendering normative power Europe: lessons of the Women, Peace and Security agenda." *International Affairs* 92 (2): 293–312. https://doi.org/10.1111/1468-2346.12555.

Guteland, Jytte. 2016. "Speech." In European Parliament (EP) 2016.

Hajer, Maarten A. 1995. *The Politics of Environmental Discourse: Ecological Modernization and the Policy Process*. Oxford: Clarendon Press.

Harms, Rebecca. 2013. "Speech." In European Parliament (EP) 2013.

Hedegaard, Connie. 2010a. "Opening Statement for the European Parliament Hearing." Accessed September 08, 2022. https://carloscoelho.eu/old_cc/dossiers/comissao_barroso_ii/Connie-Hedegaard/hedegaard_speeches_en.pdf.

Hedegaard, Connie. 2010b. "Europe's view on International Climate Policy." Accessed September 04, 2022. https://ec.europa.eu/commission/presscorner/detail/en/SPEECH_10_468.

Hedegaard, Connie. 2011. "Statement at the opening of the high-level segment of COP17." Accessed September 04, 2022. https://unfccc.int/files/meetings/durban_nov_2011/statements/application/pdf/111206_cop17_hls_european_union.pdf.

Hedegaard, Connie. 2012a. "Moving forward together." Accessed September 04, 2022. https://unfccc.int/resource/docs/cop18_cmp8_hl_statements/Statement%20on%20behalf%20of%20the%20EU%2C%20its%2027%20Member%20States%20%2C%20and%20Croatia.pdf.

Hedegaard, Connie. 2012b. "Speech." In European Parliament (EP) 2012.

Hedegaard, Connie. 2013a. "Speech." In European Parliament (EP) 2013.

Hedegaard, Connie. 2013b. "Statement for Joint High-level segment of COP19 and CMP 9." Accessed September 04, 2022. https://unfccc.int/files/meetings/warsaw_nov_2013/statements/application/pdf/cop19_hls_european_union.pdf.

Heffernan, Rose, Patrizia Heidegger, Gabriele Köhler, Anke Stock, and Katy Wiese. 2021. *A feminist European green deal: Towards an ecological and gender just transition*. Bonn: Friedrich-Ebert-Stiftung.

Heidbreder, Eva G., and Daniel Schade. 2020. "Der institutionelle Wettstreit um die Spitzenkandidaten in der Europawahl 2019." In Kaeding, Müller, and Schmälter 2020, 169–79.

Hekman, Susan J. 1990. *Gender and Knowledge: Elements of a Postmodern Feminism*. Boston: Northeastern Univ. Press.

Helmer, Roger. 2015. "Speech." In European Parliament (EP) 2015.

Herrero, Alisa, and Hanne Knaepen. 2014. "Run-up to 2015: A moment of truth for EU external climate action?" Briefing Note 67. Accessed January 13, 2023. https://ecdpm.org/application/files/3216/5546/9249/BN67-Run-up-2015-moment-truth-for-EU-external-climate-action1.pdf.

Hibner, Jolanta E. 2010. "Speech." In European Parliament (EP) 2010.

Hillje, Johannes. 2020. "Nachwort." In Kaeding, Müller, and Schmälter 2020, 437–43.

Högl, Maximilian. 2019. "Kooperationsfaktoren in den Klimaverhandlungen: Eine vergleichende Analyse der Klimagipfel in Kopenhagen 2009 und in Paris 2015." *Z Außen Sicherheitspolit* 12 (3): 357–78. https://doi.org/10.1007/s12399-019-00782-9.

Hrbek, Rudolf. 2014. "Europawahl 2014: Kontinuität und neue Facetten." *integration*, 205–27. https://doi.org/10.5771/0720-5120-2014-2-205.

Hultman, Martin. 2017. "Natures of masculinities: Conecptualising industrial, ecomodern and ecological masculinities." In *Understanding climate change through gender relations*, edited by Susan Buckingham and Virginie Le Masson, 87–103. Routledge studies in hazards, disaster risk and climate change. London: Routledge.

Hultman, Martin, and Jonas Anshelm. 2017. "Masculinites of Global Climate Change: Exploring Ecomodern, Industrial and Ecological Masculinity." In *Climate Change and Gender in Rich Countries: Work, Public Policy and Action*, edited by Marjorie G. Cohen, 19–34. Routledge studies in climate, work and society. London: Routledge.

Huyer, Sophia, Mariola Acosta, Tatiana Gumucio, and Jasmin I. J. Ilham. 2020. "Can we turn the tide? Confronting gender inequality in climate policy." *Gender & Development* 28 (3): 571–91. https://doi.org/10.1080/13552074.2020.1836817.

Institute for European Environmental Policy. 2019. "2019 European elections: Towards a more sustainable Europe? Newsletter Spring 2019." https://ieep.eu/wp-content/uploads/2022/12/IEEP-Spring-2019-Newsletter.pdf.

Integovernmental Panel on Climate Change. 2007. "Climate Change 2007. Synthesis Report." Accessed March 19, 2022. https://www.ipcc.ch/site/assets/uploads/2018/02/ar4_syr_full_report.pdf.

International Carbon Action Partnership. 2022. "Emissions Trading Worldwide." Accessed January 13, 2023. https://icapcarbonaction.com/system/files/document/220408_icap_report_rz_web.pdf.

IPCC. 2018. *Global Warming of 1.5°C:* Cambridge University Press.

Irving, Simon, and Jenny Helin. 2018. "A World for Sale? An Ecofeminist Reading of Sustainable Development Discourse." *Gender, Work & Organization* 25 (3): 264–78. https://doi.org/10.1111/gwao.12196.

Jacobs, Michael. 2016. "High pressure for low emissions: How civil society created the Paris climate agreement." *Juncture* 22 (4): 314–23. https://doi.org/10.1111/j.2050-5876.2016.00881.x.

Jadot, Yannick. 2012. "Speech." In European Parliament (EP) 2012.

Jadot, Yannick. 2017. "Speech." In European Parliament (EP) 2017.

Jazłowiecka, Danuta. 2010. "Speech." In European Parliament (EP) 2010.

Jepsen, Henrik, Magnus Lundgren, Kai Monheim, and Hayley Walker, eds. 2021. *Negotiating the Paris Agreement:* Cambridge University Press.

Johnsson-Latham, Gerd. 2007. "A study on gender equality as a prerequisite for sustainable development. What we know about the extent to which women globally live in a more sustainable way than men, leave a smaller ecological footprint and cause less climate change.".

Jordan, Andrew, Dave Huitema, Harro van Asselt, Tim Rayner, and Frans Berkhout. 2011. *Climate Change Policy in the European Union:* Cambridge University Press.

Jordan Cizelj, Romana. 2010. "Speech." In European Parliament (EP) 2010.

Jørgensen, Dan. 2010. "Speech." In European Parliament (EP) 2010.

Jørgensen, Dan. 2012. "Speech." In European Parliament (EP) 2012.

Jørgensen, Dan. 2013. "Speech." In European Parliament (EP) 2013.

Jotzo, Frank, Joanna Depledge, and Harald Winkler. 2018. "US and international climate policy under President Trump." *Climate Policy* 18 (7): 813–17. https://doi.org/10.1080/14693062.2018.1490051.

Juncker, Jean-Claude. 2014. "Mission Letter: Commissioner for Climate Action and Energy." Accessed March 12, 2021. https://ec.europa.eu/archives/juncker-commission/docs/arias-canete_en.pdf.

Juncker, Jean-Claude. 2015a. "State of the Union 2015." Accessed April 03, 2021. https://ec.europa.eu/info/sites/default/files/state_of_the_union_2015_en.pdf.

Juncker, Jean-Claude. 2015b. "A climate Agreement here and now in Paris." Accessed September 04, 2022. https://ec.europa.eu/commission/presscorner/detail/en/SPEECH_15_6211.

Juncker, Jean-Claude. 2016. "State of the Union 2016: Towards a better Europe - a Europe that Protects, Empowers and Defends." Accessed April 03, 2021. https://ec.europa.eu/commission/presscorner/api/files/attachment/93600/SOTEU%20brochure%20EN.pdf.

Juncker, Jean-Claude. 2017a. "Speech." In European Parliament (EP) 2017.

Juncker, Jean-Claude. 2017b. "State of the Union Address 2017." Accessed April 03, 2021. https://ec.europa.eu/commission/presscorner/detail/en/SPEECH_17_3165.

Kaeding, Michael, Manuel Müller, and Julia Schmälter, eds. 2020. *Die Europawahl 2019: Ringen um die Zukunft Europas*. Wiesbaden: VS Verlag für Sozialwissenschaften.

Kangasluoma, Sohvi. 2020. "Drilling for the future: Gendered justifications of the Arctic fossil fuel industry." *Polar Record* 56. https://doi.org/10.1017/S003224742000042X.

Kariņš, Krišjānis. 2015. "Speech." In European Parliament (EP) 2015.

Keating, Dave. 2013. "Marcin Korolec – Persistent Pole: Profile of Poland's minister of environment." *Politico*, November 6. Accessed July 27, 2024. https://www.politico.eu/article/persistent-pole/.

Keating, Dave. 2015. "EU finalises emissions pledge for UN climate summit." *Politico*, March 6. Accessed February 17, 2023. https://www.politico.eu/article/eu-finalises-emissions-pledge-for-un%e2%80%88climate-summit/.

Keller, Ska. 2017. "Speech." In European Parliament (EP) 2017.

Kelly, Seán. 2016. "Speech." In European Parliament (EP) 2016.

Khoury, Stefanie. 2023. "A 'lifeline out of the COVID -19 crisis'? An ecofeminist critique of the European Green Deal." *Law & Policy*. https://doi.org/10.1111/lapo.12211.

King, Jennie, Lukasz Janulewicz, and Francesca Arcostanzo. 2022. "Deny, Deceive, Delay: Documenting and Responding to Climate Disinformation at COP26 and Beyond." Accessed June 08, 2023. https://www.isdglobal.org/wp-content/uploads/2022/06/Summative-Report-COP26.pdf.

Kirchherr, Julian, Reike, Denise, and Hekkert, Marko. 2017. "Conceptualizing the circular economy: An analysis of 114 definitions." In Resources, Conservation and Recycling 127, 221–232. https://doi.org/10.1016/j.resconrec.2017.09.005.

Kirchherr, Julian, Yang, Nan-Hua Nadja, Schulze-Spüntrup, Frederik, Heerink, Maarten J., and Hartley, Kris. 2023. "Conceptualizing the Circular Economy (Revisited): An Analysis of 221 Definitions." In Resources, Conservation and Recycling 194. https://doi.org/10.1016/j.resconrec.2023.107001.

Kirton-Darling, Jude. 2015. "Speech." In European Parliament (EP) 2015.

Kokkalis, Petros. 2021. "Speech." In European Parliament (EP) 2021.

Kottasová, Ivana, Ingrid Formanek, Angela Dewan, and Rachel Ramirez. 2021. "'Historic breakthrough': 20 countries say they will stop funding fossil fuel projects abroad." *CNN*, November 4. Accessed July 19, 2023. https://edition.cnn.com/2021/11/03/world/countries-agree-to-end-fossil-fuel-financing-abroad-cop26-climate/index.html.

Kreienkamp, Julia, Tom Pegram, and David Coen. 2022. "Explaining transformative change in EU climate policy: multilevel problems, policies, and politics." *Journal of European Integration* 44 (5): 731–48. https://doi.org/10.1080/07036337.2022.2072838.

Krotký, Jan. 2022. "Debating irregular migration in the European Parliament: a 'parliament without a public' or the voice of the people?" *Journal of Contemporary European Studies*, 1–15. https://doi.org/10.1080/14782804.2022.2086108.

Krzyżanowski, Michał. 2015. "Discourse analysis, policy analysis, and the borders of EU identity." *JLP* 14 (1): 110–33. https://doi.org/10.1075/jlp.14.1.06krz.

Kurze, Kristina. 2020. "Die internationale Klimapolitik nach Paris: EU-Leadership auf dem Prüfstand." *Z Außen Sicherheitspolit* 13 (4): 357–78. https://doi.org/10.1007/s12399-020-00827-4.

Kyriazi, Anna, and Joan Miró. 2023. "Towards a socially fair green transition in the EU? An analysis of the Just Transition Fund using the Multiple Streams Framework." *Comparative European Politics* 21 (1): 112–32. https://doi.org/10.1057/s41295-022-00304-6.

La Via, Giovanni. 2015. "Speech." In European Parliament (EP) 2015.

La Via, Giovanni. 2016a. "Speech." In European Parliament (EP) 2016.

La Via, Giovanni. 2016b. "Speech." In European Parliament (EP) 2016.

La Via, Giovanni. 2017. "Speech." In European Parliament (EP) 2017.

Lamb, William F., Giulio Mattioli, Sebastian Levi, J. T. Roberts, Stuart Capstick, Felix Creutzig, Jan C. Minx, Finn Müller-Hansen, Trevor Culhane, and Julia K. Steinberger. 2020. "Discourses of climate delay." *Glob. Sustain.* 3. https://doi.org/10.1017/sus.2020.13.

Langbakk, Virginija. 2012. "Foreword." In *Review of the Implementation in the EU of area K of the Beijing Platform for Action: Women and the Environment Gender Equality and Climate Change – Report*, edited by European Institute for Gender Equality (EIGE), 3. Luxembourg.

Latvia and the European Commission. 2015. "Intended Nationally Determined Contribution of the EU and its Member States." Accessed April 29, 2022. https://www4.unfccc.int/sites/ndcstaging/PublishedDocuments/European%20Union%20First/LV-03-06-EU%20INDC(Archived).pdf.

Legislative Observatory European Parliament. 2019. "Results of Vote in Parliament: Statistics - 2019/2582(RSP) | B8–0195/2019." Accessed May 22, 2023. https://oeil.secure.europarl.europa.eu/oeil/popups/sda.do?id=32471&l=en.

Leinen, Jo. 2010. "Speech." In *Debate: Outcome of the Copenhagen Summit on climate change. 20.01.2010*, edited by European Parliament (EP).

Leinen, Jo. 2015. "Speech." In European Parliament (EP) 2015.

Leinen, Jo. 2016. "Speech." In European Parliament (EP) 2016.

Leinen, Jo. 2017. "Speech." In European Parliament (EP) 2017.

Leipold, Sina. 2021. "Transforming ecological modernization 'from within' or perpetuating it? The circular economy as EU environmental policy narrative." *Environmental Politics* 30 (6): 1045–67. https://doi.org/10.1080/09644016.2020.1868863.

Lenaerts, Klaas, Simone Tagliapietra, and Guntram Wolff. 2022. "How can the European Union adapt to climate change while avoiding a new fault line?" Policy Contribution. Accessed July 24, 2023. https://www.bruegel.org/sites/default/files/2022-06/PC%2011%202022.pdf.

Leppänen, Taru, and Duncan Liefferink. 2022. "Agenda-setting, policy formulation, and the EU institutional context: The case of the Just Transition Fund." *European Policy Analysis* 8 (1): 51–67. https://doi.org/10.1002/epa2.1136.

Liefferink, Duncan, and Rüdiger K. Wurzel. 2017. "Environmental leaders and pioneers: agents of change?" *Journal of European Public Policy* 24 (7): 951–68. https://doi.org/10.1080/13501763.2016.1161657.

Liese, Peter. 2015. "Speech." In European Parliament (EP) 2015.

Liese, Peter. 2019. "Speech." In European Parliament 2019.

Littig, Beate. 2017. "Good Green Jobs for Whom? A feminist critique of the green economy." In MacGregor 2017, 318–30.

López, Javi. 2021. "Speech." In European Parliament (EP) 2021.

Lorek, Sylvia, and Doris Fuchs. 2013. "Strong sustainable consumption governance – precondition for a degrowth path?" *Journal of Cleaner Production* 38: 36–43. https://doi.org/10.1016/j.jclepro.2011.08.008.

Lütz, Susanne, Tobias Leeg, Daniel Otto, and Vincent Woyames Dreher. 2021. "Climate Policy." In *The European Union as a Global Actor*, edited by Susanne Lütz, Tobias Leeg, Daniel Otto, and Vincent Woyames Dreher, 169–214. Springer Texts in Political Science and International Relations. Cham: Springer International Publishing.

MacGregor, Sherilyn. 2010. "'Gender and climate change': from impacts to discourses." *Journal of the Indian Ocean Region* 6 (2): 223–38.

MacGregor, Sherilyn. 2014. "Only Resist: Feminist Ecological Citizenship and the Post-politics of Climate Change." *Hypatia* 29 (3): 617–33. https://doi.org/10.1111/hypa.12065.

MacGregor, Sherilyn, ed. 2017. *Routledge international handbook of gender and environment.* Routledge international handbooks. London: Routledge.

Machin, Amanda. 2019. "Changing the story? The discourse of ecological modernisation in the European Union." *Environmental Politics* 28 (2): 208–27. https://doi.org/10.1080/09644016.2019.1549780.

Magnusdottir, Gunnhildur L., and Annica Kronsell. 2015. "The (In)Visibility of Gender in Scandinavian Climate Policy-Making." *International Feminist Journal of Politics* 17 (2): 308–26. https://doi.org/10.1080/14616742.2014.896661.

Marcinkiewicz, Kamil, and Jale Tosun. 2015. "Contesting climate change: mapping the political debate in Poland." *East European Politics* 31 (2): 1–21. https://doi.org/10.1080/21599165.2015.1022648.

Marquardt, Jens. 2020. "Fridays for Future's Disruptive Potential: An Inconvenient Youth Between Moderate and Radical Ideas." *Frontiers in Communication* 5. https://doi.org/10.3389/fcomm.2020.00048.

Mathiesen, Karl. 2016. "Uncertainty on Trump climate stance overshadows UN talks." *Climate Home News*, November 9. Accessed April 16, 2021. https://www.climatechangenews.com/2016/11/09/uncertainty-over-trump-climate-stance-overshadows-un-talks/.

Mazzoncini, Renato, Marta Cañada, Jean-Marc Ollagnier, and José M. Entrecanales Domecq. 2020. "CEO letter on EU 2030 targets." Accessed July 11, 2023. https://www.corporateleadersgroup.com/system/files/documents/business-and-investor-ceo-letter-on-eu-2030-ghg.pdf.

Michael, Kavya, Manish K. Shrivastava, Arunima Hakhu, and Kavya Bajaj. 2020. "A two-step approach to integrating gender justice into mitigation policy: examples from India." *Climate Policy* 20 (7): 800–814. https://doi.org/10.1080/14693062.2019.1676688.

Modig, Silvia. 2021. "Speech." In European Parliament (EP) 2021.

Moesker, Karen, and Udo Pesch. 2022. "The just transition fund – Did the European Union learn from Europe's past transition experiences?" *Energy Research & Social Science* 91: 102750. https://doi.org/10.1016/j.erss.2022.102750.

Morgan, Jennifer. 2021. "The Power of Civil Society." In Jepsen, Lundgren, Monheim, and Walker 2021, 245–64.

Morozov, Evgeny. 2013. *To Save Everything, Click Here: Technology, Solutionism and the Urge to Fix Problems That Don't Exist.* London: Allen Lane.

Müller Gómez, Johannes, and Alina Thieme. 2020. "The Appointment of the President of the European Commission 2019: A Toothless eruopean Parliament?" In Kaeding, Müller, and Schmälter 2020, 181–90.

Müller-Brandeck-Bocquet, Gisela. 2022. *Germany and the European Union.* Cham: Springer International Publishing.

Müller-Brandeck-Bocquet, Gisela, Manuel Pietzko, and Anja Zürn. 2018. "Die Zukunft der Europäischen Union: Potenziale besser nutzen." Würzburger Jean-Monnet-Papers 1.

Müller-Brandeck-Bocquet, Gisela, and Carolin Rüger. 2015. *Die Außenpolitik der EU.* Berlin: De Gruyter Oldenbourg.

Munić, Jagoda. 2019. "EU Green Deal: fails to slam on the brakes." https://friendsoftheearth.eu/press-release/eu-green-deal-fails-to-slam-on-the-brakes/.

Nachtwey, Oliver, and Timo Seidl. 2017. "Die Ethik der Solution und der Geist des digitalen Kapitalismus." IFS WORKING PAPER 11. Accessed August 19, 2023. https://www.ifs.uni-frankfurt.de/publikationsdetails/ifs-oliver-nachtwey-und-timo-seidl-die-ethik-der-solution-und-der-geist-des-digitalen-kapitalismus.html?file=files%2FContent%2FPublikationen%2FIfS+Working+Papers%2FIfS-WP-11.pdf&fileKey=46023a8ac8ec164c8bbafc7b91b2c8bb.

Nae, Tamara-Maria, and Narcisa-Alexandra Panie. 2021. "European Green Deal: The Recovery Strategy Addressing Inequalities." *JEERBE*, 1–12. https://doi.org/10.5171/2021.887980.

Nelson, Julie A. 2012. "Is dismissing the precautionary principle the manly thing to do? Gender and the economics of climate change." Global Development And Environmetn Institute Working Paper 12–04. Accessed December 25, 2023. https://www.bu.edu/eci/files/2020/01/12-04NelsonPrecautionManly.pdf.

Nitoiu, Cristian. 2015. "Supporting the EU's Approach to Climate Change: The Discourse of the Transnational Media Within the 'Brussels Bubble'." *Journal of European Integration* 37 (5): 535–52. https://doi.org/10.1080/07036337.2015.1019879.

Nonhoff, Martin. 2006. *Politischer Diskurs und Hegemonie: Das Projekt »Soziale Marktwirtschaft«.* Bielefeld: transcript Verlag.

Nonhoff, Martin. 2010. "Hegemonieanalyse: Theorie, Methode und Forschungspraxis." In *Handbuch sozialwissenschaftliche Diskursanalyse*, edited by Reiner Keller. 4. Aufl., 299–331. Wiesbaden: VS Verlag für Sozialwissenschaften.

Obergassel, Wolfgang, Christof Arens, Lukas Hermwille, Nicolas Kreibich, Florian Mersmann, Hermann E. Ott, and Hanna Wang-Helmreich. 2016. "Good signs, bad signs. An Initial Assessment of the Marrakech Climate Conference." Accessed April 18, 2021.

https://wupperinst.org/fa/redaktion/downloads/publications/COP22_First_Assessment_en.pdf.
Oberthür, Sebastian, and Claire Dupont. 2021. "The European Union's international climate leadership: towards a grand climate strategy?" *Journal of European Public Policy* 28 (7): 1095–1114. https://doi.org/10.1080/13501763.2021.1918218.
Oberthür, Sebastian, and Lisanne Groen. 2015. "The Effectiveness Dimension of the EU's Performance in International Institutions: Toward a More Comprehensive Assessment Framework." *JCMS: Journal of Common Market Studies* 53 (6): 1319–35. https://doi.org/10.1111/jcms.12279.
Oberthür, Sebastian, and Lisanne Groen. 2017. "The European Union and the Paris Agreement: leader, mediator, or bystander?" *WIREs Clim Change* 8 (1). https://doi.org/10.1002/wcc.445.
Oberthür, Sebastian, and Lisanne Groen. 2018. "Explaining goal achievement in international negotiations: the EU and the Paris Agreement on climate change." *Journal of European Public Policy* 25 (5): 708–27. https://doi.org/10.1080/13501763.2017.1291708.
Oberthür, Sebastian, and Ingmar von Homeyer. 2023. "From emissions trading to the European Green Deal: the evolution of the climate policy mix and climate policy integration in the EU." *Journal of European Public Policy* 30 (3): 445–68. https://doi.org/10.1080/13501763.2022.2120528.
Oberthür, Sebastian, and Claire Roche Kelly. 2008. "EU Leadership in International Climate Policy: Achievements and Challenges." *The International Spectator* 43 (3): 35–50. https://doi.org/10.1080/03932720802280594.
Oels, Angela. 2005. "Rendering climate change governable: From biopower to advanced liberal government?" *Journal of Environmental Policy & Planning* 7 (3): 185–207. https://doi.org/10.1080/15239080500339661.
Optenhögel, Uwe, Marcel Humuza, and Stephan Thalhofer. 2014. "Europa hat gewählt: Kurze Analysen der Wahlen zum Europäischen Parlament 2014 aus den Büros der Friedrich-Ebert-Stiftung." In *Europa hat gewählt: Kurze Analysen der Wahlen zum Europäischen Parlament 2014 aus den Büros der Friedrich-Ebert-Stiftung*, edited by Reinhard Krumm and Anne Seyfferth. Perspektive / Friedrich-Ebert-Stiftung. Berlin.
Oroschakoff, Kalina, and Karl Mathiesen. 2020. "How the EU's Green Deal survived the coronavirus pandemic." *Politico*, December 17. Accessed June 29, 2023. https://www.politico.eu/article/how-eu-climate-change-promises-survived-the-coronavirus-plague/.
Paleari, Susanna. 2022. "The Impact of the European Green Deal on EU Environmental Policy." *The Journal of Environment & Development* 31 (2): 196–220. https://doi.org/10.1177/10704965221082222.
Pallemaerts, Marc, and Rhiannon Williams. 2007. "Climate Change: the International and European Policy Framework." In Peeters and Deketelaere 2007, 22–50.
Panayotov, Vladko. 2012. "Speech." In European Parliament (EP) 2012.
Pansardi, Pamela, and Francesco Battegazzorre. 2018. "The discursive legitimation strategies of the president of the commission: a qualitative content analysis of the State of the Union Addresses (SOTEU)." *Journal of European Integration* 40 (7): 853–71. https://doi.org/10.1080/07036337.2018.1482286.
Parker, Charles F., and Christer Karlsson. 2018. "The UN climate change negotiations and the role of the United States: assessing American leadership from Copenhagen to Paris." *Environmental Politics* 27 (3): 519–40. https://doi.org/10.1080/09644016.2018.1442388.

Parker, Charles F., Christer Karlsson, and Mattias Hjerpe. 2017. "Assessing the European Union's global climate change leadership: from Copenhagen to the Paris Agreement." *Journal of European Integration* 39 (2): 239–52. https://doi.org/10.1080/07036337.2016.1275608.

Party of European Socialists. 2014. "Towards a New Europe: PES Manifesto." Accessed March 12, 2021. https://d3n8a8pro7vhmx.cloudfront.net/partyofeuropeansocialists/pages/1/attachments/original/1397230958/110001306_PES_Manifesto_UK.pdf?1397230958.

Party of European Socialists. 2019. "A new Social Contract for Europe: Party of European Socialists 2019 Manifesto." Accessed March 14, 2021. https://pes.eu/wp-content/uploads/2022/07/PES-Manifesto-2019_EN.pdf_2063069299.pdf.

Pearson, Mitya, and Wolfgang Rüdig. 2020. "The Greens in the 2019 European elections." *Environmental Politics* 29 (2): 336–43. https://doi.org/10.1080/09644016.2019.1709252.

Peeters, Marjan, and Kurt Deketelaere, eds. 2007. *EU Climate Change Policy: The Challenge of New Regulatory Initiatives*. Cheltenham: Elgar.

Pereira, Lídia. 2021. "Speech." In European Parliament (EP) 2021.

Peterson, V. S. 2000. "Rereading Public and Private: The Dichotomy that is Not One." *SAIS Review* 20 (2): 11–29. https://doi.org/10.1353/sais.2000.0048.

Petri, Franziska, Veronika Zapletalová, and Katja Biedenkopf. 2023. "The Euroepean Parliament: a strong internal actor with external ambitions." In Rayner, Szulecki, Jordan, and Oberthür 2023, 53–67.

Pianta, Silvia, and Matthew R. Sisco. 2020. "A hot topic in hot times: how media coverage of climate change is affected by temperature abnormalities." *Environ. Res. Lett.* 15 (11). https://doi.org/10.1088/1748-9326/abb732.

Pickering, Jonathan, Jeffrey S. McGee, Tim Stephens, and Sylvia I. Karlsson-Vinkhuyzen. 2018. "The impact of the US retreat from the Paris Agreement: Kyoto revisited?" *Climate Policy* 18 (7): 818–27. https://doi.org/10.1080/14693062.2017.1412934.

Platzer, Hans-Wolfgang. 2017. "Europäische Arbeitgeber- und Wirtschaftsverbände." In *Handbuch Arbeitgeber- und Wirtschaftsverbände in Deutschland*, edited by Wolfgang Schroeder and Bernhard Weßels, 589–615. Wiesbaden: Springer Fachmedien Wiesbaden.

Plumb, Rovana. 2010. "Speech." In European Parliament (EP) 2010.

Plumwood, Val. 1993. *Feminism and the Mastery of Nature*. Reprinted 1997. Transferred to digital print 2003. Feminism for today. London: Routledge.

Politico. 2010. "Climate-change diplomacy begins at home: The EU can do a great deal to build upon the modest success for climate diplomacy in Cancún." *Politico*, December 15. Accessed March 14, 2021. https://www.politico.eu/article/climate-change-diplomacy-begins-at-home/#.

Prügl, Elisabeth. 2021. "Untenable dichotomies: de-gendering political economy." *Review of International Political Economy* 28 (2): 295–306. https://doi.org/10.1080/09692290.2020.1830834.

Rankin, Jennifer. 2011a. "A year in the hot seat: Commissioner vows to push for climate deal and praises 'balanced' agreement at Cancún." *Politico*, January 5. Accessed January 12, 2023. https://www.politico.eu/article/a-year-in-the-hot-seat/.

Rankin, Jennifer. 2011b. "Commission plans for a low-carbon future: 'Roadmap' says EU should cut greenhouse-gas emissions by up to 44% by 2030." March 8. Accessed January 12, 2023. https://www.politico.eu/article/commission-plans-for-a-low-carbon-future/.

Rankin, Jennifer. 2019. "European Green Deal to press ahead despite Polish targets opt-out." *The Guardian*, December 13. Accessed July 06, 2023. https://www.theguardian.com/environment/2019/dec/13/european-green-deal-to-press-ahead-despite-polish-targets-opt-out.

Rao, Aruna, Joanne Sandler, David Kelleher, and Carol Miller. 2016. *Gender at Work: Theory and Practice for 21st Century Organizations.* London: Routledge.

Rask, Nanna. 2022. "An intersectional reading of circular economy policies: towards just and sufficiency-driven sustainabilities." *Local Environment* 27 (10-11): 1287–1303. https://doi.org/10.1080/13549839.2022.2040467.

Rayner, Tim, and Andrew Jordan. 2016. "Climate Change Policy in the European Union." In *Oxford Research Encyclopedia of Climate Science*, edited by Tim Rayner and Andrew Jordan, 1–25: Oxford University Press.

Rayner, Tim, Kacper Szulecki, Andrew J. Jordan, and Sebastian Oberthür, eds. 2023. *Handbook on European Union Climate Change Policy and Politics.* Cheltenham: Edward Elgar.

Rayner, Tim, Kacper Szulecki, Andrew J. Jordan, and Sebastian Oberthür. 2023. "The global importance of EU climate policy: an introduction." In Rayner, Szulecki, Jordan, and Oberthür 2023, 2–21.

Rietig, Katharina, and Claire Dupont. 2021. "Presidential leadership styles and institutional capacity for climate policy integration in the European Commission." *Policy and Society* 40 (1): 19–36. https://doi.org/10.1080/14494035.2021.1936913.

Risse, Thomas. 2010. *A Community of Europeans? Transnational Identities and Public Spheres.* Ithaca, N.Y. Cornell Univ. Press.

Robert, Aline. 2014a. "Election of Donald Tusk puts EU climate position in doubt." *Euractiv*, September 9. Accessed March 12, 2021. https://www.euractiv.com/section/sustainable-dev/news/election-of-donald-tusk-puts-eu-climate-position-in-doubt/.

Robert, Aline. 2014b. "The climate: biggest loser of the new Commission?" *Euractiv*, September 12. Accessed March 12, 2021. https://www.euractiv.com/section/climate-environment/news/the-climate-biggest-loser-of-the-new-commission/.

Rodríguez Zapatero, José L. 2010. "Presentation of the programme of the Spanish Presidency." Accessed September 05, 2022. https://www.europarl.europa.eu/doceo/document/CRE-7-2010-01-20-INT-3-008_EN.html (Rodríguez Zapatero 2010 Statement Presidency's Programme, Pos. 1).

Rüger, Carolin. 2013. *Europäische Außen- und Sicherheitspolitik - (k)ein Thema für die Öffentlichkeit? Die außen- und sicherheitspolitische Rolle der EU im Blickwinkel von öffentlicher Meinung und Medien* 5. Baden-Baden: Nomos.

Sachverständigenrat für Umweltfragen. 2020. "Für eine entschlossene Umweltpolitik in Deutschland und Europa: Umweltgutachten 2020." Accessed June 27, 2023. https://www.umweltrat.de/SharedDocs/Downloads/DE/01_Umweltgutachten/2016_2020/2020_Umweltgutachten_Entschlossene_Umweltpolitik.pdf?__blob=publicationFile&v=2.

Sarkki, Simo, Alice Ludvig, Maria Nijnik, and Serhiy Kopiy. 2022. "Embracing policy paradoxes: EU's Just Transition Fund and the aim 'to leave no one behind'." *Int Environ Agreements* 22 (4): 761–92. https://doi.org/10.1007/s10784-022-09584-5.

Savaresi, Annalisa. 2016. "The Paris Agreement: a new beginning?" *Journal of Energy & Natural Resources Law* 34 (1): 16–26. https://doi.org/10.1080/02646811.2016.1133983.

Scherbarth, Susann. 2017. "Statement." In Climate Action Network 2017.

Schoenfeld, Jonas, Brendan Moore, and Viviane Gravey. 2014. "From Hedegaard to Cañete: Still a Coherent EU Voice on Climate Change?" Accessed August 14, 2022. https://psaenvironment.wordpress.com/2014/12/01/from-hedegaard-to-canete-still-a-coherent-eu-voice-on-climate-change/.

Schreurs, Miranda A. 2016. "The Paris Climate Agreement and the Three Largest Emitters: China, the United States, and the European Union." *Politics and Governance* 4 (3): 219–23.

Schreurs, Miranda A., and Yves Tiberghien. 2010. "European Union Leadership in Climate Change: Mitigation through Multilevel Reinforcement." In *Global Commons Domestic Decisions: The Comparative Politics of Climate Change*, edited by Kathryn Harrison and Lisa McIntosh Sundstrom, 23–66. Cambridge: MIT Press.

Schulz, Florence. 2020. "Timmermans promises green recovery to EU lawmakers." *Euractiv*, April 22. https://www.euractiv.com/section/energy-environment/news/timmermans-promises-green-recovery-to-eu-lawmakers/.

Schunz, Simon. 2015. "The European Union's Climate Change Diplomacy." In *The European Union as a Diplomatic Actor*, edited by Joachim A. Koops and Gjovalin Macaj, 178–201. London: Palgrave Macmillan.

Seeber, Richard. 2010. "Speech." In European Parliament (EP) 2010.

Seeber, Richard. 2012. "Speech." In European Parliament (EP) 2012.

Skjærseth, Jon B. 2017. "The Commission's Shifting Climate Leadership: From Emissions Trading to Energy Union." In Wurzel, Connelly, and Liefferink 2017, 37–51.

Spring Alliance. 2014. "Spring Alliance Letter to President Barroso and EU Commissioners." Accessed March 12, 2021. https://concordeurope.org/2014/01/13/spring-alliance-letter-to-president-barroso-and-eu-commissioners/.

Spyraki, Maria. 2017. "Speech." In European Parliament (EP) 2017.

Stam, Claire. 2018. "EU environment ministers to target 1.5C temperature rise, hints leaked paper." *Euractiv*, October 3. Accessed May 03, 2023. https://www.euractiv.com/section/climate-environment/news/eu-environment-ministers-to-target-1-5c-temperature-rise-hints-leaked-paper/.

Stam, Claire, and Florence Schulz. 2018. "Greens to push industrial and social agenda for EU elections." November 26. Accessed May 25, 2023. https://www.euractiv.com/section/eu-elections-2019/news/greens-to-push-industrial-and-social-agenda-for-eu-election/.

Stam, Claire, and Frédéric Simon. 2018. "Europe stuck on climate ambition at COP24." *Euractiv*, December 11. Accessed May 04, 2023. https://www.euractiv.com/section/climate-environment/news/europe-stuck-on-climate-ambition-at-cop24/.

Stefanini, Sara. 2016. "Marrakech primer: 5 things to know about the next climate summit." *Politico*, October 20. Accessed June 14, 2022. https://www.politico.eu/article/marrakech-primer-5-things-to-know-about-the-next-climate-summit/.

Streck, Charlotte, Paul Keenlyside, and Moritz von Unger. 2016. "The Paris Agreement: A New Beginning." *J Eur Environ Plan Law* 13 (1): 3–29. https://doi.org/10.1163/18760104-01301002.

Stuger, Olaf. 2015. "Speech." In European Parliament (EP) 2015.

Süddeutsche Zeitung. 2023. "Klimabeirat: EU sollte Emissionen schon bis 2040 drastisch senken." *Süddeutsche Zeitung*, June 15. Accessed June 20, 2023. https://www.sueddeutsche.de/politik/klima-eu-klimabeirat-2040-emissionen-1.5932896.

Tabajdi, Csaba S. 2010. "Speech." In European Parliament (EP) 2010.
Tendero, Marjorie. 2022. "The Time Has Come! the Development of the European Environmental Conscience: Evidence from the Eurobarometer Surveys from 1974 to 2020." In *The European Environmental Conscience in EU Politics: A Developing Ideology*, edited by Thomas Hörber and Gabriel Weber, 156–84. London: Routledge Taylor & Francis Group.
The Economist. 2021. "The power of lobbyists is growing in Brussels and Berlin: Europe's politics are getting more swampy." *The Economist*, May 13. Accessed August 15, 2022. https://www.economist.com/business/2021/05/13/the-power-of-lobbyists-is-growing-in-brussels-and-berlin.
Thunberg, Greta. 2019. "Speech." Accessed April 03, 2021. https://www.europarl.europa.eu/resources/library/media/20190416RES41665/20190416RES41665.pdf.
Thunberg, Greta. 2020. "Opening Statement." March 4. Accessed June 26, 2023. https://www.youtube.com/watch?v=pxh6X4CW38Y.
Tickner, J. A. 1993. "States and Markets: An Ecofeminist Perspective on International Political Economy." *International Political Science Review* 14 (1): 59–69. https://doi.org/10.1177/019251219301400104.
Tobin, Paul, and Nicole M. Schmidt. 2021. "European Union Leadership Before, During and After the Paris Conference of the Parties." In *Climate Governance Across the Globe: Pioneers, Leaders and Followers*, edited by Rüdiger Wurzel, Mikael S. Andersen, and Paul Tobin, 142–59. London: Routledge Taylor & Francis Group.
Tomaselli, M. F., R. Hajjar, A. E. Ramón-Hidalgo, and A. M. Vásquez-Fernández. 2017. "The problematic old roots of the new green economy narrative: how far can it take us in re-imagining sustainability in forestry?" *int. forest. rev.* 19 (1): 139–51. https://doi.org/10.1505/146554817822407376.
Torney, Diarmuid. 2019. "Follow the leader? Conceptualising the relationship between leaders and followers in polycentric climate governance." *Environmental Politics* 28 (1): 167–86. https://doi.org/10.1080/09644016.2019.1522029.
Trio, Wendel. 2017. "Statement." In Climate Action Network 2017.
Tröltzsch, Jenny. 2017. "Die Europäische Union." In *Handbuch globale Klimapolitik*, edited by Georg Simonis, 302–33 8672. Paderborn: Ferdinand Schöningh.
Turmes, Claude. 2016. "Speech." In European Parliament (EP) 2016.
Umweltbundesamt. 2018. "Fact Sheet: EU 2050 strategic vision 'A Clean Planet for All': Brief Summary of the European Commission proposal." Accessed April 12, 2023. https://www.umweltbundesamt.de/sites/default/files/medien/376/publikationen/eu_2050_strategic_vision_a_clean_planet_for_all.pdf.
UNFCCC. 2012. "Promoting gender balance and improving the participation of women in UNFCCC negotiations and in the representation of Parties in bodies established pursuant to the Convention or the Kyoto Protocol." https://unfccc.int/sites/default/files/cop18_gender_balance.pdf.
United Kingdom. 2021. "Statement on International Public Support for the Clean Energy Transition." Accessed July 19, 2023. https://webarchive.nationalarchives.gov.uk/ukgwa/20230313124743/https://ukcop26.org/statement-on-international-public-support-for-the-clean-energy-transition/.
United Nations Environment Programme. 2011. "Towards a green economy. Pathways to Sustainable Development and Poverty Eradication." https://sustainabledevelopment.un.org/content/documents/126GER_synthesis_en.pdf.

United Nations Framework Convention on Climate Change. 1998. "Kyoto Protocol." Accessed May 04, 2020. https://unfccc.int/resource/docs/convkp/kpeng.pdf.
United Nations Framework Convention on Climate Change. 2000. "List of Participants: Conference of the Parties 6." Accessed May 10, 2022. https://unfccc.int/sites/default/files/resource/docs/cop6/inf02.pdf.
United Nations Framework Convention on Climate Change. 2001. "List of Participants: Conference of the Parties 7." Accessed May 10, 2022. https://unfccc.int/sites/default/files/resource/docs/cop7/inf04.pdf.
United Nations Framework Convention on Climate Change. 2002. "List of Participants: Conference of the Parties 8." Accessed May 10, 2022. https://unfccc.int/sites/default/files/resource/docs/cop8/inf02.pdf.
United Nations Framework Convention on Climate Change. 2003. "List of Participants: Conference of the Parties 9." Accessed May 10, 2022. https://unfccc.int/sites/default/files/resource/docs/cop9/inf01.pdf.
United Nations Framework Convention on Climate Change. 2004. "List of Participants: Conference of the Parties 10." https://unfccc.int/sites/default/files/resource/docs/cop10/inf03.pdf.
United Nations Framework Convention on Climate Change. 2005. "List of Participants: Conference of the Paries 11." Accessed May 10, 2022. https://unfccc.int/sites/default/files/resource/docs/2005/cop11/eng/inf02p01.pdf.
United Nations Framework Convention on Climate Change. 2006. "List of Participants: Conference of the Parties 12." Accessed May 10, 2022. https://unfccc.int/sites/default/files/resource/docs/2006/cop12/eng/inf01.pdf.
United Nations Framework Convention on Climate Change. 2007. "List of Participants: Conference of the Parties 13." Accessed May 10, 2022. https://unfccc.int/sites/default/files/resource/docs/2007/cop13/eng/inf01p01.pdf.
United Nations Framework Convention on Climate Change. 2008. "List of Participants: Conference of the Parties 14." Accessed May 10, 2022. https://unfccc.int/sites/default/files/resource/docs/2008/cop14/eng/inf01p01.pdf.
United Nations Framework Convention on Climate Change. 2009. "List of Participants: Conference of the Parties 15." Accessed May 10, 2022. https://unfccc.int/sites/default/files/resource/docs/2009/cop15/eng/inf01p01.pdf.
United Nations Framework Convention on Climate Change. 2010. "List of Participants: Conference of the Parties 16." Accessed May 10, 2022. https://unfccc.int/sites/default/files/resource/docs/2010/cop16/eng/inf01p01.pdf.
United Nations Framework Convention on Climate Change. 2011. "List of Participants: Conference of the Parties 17." Accessed May 10, 2022. https://unfccc.int/sites/default/files/resource/docs/2011/cop17/eng/inf03p01.pdf.
United Nations Framework Convention on Climate Change. 2012. "List of Participants: Conference of the Parties 18." Accessed May 10, 2022. https://unfccc.int/sites/default/files/resource/docs/2012/cop18/eng/inf02.pdf.
United Nations Framework Convention on Climate Change. 2013. "List of Paricipants: Conference of the Parties 19." Accessed May 10, 2022. https://digitallibrary.un.org/record/768925?ln=en.

United Nations Framework Convention on Climate Change. 2014. "List of Participants: Conference of the Parties 20." Accessed May 10, 2022. https://unfccc.int/sites/default/files/resource/docs/2014/cop20/eng/misc01p01.pdf.

United Nations Framework Convention on Climate Change. 2015a. "List of Participants: Conference of the Parties 21." Accessed May 10, 2022. https://unfccc.int/sites/default/files/resource/docs/2015/cop21/eng/inf03p01.pdf.

United Nations Framework Convention on Climate Change. 2015b. "Decisions adopted by the Conference of the Parties." Accessed February 19, 2023. https://unfccc.int/resource/docs/2014/cop20/eng/10a01.pdf.

United Nations Framework Convention on Climate Change. 2016. "List of Participants: Conference of the Parties 22." Accessed May 10, 2022. https://unfccc.int/resource/docs/2016/cop22/eng/inf03p01.pdf.

United Nations Framework Convention on Climate Change. 2017. "List of Participants: Conference of the Parties 23." Accessed May 10, 2022. https://unfccc.int/sites/default/files/resource/inf04_0.pdf.

United Nations Framework Convention on Climate Change. 2018. "List of Participants: Conference of the Parties 24." Accessed May 10, 2022. https://unfccc.int/sites/default/files/resource/i03.pdf.

United Nations Framework Convention on Climate Change. 2019. "List of Participants: Conference of the Parties 25." Accessed May 10, 2022. https://unfccc.int/sites/default/files/resource/cp_inf4.pdf.

United Nations Framework Convention on Climate Change. 2021a. "Glasgow Climate Pact." Accessed June 28, 2022. https://unfccc.int/sites/default/files/resource/cop26_auv_2f_cover_decision.pdf.

United Nations Framework Convention on Climate Change. 2021b. "List of Participants: Conference of the Parties 26." Accessed May 10, 2022. https://unfccc.int/sites/default/files/resource/cp2021_inf03p01.pdf.

United Nations Framework Convention on Climate Change. 2023a. "INDCs as communicated by Parties." Accessed February 16, 2023. https://www4.unfccc.int/sites/submissions/indc/Submission%20Pages/submissions.aspx.

United Nations Framework Convention on Climate Change. 2023b. "The Doha Climate Gateway." Accessed January 24, 2023. https://unfccc.int/process/conferences/the-big-picture/milestones/the-doha-climate-gateway.

Unmüßig, Barbara. 2012. "Grüne Ökonomie – die neue Zauberformel?" *Vereinte Nationen* 60 (1): 3–9.

Urbansky, Frank. 2023. "CCS - nur beschränkt nutzbar, doch als Retter gepriesen." *Z Energiewirtsch* 47 (1): 14–19. https://doi.org/10.1007/s12398-023-0901-9.

van Brempt, Kathleen. 2015. "Speech." In European Parliament (EP) 2015.

van Brempt, Kathleen. 2016. "Speech." In European Parliament (EP) 2016.

van Brempt, Kathleen. 2017. "Speech." In European Parliament (EP) 2017.

van Brempt, Kathleen. 2018. "Speech." In *Debate: Strategy for long-term EU greenhouse gas emissions reduction in accordance with the Paris agreement*, edited by European Parliament (EP).

Van Schaik, Louise. 2013. "The European Union and the Climate Change Regime." In *Routledge Handbook on the European Union and International Institutions: Performance,*

Policy, Power, edited by Knud E. Jørgensen, 357–70. Routledge handbooks. London: Routledge.

Van Schaik, Louise, and Simon Schunz. 2012. "Explaining EU Activism and Impact in Global Climate Politics: Is the Union a Norm- or Interest-Driven Actor?" *JCMS: Journal of Common Market Studies* 50 (1): 169–86. https://doi.org/10.1111/j.1468-5965.2011.02214.x.

Varoufakis, Yanis, and David Adler. 2020. "The EU's green deal is a colossal exercise in greenwashing." *The Guardian*, February 7. Accessed June 10, 2023. https://www.theguardian.com/commentisfree/2020/feb/07/eu-green-deal-greenwash-ursula-von-der-leyen-climate.

Vida, Bianka. 2021. "Policy framing and resistance: Gender mainstreaming in Horizon 2020." *European Journal of Women's Studies* 28 (1): 26–41. https://doi.org/10.1177/1350506820935495.

Vidal, John, and Fiona Harvey. 2013. "Green groups walk out of UN climate talks." *The Guardian*, November 21. Accessed March 05, 2021. https://www.theguardian.com/environment/2013/nov/21/mass-walk-out-un-climate-talks-warsaw.

Visegrád Group, Republic of Bulgaria, and Romania. 2020. "Joint letter from the the Visegrad Group /V4/, Republic of Bulgaria and Romania Environment and Climate Ministers to Mr. Frans Timmermans, Executive Vice-President of the European Commission on Impact Assessment (IA) for the European Union's 2030 climate ambition and the action plan." Accessed February 18, 2023. https://www.euractiv.com/wp-content/uploads/sites/2/2020/08/V4BGRO-letter-IA-2030_20200712.pdf.

Von der Leyen, Ursula. 2019a. "A Union that strives for more My agenda for Europe." Accessed October 27, 2022. https://ec.europa.eu/info/sites/default/files/political-guidelines-next-commission_en_0.pdf.

Von der Leyen, Ursula. 2019b. "Press remarks by President von der Leyen on the occasion of the adoption of the European Green Deal Communication." Accessed November 14, 2023. https://ec.europa.eu/commission/presscorner/api/files/document/print/en/speech_19_6749/SPEECH_19_6749_EN.pdf.

Von der Leyen, Ursula. *Opening Statement*, 2020a. Accessed June 29, 2023. https://multimedia.europarl.europa.eu/fr/video/eu-coordinated-action-to-combat-the-covid-19-pandemic-and-its-consequences-opening-statements-by-council-and-commission_I189086.

Von der Leyen, Ursula. 2020b. "State of the Union 2020: Building the World We want to live in: A Union of Vitality in a World of Fragility." Accessed April 03, 2021. https://ec.europa.eu/info/sites/default/files/soteu_2020_en.pdf.

von Lucke, Franziskus, Thomas Diez, Solveig Aamodt, and Bettina Ahrens. 2021. *The EU and Global Climate Justice: Normative Power Caught in Normative Battles.* Routledge studies on the European Union and global order. London: Routledge. https://www.taylorfrancis.com/books/9781003081517.

Voßwinkel, Jan, and Götz Reichert. 2011. "EU Communication Low Carbon Economy in 2050." cepPolicy Brief. Accessed January 16, 2023. https://www.cep.eu/fileadmin/user_upload/cep.eu/Analysen_KOM/KOM_2011_112_CO2-arme_Wirtschaft/cepPolicyBrief_KOM_2011_112_Low_Carbon_Economy.pdf.

Wachsmuth, Jakob, Vicki Duscha, Johannes Eckstein, Matthias Duwe, Nick Evans, Mona Freundt, Katharina Umpfenbach, Kjell Bettgenhäuser, and Andreas Hermelink. 2022. "The European Commission's 2050 Vision 'A clean planet for all' – Implications for

Sector Strategies and Climate Governance." Accessed February 13, 2023. https://www.ecologic.eu/sites/default/files/publication/2022/3550-the-european-commissions-2050-vision-web.pdf.

Wæver, Ole. 1996. "European Security Identities." *JCMS: Journal of Common Market Studies* 34 (1): 103–32. https://doi.org/10.1111/j.1468-5965.1996.tb00562.x.

Wandler, Reiner. 2014. "'Die schlechtestmögliche Wahl'." *taz die tageszeitung*, September 15. Accessed March 12, 2021. https://taz.de/Kritik-am-neuen-EU-Klimakommissar/!5033170/.

Wanner, Thomas. 2015. "The New 'Passive Revolution' of the Green Economy and Growth Discourse: Maintaining the 'Sustainable Development' of Neoliberal Capitalism." *New Political Economy* 20 (1): 21–41. https://doi.org/10.1080/13563467.2013.866081.

Wendler, Frank. 2022a. "Contesting the European Union in a changing climate: policy narratives and the justification of supranational governance." *Journal of Contemporary European Studies* 30 (1): 67–83. https://doi.org/10.1080/14782804.2021.1882107.

Wendler, Frank. 2022b. *Framing Climate Change in the EU and US After the Paris Agreement.* Cham: Springer International Publishing AG.

Weyers, Till, Alexandra Goritz, Katharina Lange, and Kerstin Mohr. 2020. "Die Bedeutung von Klimapolitik während der Europawahl 2019: Wie das Thema die Agenda prägte." In Kaeding, Müller, and Schmälter 2020, 289–99.

Wils, Sabine. 2010. "Speech." In European Parliament (EP) 2010.

Wilson, Joanna, and Eric Chu. 2020. "The embodied politics of climate change: analysing the gendered division of environmental labour in the UK." *Environmental Politics* 29 (6): 1085–1104. https://doi.org/10.1080/09644016.2019.1629170.

Wurzel, Rüdiger, and James Connelly, eds. 2011. *The European Union as a Leader in International Climate Change Politics.* With the assistance of J. Connelly and R. Wurzel 15. Abingdon: Routledge.

Wurzel, Rüdiger, James Connelly, and J. D. Liefferink, eds. 2017. *The European Union in International Climate Change Politics: Still Taking a Lead?* London: Routledge Taylor & Francis Group.

Wurzel, Rüdiger K., James Connelly, and Elizabeth Monaghan. 2017. "Environmental NGOs: Pushing for Leadership." In Wurzel, Connelly, and Liefferink 2017, 221–36.

Wurzel, Rüdiger K., Maurizio Di Lullo, and Duncan Liefferink. 2023. "The European Council, Council and member States: jostling for influence." In Rayner, Szulecki, Jordan, and Oberthür 2023, 38–52.

Wurzel, Rüdiger K., Duncan Liefferink, and James Connelly. 2017. "Introduction. European Union Climate Leadership." In Wurzel, Connelly, and Liefferink 2017, 3–19.

WWF European Policy Office. 2011. "Make Space for Nature: Giving a Voice to Nature in the European Union." Accessed September 07, 2022. http://awsassets.panda.org/downloads/wwfepo_annualreport_2010.pdf.

WWF European Policy Office. 2012. "Policy Works! WWF in Brussels 2011." Accessed September 07, 2022. https://wwfeu.awsassets.panda.org/downloads/wwf_epo_annual_review_2011.pdf.

WWF European Policy Office. 2014. "Campaigning for Policy Wins: WWF in Brussels 2013." Accessed September 07, 2022. https://wwfeu.awsassets.panda.org/downloads/web_wwf_ar_2013.pdf.

WWF European Policy Office. 2015. "Annual Review 2014." Accessed September 05, 2022. https://wwfeu.awsassets.panda.org/downloads/wwf_epo_ar_2014_web.pdf.

WWF European Policy Office. 2016a. "Annual Review 2015." Accessed September 05, 2022. https://wwfeu.awsassets.panda.org/downloads/wwf_annual_review_2015_lowres.pdf.

WWF European Policy Office. 2016b. "EU faces credibility test on climate action." https://www.wwf.mg/?278630/EU%2Dfaces%2Dcredibility%2Dtest%2Don%2Dclimate%2Daction.

WWF European Policy Office. 2018. "The EU's long term climate strategy." Accessed April 11, 2023. https://wwfeu.awsassets.panda.org/downloads/eu_long_term_climate_strategy___wwf_position___july_2018_final.pdf.

WWF European Policy Office. 2019. "Annual Review 2018." https://wwfeu.awsassets.panda.org/downloads/wwf_epo_annual_review_2018_1.pdf.

WWF European Policy Office. 2020. "Annual Reviw 2019." Accessed September 05, 2022. https://wwfeu.awsassets.panda.org/downloads/wwf_epo_annual_review_2019.pdf.

WWF European Policy Office. 2021. "Annual Review 2020." Accessed September 05, 2022. https://wwfeu.awsassets.panda.org/downloads/wwf_epo_2020_annual_report_spread.pdf.

WWF European Policy Office. 2022. "Annual Review 2021." Accessed September 06, 2022. https://wwfeu.awsassets.panda.org/downloads/wwf_epo_2021_annual_report_spreads.pdf.

WWF European Policy Office, European Environmental Bureau, Carbon Market Watch, Bird Life, NABU, Clima Action Network Europe, Greenpeace Europe, and Transport & Environment. 2020. "The EU's 2030 climate target: NGO views on the European Commission's proposals." Accessed June 14, 2021. https://eeb.org/wp-content/uploads/2020/09/The-EU-2030-climate-target-_NGO-views.pdf.

Yamin, Farhana. 2021. "The Hig Ambition Coalition." In Jepsen, Lundgren, Monheim, and Walker 2021, 216–44.

Zeigermann, Ulrike. 2020. *Transnational Policy Entrepreneurs*. Cham: Springer International Publishing.

Zürn, Anja. 2017. "COP 23: Fortschritte zur Umsetzung des Pariser Klimaabkommens." Accessed March 12, 2021. https://maineuropa.de/cop-23-bonn-fortschritte-zur-umsetzung-des-pariser-klimaabkommens-talanoa.

Zürn, Anja. 2019. "COP 24: Die Grundsteinlegung zur Umsetzung des Pariser Klimaabkommens?" Accessed March 12, 2021. https://maineuropa.de/cop-24-kattowitz-klima-konferenz-klimakonferenz-umsetzung-des-pariser-klimaabkommens-paris.

Zürn, Anja. 2020. "From Sex and Gender to Intersectional Approaches? UN-Written Identities of Local Women in Participation and Protection Discourses." In *Gender Roles in Peace and Security*, edited by Manuela Scheuermann and Anja Zürn, 11–33. Cham: Springer International Publishing.

Zürn, Anja. 2023. "Die Europäische Union im Anthropozän – Potenziale einer gendersensiblen Klimaakteurin." In Gieg, Lowinger, Pietzko, Rüger, Scheuermann, and Zürn 2023, 229–52.

5 Conclusion

After reconstructing and discussing the climate policy identities of India and the EU, applying ecofeminist hegemony analysis, I will now scrutinise the general findings, synoptically answer the research questions and give a final outlook. This will be done in three stages.

Section 5.1 will contrast the results of the case studies of India and the EU with each other. While they exhibit major differences, they nevertheless have in common that their climate policy identities are very strongly masculinised with a similar yet not identical significance for the respective climate policy. However, this chapter should not be understood as a systematic comparison of the Indian and EU discourse and the resulting gendered identities, either in the sense of a most similar cases design, or in the sense of its inverted form, a most different case design. Rather, the juxtaposition is intended to succinctly summarise, as part of the process, exactly what answers have been discovered regarding the research questions, having been derived through the thorough in-depth analysis in Chaps. 3 and 4—and to prepare the ground for the considerations in the following two chapters.

Reflections on the theory and methodology used will be the core of Sect. 5.2. The amalgam of poststructuralist hegemony analysis and ecofeminist, normatively informed theoretical background chosen here will be evaluated in two key respects: First, the feasibility of applying this novel conceptual framework to the two cases, which were analysed here for the first time using this unique combination; second, the assumptions developed in the introduction. This will be done against the backdrop of the concise juxtaposition of both case studies in the preceding chapter, which now allows us to ask whether these case studies can serve as blueprints for analysing further climate discourses through a comprehensive hegemony analysis with an ecofeminist focus. In doing so, this study is also

A. Zürn, *Of Climate and Gender*, https://doi.org/10.1007/978-3-658-46774-6_5

located in the wider field of International Relations (IR), climate policy research and the increasing number of feminist studies—thereby reflecting my contribution to the research gaps identified in the introduction. Beyond that, I offer an insight into both the necessity for (and possible strands of) further research in the nascent field of climate policy identities examined through the lens of ecofeminist hegemony analysis.

In the final Sect., 5.3, the study will be completed. The final point for reflection concerns the implications of this study's varied findings at the bilateral and global levels. While the previous chapter, inter alia, asked from a theoretical and methodological perspective whether the cases of India and EU can serve as blueprints, Sect. 5.3 takes a more empirical direction: What is the significance of the EU's and India's gendered identities at bilateral and global levels? This includes examining—from an ecofeminist perspective critical of the prevailing economic model and the insufficient recognition of global environmental stewardship—potential starting points for a global strong sustainability approach less affected by the outdated gender dichotomies that currently define both our society and, consequently, also climate policy.

5.1 The Gendered Climate Policy Identities of India and the EU

This study has shown for India and the EU that the hegemony of notions of masculinity and femininity leads to a specific understanding by the actors what the resulting climate policy identity should look like. The preferred responses to the climate crisis are derived from this identity discursively produced and represented by the respective society. In examining both the Indian and EU discourses, the focus has been on assessing contributions from key entities such as the government, parliament and society. Diverse actors from the realms of business and non-governmental organisations have been included in the analysis. Additionally, media perspectives have been analysed in order to offer a comprehensive view of societal preferences in climate policy. While the actors shaping the Indian and EU climate discourses used both feminised and masculinised discourse patterns, the dominating structures have been revealed: Masculinised demands are clearly at the forefront, pushing economic growth through a focus on innovation, technology and the necessary political framework as the primary strategies to combat climate change (cf. Sects. 3.3 and 4.3). In other words, in both cases the actors establish a hegemony of ecological modernisation, positioning the triad of innovation, technology and economic responses as the solution to climate change, a perspective stemming from masculinised identities. It can therefore also be specified that both the Indian and the EU discourses are characterised by ecomodern masculinity (Hultman 2017b) reproducing technological responses and not seeing the solution as lying in sufficiency approaches or major changes within the economic system. It is almost an irony that the solution to a predicament spawned by technological progress is now being sought through further technological innovation (Hultman and Anshelm 2017, 26).

If we take a closer look at the findings of Chaps. 3 and 4 and go beyond the common ground of the actors India and the EU, it quickly becomes evident that the cases vary significantly in other aspects. Some of these already discussed are the respective disparate framework conditions such as the different configuration of actors shaping the respective climate policy, the fundamentally distinct institutional structures, economic situation and historical contexts. In the case of India, this study has shown that its identity changed several times during the period under investigation, from 2010 to 2021. The starting point of the analysis was the post-Copenhagen period, when it was up to the Singh government to change the old certainties of the Indian climate discourse as an advocate of the Global South demanding the Global North to live up to its historical responsibility. This identity underwent a transformation after Copenhagen, driven not only by pressure from particularly vulnerable states but also by the desire to generate co-benefits through proactive climate policies. Accordingly, India can be categorised as a state that adopted *Sustainable Development* as its identity in climate policy in the period from 2010 to 2014. The Indian identity changed with the new Prime Minister in office—albeit with one intermediate step—towards a *significant role culturally-rooted close to nature* in the specific interpretation of Modi, who portrays India as a state-that-cannot-be-ignored debating global climate policy. The Indian discourse reproduced both masculinised and feminised demands, but as in the case of the EU, the masculinised patterns are to be seen as hegemonic in predominantly shaping the identity.

Reflecting on Singh's tenure until 2014, it becomes apparent that India was indeed pursuing ecological modernisation, particularly underpinned by masculinised patterns (cf. Sect. 3.3.1). The ecofeminist lens opened up the perspective that the concept of sustainable development often portrays growth as the solution rather than the problem, especially due to society's firmly-anchored gendered values (Irving and Helin 2018, 269). While it could be argued that Modi later employed this concept as a veneer, Singh utilised SD to rationalise growth—albeit growth infused with sustainability, that is, with feminised aspects such as safeguarding vulnerable groups and non-conditional environmental protection. His administration laid the groundwork for the later substantial expansion of solar energy and other climate protection measures, which are now integral to India's contemporary, though still largely masculinised climate policy identity.

Conversely, Modi has not only challenged the limited use of feminised discourse patterns but has actively instated new ones—notably privatisation of climate protection (cf. Sects. 3.2.3 and 3.3.2), putting the responsibility of protecting the climate on women's shoulders in order to further advance his hegemonic project. Moreover, his administration has leveraged the cultural value in India of people's closeness with nature to significantly diminish its accountability for mitigation efforts. In summary, the Indian case affirms the premise that gendered patterns serve as justifications for climate policy, with formulated solutions drawing upon deeply ingrained and centuries-old gender norms. The discourse reconstruction has underscored that economic growth and the will to be recognised as a significant global actor are well supported within the Indian society and only face minimal critical scrutiny, while the notion of care (encompassing environmental concerns) is positioned in juxtaposition to economic growth—and subordinated to it.

The EU has also relied on masculinised discourse patterns to shape its identity. Admittedly, as with India, this is most likely unconscious and implicit. Nevertheless, the EU's climate policy identity, again as in the Indian case in the wake of Copenhagen 2009, has developed from playing the role of a climate policy leader towards that of leader by example, with a brief episode of this identity being contested by the newly popularised concept of the green economy. The analysis has shown that the characteristics of the leader varied depending on the EU-internal and global contexts. For example, the identity notably underwent a transformation in response to the shift in the 'radical other' when Trump declared the USA's withdrawal from the Paris Agreement (cf. Sect. 4.2.5). The result was a heightened emphasis on masculinised patterns, leaning more towards the masculinised demands for strength and leadership and less towards encouraging others to follow and emulate the positive example of the EU. While the exact gendered structures have already been discussed in detail, it shall still be evinced once more that the European Green Deal introduced some feminised discourse patterns. However, these were used, as in the case of India, mainly to support economic growth without questioning the old answers to climate change—technological innovation and its economic dissemination.

If we juxtapose the results once again, it becomes clear that in addition to the primacy of economic growth, the striving for an important role also links the two cases. But here again, the different context becomes apparent: India is not striving for a leadership role closely associated with setting an example. Unlike the EU's ambitious targets and innovative policy instruments such as the ETS or CBAM, India is concerned with assuming a significant role that emphasises the new importance of Emerging Power India as a decisive player that cannot be ignored. Even though the configurations differ slightly on closer inspection, both actors use very strongly masculinised patterns to generate identity. What's more, both actors underpin these with feminised patterns wherever they are advantageous. But the application of this dichotomised order results in an encrustation of the same—and not in a new dynamic in the fight against climate change.

5.2 Understanding Climate Policy Identities through an Ecofeminist Hegemony Analysis

Now, we will reflect the importance of comprehending these identities using an ecofeminist hegemony analysis. This aims not only to discuss this study's contribution to the research gaps identified in the introduction (cf. Sect. 1.3), but also to assess the feasibility of applying the conceptual framework to the two cases. After all, the unique combination of poststructuralist hegemony analysis and ecofeminism, complemented by the theory of subjectivation, has been used for the first time in this study. To evaluate the practicality necessitates, inter alia, identifying limitations of the study's framework, but it also involves determining the potential added value generated. Therefore, this chapter identifies future trajectories in the nascent yet dynamic research on climate policy identities that the ecofeminist hegemony analysis conducted in this study has helped to lay bare.

To elucidate the intricacies of identity construction, an exhaustive hegemony analysis was imperative and has allowed us to gain a comprehensive understanding of both the broad lines and the small nuances in the discursive negotiation of the climate identities of the EU and India. Being able to also grasp the gendered nature of the discourses, that is, the fusion of Nonhoff's hegemony analysis with Judith Butler's element of subjectivation, alongside manifold ecofeminist insights, has proven exceptionally fruitful for this undertaking. Although the first dimension of gendered discourses and thus Butler's theoretical expansion has played only a subordinate role due to the nature of the discourses, the second dimension of gendered discourses was able to be reconstructed in tremendous detail which has yielded valuable and highly nuanced insight into the actors' reasoning with regard to their climate policy. The hegemony analysis, accomplished through the reconstruction of various strategemes, has facilitated an understanding of both the content of the Indian and EU climate discourses and the strategic components employed to follow the actors' hegemonic projects. This approach has allowed for the reconstruction not only of the identity of the entities but also of the structure of the discourse and the underlying hegemony. The gendered dichotomies woven into it have been recognised and identified due to the ecofeminist theory and critique.

As previously affirmed for the empirical findings, it can also be asserted from a theoretical perspective that the conceptual framework of ecofeminist hegemony analysis is applicable to both nation states and regional organisations. Initial caveats that the ecofeminist approach was predestined in the analysis of the Indian discourse but would only work to a limited extent for the EU's case (cf. Sect. 1.2) can be negated. Rather, "the concept of dualisms is an illuminating means of understanding and deconstructing rhetorical strategies in climate politics" (Flavell 2023, 58). Furthermore, the study has shown that the theoretical derivation of two dimensions of gendered discourse patterns (cf. Sect. 2.3.3) has been important in this respect, as it has enabled the discourse of the EU to be analysed together with its implicit gendered structure. Indeed, the second gendered discourse dimension has also been thoroughly reconstructed within the Indian discourse, whereas previous case studies had suggested that the first dimension might be more common.

Undoubtedly, there were instances of tension encountered along the way. The empirical cases of India and EU have prompted nuanced considerations within this approach. India's climate discourse is metaphorically speaking almost insular, facilitating the effective implementation of the devised set of strategemes. Notwithstanding, an extension of the set proved necessary in order to grasp the specifics of the Indian discourse, incorporating the demand from the cultural discourse to be *culturally rooted*. This has been done by introducing Strategeme X: Opening the border to import a particular demand of another thematical discourse has been conceptualised as a tactical discursive device (cf. Sect. 3.2.8). In the EU, the attribution of Strategeme X transpired early in the discourse, and I have further refined it. Notably, the EU *exports* individual demands beyond the borders of its climate discourse, which requires a more frequent opening of the discourse's boundaries and thus further undermines the conditions formulated by Nonhoff for the application of his approach. This deviation is particularly evident due to the inherently more open nature of the EU climate discourse compared to that of the Indian Union. The key factor contributing to this distinction is that the EU climate policy operates within a cross-cutting policy field, aligning seamlessly with various other policy fields—a factor reinforced by the European Green Deal.

Another potential limitation in the application of the theory emerged during the course of the study period. Hegemony analysis is designed for use in liberal democracies (Nonhoff 2010, 306), assuming that through an offensive hegemonic strategy, other significant actors can be won over for one's own hegemonic project using legitimate discursive means. However, with the election of BJP politician Narendra Modi as Prime Minister in 2014, a party came to power that had already come to attention in various states due to practices restricting democracy. Indeed, India's democracy has become "increasingly deficient" (Gieg and Müller-Brandeck-Bocquet 2024), from 2018 at the latest. As the elections approached, various initiatives were observed that restricted democracy in India. In the context of our example of climate policy, first indications were visible as early as 2015 with the restriction of environmental NGOs (cf. Sect. 3.2.4) and further civil society initiatives (Lowinger et al. 2021). Democracy indices also support this assessment, stating that democracy in India is declining (Demokratiematrix 2023). In 2019, the V-Dem Institute (2021, 14) downgraded India from a democracy to an "electoral autocracy". As these new circumstances concern 'only' some of the twelve years studied here, the theoretical framework has been applied—although the results of the analysis must be interpreted against this backdrop, as done in this project.

Through the ecofeminist hegemony analysis conducted, and in spite of the highlighted challenges, I provide a comprehensive understanding of the gendered discursively constructed identities of India and the EU. Additionally, the outcome of the analysis goes beyond the reconstructed and critically discussed identities by also encompassing the comprehensive reconstruction of the climate discourse of both entities—a result that has not been presented in this form to date, either in that level of detail, or based on this novel and uniquely configured analytical framework.

Which of the additional lacunae identified in the state of research (cf. Sect. 1.3) have been addressed? Firstly, it must be emphasised that this work presents an unprecedented analysis of the Indian and EU climate discourses beyond the events of the Paris Agreement into the initial impact of the global pandemic. In addition to presenting the first ecofeminist hegemony analysis spanning twelve years of the key global climate policy actors, India and the EU, this work also contributes to developing an understanding of the EU identity in its climate policy, reconstructing in detail the debate about whether and in what form the EU is a leader, but also deconstructing further aspects of its identity. It is particularly noteworthy that the analysis succeeded in addressing the hegemonic struggles surrounding the identity formation, such as individual actors' efforts to construct green economy as the empty signifier. Furthermore, I was able to deconstruct the significant contribution of implicit gender structures to the construction and stabilisation of the EU's identity as a leader. This perspective should be seen as an innovative contribution to EU climate policy research and in particular to the scholarship on EU climate leadership.

Through the examination of the Indian case study, the research gap of (eco-) feminist engagement with India's foreign climate policy has been addressed. The study at hand is the first of its kind to scrutinise gender structures at the level of identity formation of the Indian state, going beyond studies that have explored the first dimension of gendered discourses through an ecofeminist lens.

Furthermore, by analysing the climate discourses of two different entities within one project, the gap identified by Sina Leipold and her colleagues (2019, 456) in their review of the broad field of discourse analysis of environmental and climate policy was also addressed. The added value of undertaking two case studies has already been touched upon, first, to examine the applicability of ecofeminist hegemony analysis to different entities and second, to gain insights into the pursuit of the same goals, namely ecological modernisation with sometimes very different patterns of gendered argumentation. And even if it was not part of the description of the research gaps; impressively, the coupling of the two cases was also able to show that it is not only in isolated cases that gendered patterns play an important role—which might have been the first impression when only focusing on India's climate discourse.

The last, more conceptual research gap addressed in this work is the hitherto inadequate delineation of the concept of power. While I did not undertake a comprehensive elaboration of power within the scope of this study, I contribute a crucial piece to the mosaic of the too-often only vaguely-employed concept, which more often than not seems like an empty signifier, void of any precise meaning. Through a decisive focus and detailed ecofeminist analysis, a crucial yet mostly neglected dimension of the power structures in climate policy could be analysed: Its gendered nature.

Of course, the encompassing conceptualisation of power ultimately remains an unfinished construction site to be elaborated by other scholars. Moreover, further research needs out of this study are evident in other respects as well. Firstly, other case studies exploring the gendered identity of state entities in climate policy are essential. Applying the developed analytical framework to other entities would provide a broader understanding of the significance of gendered structures discursively constructing climate identities but also trying to formulate answers on how to deal with the consequences of gendered natures. Secondly, it would be an important subsequent endeavour to focus more on how the existing alternative responses to climate change, such as sufficiency and precautionary principle (which were also an issue in the EU and Indian discourses, but could not prevail), are repeatedly hegemonically overformed, where the counter-hegemonic spaces lie and how these can be strengthened in the interest of strong sustainability. Thirdly, it seems worthwhile to comprehensively investigate the key actors identified in this project, such as the EU Climate Commissioners or prominent figures in the Indian climate discourse, focusing particularly on the first dimension of gendered discourses. For instance, building on my findings, one logical next step would be to examine to what extent Arias Cañete can be considered a representative of "petro-masculinity" (Daggett 2018) or "industrial masculinity" (Hultman 2017a; Hultman and Pulé 2019) and whether this was reflected in his policies supporting fossil fuels. Furthermore, as already indicated in Sect. 4.2.7, the Conference on the Future of Europe (CoFE) could provide an excellent case study. Analysing the CoFE process and its recommendations (Greubel 2023) in the framework of an ecofeminist hegemony analysis would certainly promote an understanding of the role of the public in shaping the EU's climate identity and the gendered patterns at play. What's more, this approach could further explore the interplay between the first and second dimensions of gendered discourses—further elucidating the potential lying in the nascent field of ecofeminist hegemony research.

Sherilyn MacGregor (2010, 229) generally identified a great need to explore the "interrogation of how (…) the politics of climate change are gendered" as an "important addition to—and perhaps as a corrective to—the existing 'gender impacts' research"; this gap that existed until the present project was addressed by analysing both climate identities of the EU and India. These two case studies can serve as an important impetus to systematically deal with this form of gender structures in climate policy and identity. After all, they can be understood as a mandate to close this gap also with regard to other important climate policy actors. This of course raises the question: Can the two case studies serve as a blueprint for further investigation of different actors in global climate policy? As the analysis has shown, gender-specific structures play a significant role in the discursively produced identity of both actors. This general finding can also be applied to other states and regional organisations; after all, no democratic entity can create its identity outside of society or in disregard of its values, and given that we still live in a time of patriarchy, gender is a veritable part of these structures. However, the analysis has also shown the subtle differences that exist. From the examination of a nation state from the Global South and a regional organisation from the Global North in the study at hand, we know that the context of the respective identity construction plays an important role, so that further studies must follow to generate a more comprehensive picture.

For example, the dichotomies that characterise society and its discourse are sometimes gendered differently from what ecofeminism assumes based on well-established Western structures from the time of enlightenment, as the cultural reference in the Indian discourse has exemplified. As already pointed out in Sect. 3.3.2, this example necessitated an adoption of the theoretical foundation. Such an exercise does not limit the suitability of the conceptual framework developed in this study; it does, however, call for prudent application and constant scrutiny of the implicit or explicit theoretical or methodological assumptions, which must not be blindly followed without regard to the circumstances of the case. Moreover, when applying the analytical framework, further studies must consider the overall features of the respective political system, because to employ the set of strategemes, the entity under investigation must be largely democratic, as already mentioned above. If not, it must be questioned whether, for example, a 'critical mass' is actually necessary in order to support a hegemonic project or whether an autocratic state can enforce its respective interests in other ways—further aspects a circumspect ecofeminist hegemony analysis must take into account when building on the novel and innovative theoretical and methodological framework developed in this study. With this framework, the project at hand—operating at the intersection of poststructuralist discourse analysis, ecofeminism and International Relations (IR)—has provided a valuable contribution to the field of poststructuralist foreign policy analysis which may inspire further research.

5.3 Implications of Gendered Climate Policy Identities at Bilateral and Global Levels

The final chapter now turns to the implications of this study's varied findings at bilateral and global levels, for instance regarding the global discourse of climate policy or the India-EU Strategic Partnership.

The findings of the study underscore the presence of strongly gendered identities that underpin and justify the respective climate policies. Given the importance of the third and fourth greatest emitters, India and the EU, in global climate affairs, the structuring of their climate policies holds profound implications for global climate governance. The negotiations within the UNFCCC not only impact the discourse at the global level (Aykut and Maertens 2021) but also exert a reciprocal influence, shaping the climate discourse of individual actors, as multiple times illustrated through the cases of India and the EU in this study.

Crucially, the identities of India and the EU do not emerge in isolation but are shaped in dialogue and in response to third parties—including the 'radical other'—who are integral to the group negotiating global climate governance. The results of the project highlight that for the EU to maintain its leadership role, it is imperative to assert its concerns in global negotiations while also inspiring others to follow its example as a "normative power" (Manners 2002). Simultaneously, it became evident that the EU adopts and supports demands from partners to achieve its own objectives in line with its identity. This was exemplified when the EU embraced the 1.5-degree target advocated by the Marshall Islands and other small island states, establishing a robust alliance within the High Ambition Coalition (cf. Sect. 4.2.4). It was through such collaborative efforts that the EU solidified its identity as a leader within the framework of the Paris Agreement.

Similarly, India relies on the resonance of global climate negotiations for its identity construction, as exemplified by its impactful role in altering the concluding text during COP26 in 2021, emphasising its significant part in the process (cf. Sect. 3.2.7).

Addressing the implications for global climate governance, the gendered identity, with its unique nuances, plays a pivotal role in determining the global demands put forth by these actors, and thus in co-determining the global discourse. Over time, shifts in the gendered identity, such as India's transition from emphasising historical responsibility to actively engaging in renewable energy, signal potential alterations in global responses to climate change. As gendered identities evolve, moving away from the prevailing emphasis on ecological modernisation that predominantly equates to economic growth, alternative global responses to climate change become conceivable. Notably, feminised responses, including loss and damage or general adaptation, often face challenges to be given the necessary attention compared to mitigation through technological innovation on the global stage. Observing the trajectory of the loss and damage initiative reveals the struggle for recognition and political action. The initiative, however, has not historically held priority among powerful states, prompting further inquiry into whether states prioritising it possess a differently gendered identity.

If the identities of the actors India and EU underwent changes, prioritising the idea of caring for environment and climate, the global emphasis in formulating responses to the climate crisis could also shift. Given that both entities are inclined to generate recognition and to some extent followers, there are potential avenues here to transform global responses to the climate crisis towards strong sustainability.

This capability is highly relevant for the somewhat cumbersome Strategic Partnership between the two entities. There exists a clear untapped potential, as India and the EU could, based on the insights gained in the study, contribute not only to greater mutual understanding but also to fruitful and far-reaching climate cooperation. The realisation emerges that the EU links its identity with the nodal point to *lead by example*, and India demands to be recognised as a *significant actor culturally-rooted close to nature*, forms a duo capable of generating high impact regarding both the global high table and the bilateral cooperation. To add one more potential strand of future research to the issues already presented in the previous chapter: Future studies could explore how exactly respective gendered identities in climate policy influence pertinent bilateral cooperation between two actors beyond the often-discussed aspect of the 'radical other'. In the case of the India-EU Strategic Partnership, such research could shed light on the role of climate policy in the overall framework of bilateral relations. Indeed, this policy undoubtedly harbours positive potential for closer and more substantial bilateral relations; at the same time, however, it also has the potential to serve as (yet another) roadblock for the partnership—as has been proven several times in the past (Jörgensen 2021, 187–88). At this point in this study, it can certainly be established that if India's and the EU's identities develop even more strongly in the direction of cooperation, climate policy could indeed emerge as a significant cornerstone of the India-EU Strategic Partnership (Jayaram 2021, 221–22; Jörgensen 2021, 195)—which undoubtedly is in dire need of new impulses (Müller-Brandeck-Bocquet et al. 2021, 3–4). But regarding the current state of their identities, there is of course still a danger that the traditional responses to the climate catastrophe will continue to be formulated based on gender stereotypes, so the need to change this is even greater. The ways in which this could be achieved within the gendered dichotomous structuring of discourses and identity construction are elucidated towards the conclusion of this chapter.

Even though there are potential points of intervention suggesting that global climate policy could transform towards greater sustainability due to changing or dissolving gender stereotypes, the current observable trend is the stabilisation of the gendered order. This is evident, for instance, in the recent developments at COP28 held in Dubai in November/December 2023, where the prospective phase-out of fossil fuel combustion did not materialise. Here we can see how inter alia gendered identities influence global climate governance.

A further aspect of the significance of the results for global climate governance lies in the interaction between the actors India and the EU and their respective 'radical other'. The construction of each identity occurs in contrast to this radical other, creating a dynamic interplay that can influence the global stage. India's radical other includes both industrialised states and China, whose policies can serve as stimuli for India's climate policy identity. The EU's radical other is the USA, and recent developments illustrate how the EU's identity adapts in response to stimuli from across the Atlantic. US President Joe Biden's initiation of the Inflation Reduction Act, which is actually a green growth package with substantial implications completely outshining the EGD, placed considerable economic and climate policy pressure on the EU (Kleimann et al. 2023). Consequently, the EU responded by sharpening its own existing programmes, though not to the same extent (Scheinert 2023). Nevertheless, this instance highlights the consolidation of the EU's identity, persistently emphasising green innovation in pursuit of leadership. The question of whether the EU's identity will undergo further transformations remains uncertain, as does the trajectory for India. However, this analysis demonstrates that any changes in gendered identity hold significance for global climate governance. The specific mechanisms through which alterations in gendered identity could be brought about warrant consideration. Normatively, it is evident that the current configurations of both identities neither account for gender equality nor ensure strong sustainability.

While the study unequivocally illustrates that the climate policy decisions of actors are intricately linked to an implicit gendered order, this structure significantly influences which responses to climate change are deemed appropriate. The analysis of the cases of India and the EU underscores that the formulated approaches are primarily masculinised, leading to a depoliticised climate discourse and, beyond that, many untapped potentials lying idle. A substantial portion of other (ecofeminist) solutions to the climate crisis, such as a social order based on care (MacGregor, Arora-Jonsson, and Cohen 2022; Bauhardt 2019), strong sustainability within planetary boundaries (Neumayer 2013) or alternative economic forms that put the common good balanced with nature at the centre (Alber, Cahoon, and Röhr 2017; McAfee 2016, 347–48; Fukuda-Parr, Heintz, and Seguino 2013; Wichterich 2012; Paech 2012; Escobar 2010), are therefore either not being discussed or not earnestly considered. This became clear, for example, when non-hegemonic actors in the EU discourse called for increased precautionary measures and more engagement in the global loss and damage initiative. Similar results were seen in the Indian discourse when the ENGOs and the opposition failed to push through climate protection in tribal areas instead of environmentally destructive economic projects. This situation could prompt us to call for the promotion and implementation of feminised aspects instead of masculinised. But through exclusively feminised measures certain perspectives on climate change, such as technological ones, would then be overlooked— perspectives that also have their justification as long as they are not pursued exclusively and suppress other points of view. Furthermore, this study has affirmed the findings of other researchers (Alston 2014) that introducing feminised patterns into the discourse without explicitly addressing and dismantling the underlying power structures would render them merely supportive of the neoliberal agenda.

Therefore, the study's conclusion is unmistakable: The system of the discursively produced dichotomies must be transcended and disrupted (Irving and Helin 2018, 275), as otherwise we are "unable to envision more than two opposing alternatives " (Peterson 2000, 11,12). Above all, the dichotomy of economy and care must be dissolved—while repoliticising the climate discourse (MacGregor 2014, 629) and ensuring that all potential solutions to the climate catastrophe are adequately deliberated on and the results taken into account. Current climate policy starkly demonstrates that limiting ourselves primarily to masculinised responses based on green growth and technological progress is highly insufficient for the purpose of meeting the 1.5-degree target. There are already numerous ideas and approaches that go beyond conventional ecological modernisation strategies, such as the valorisation of care work in comparison to the environment (Fraser 2023), or a Green New Deal which prioritises the common good (European Environmental Bureau and Women Engage for a Common Future 2021; Cohen and MacGregor 2020).

However, these can only become part of the climate policy via an actor's identity if these solutions are recognised, conceivable and feasible within our societal norms. The long-established dichotomies must be overcome. This is complicated but not utopian. Both my own and other studies have shown that these gendered "categories are always culturally specific and change over time" (Kangasluoma 2020, 3). This is precisely where the untapped potential of applying ecofeminist thought lies in practice. To tap this practical potential, it is vital to significantly empower those actors engaged in counter-hegemonic projects, who are presently marginalised by the dominant actors and their hegemonic projects.

Overall, a new narrative that goes beyond the idea of combating climate change via technological innovation, rethinks power and prioritises justice and care, is necessary. No less than a shift of the global economic order is needed (Dankelman and Naidu 2020, 450). It is imperative to overcome androcentrism and patriarchy to successfully address the global climate crisis via strong sustainability—and to achieve justice and 'The Good Life for All'.

References

Alber, Gotelind, Kate Cahoon, and Ulrike Röhr. 2017. "Gender and urban climate change policy: Tackling cross-cutting issues towards equitable, sustainable cities." In Buckingham and Le Masson 2017, 64–86.

Alston, Margaret. 2014. "Gender mainstreaming and climate change." *Women's Studies International Forum* 47: 287–94. https://doi.org/10.1016/j.wsif.2013.01.016.

Aykut, Stefan C., and Lucile Maertens. 2021. "The climatization of global politics: introduction to the special issue." *International Politics* 58 (4): 501–18. https://doi.org/10.1057/s41311-021-00325-0.

Bauhardt, Christine. 2019. "Nature, Care and Gender: Feminist Dilemmas." In *Feminist Political Ecology and the Economics of Care: In Search of Economic Alternatives*, edited by Christine Bauhardt and Wendy Harcourt, 16–35. Routledge Studies in Ecological Economics Ser. Milton: Routledge.

Buckingham, Susan, and Virginie Le Masson, eds. 2017. *Understanding climate change through gender relations*. Routledge studies in hazards, disaster risk and climate change. London: Routledge.

Cohen, Maeve, and Sherilyn MacGregor. 2020. "Towards a Feminist Green New Deal for the UK." Accessed January 04, 2023. https://genderclimatetracker.org/sites/default/files/Resources/FEM-GND-POLICY-PAPER-2.pdf.

Daggett, Cara. 2018. "Petro-masculinity: Fossil Fuels and Authoritarian Desire." *Millennium* 47 (1): 25–44. https://doi.org/10.1177/0305829818775817.

Dankelman, Irene, and Kavita Naidu. 2020. "Introduction: Gender, development, and the climate crisis." *Gender & Development* 28 (3): 447–57. https://doi.org/10.1080/13552074.2020.1843830.

Demokratiematrix. 2023. "Ranking der Länger anhand der Demokratiequalität." https://www.demokratiematrix.de/ranking.

Escobar, Arturo. 2010. "Latin America at a Crossroads." *Cultural Studies* 24 (1): 1–65. https://doi.org/10.1080/09502380903424208.

European Environmental Bureau, and Women Engage for a Common Future. 2021. "Why the European Green Deal Needs Ecofoeminism: Moving from gender-blind to gender-transformatie environmental policies." https://eeb.org/wp-content/uploads/2021/07/Report-16.pdf.

Flavell, Joanna. 2023. *Mainstreaming Gender in Global Climate Governance*. London: Routledge.

Fraser, Nancy. 2023. *Cannibal capitalism: How our system is devouring democracy, care, and the planet - and what we can do about it*. London: Verso.

Fukuda-Parr, Sakiko, James Heintz, and Stephanie Seguino. 2013. "Critical Perspectives on Financial and Economic Crises: Heterodox Macroeconomics Meets Feminist Economics." *Feminist Economics* 19 (3): 4–31. https://doi.org/10.1080/13545701.2013.806990.

Gieg, Philipp, Timo Lowinger, Manuel Pietzko, Anja Zürn, Ummu S. Bava, and Gisela Müller-Brandeck-Bocquet, eds. 2021. *EU-India Relations*. Cham: Springer International Publishing.

Gieg, Philipp, and Gisela Müller-Brandeck-Bocquet. 2024. "Navigating Norms and Pragmatism in EU-India Relations. Multilateralism, Connectivity and the Future of the Strategic Partnership." In *The European Union's Geopolitics. The Lackluster World Power*, edited by Mathias Jopp and Johannes Pollak. Cham: Springer.

Greubel, Johannes. 2023. "Neue Impulse für eine bürgernähere Europäische Union – Die Potenziale und Grenzen der Konferenz zur Zukunft Europas." In *Jenseits der Krisen: Potenziale der europäischen Integration im 21. Jahrhundert*, edited by Philipp Gieg, Timo Lowinger, Manuel Pietzko, Carolin Rüger, Manuela Scheuermann, and Anja Zürn, 365–87. Wiesbaden: Springer VS.

Hultman, Maritin, and Paul M. Pulé. 2019. *Ecological Masculinities: Theoretical foundations and practical guidance.* London: Routledge.

Hultman, Martin. 2017a. "Exploring Industrial, Ecomodern, and Ecological Masculinities." In *Routledge international handbook of gender and environment*, edited by Sherilyn MacGregor, 239–52. Routledge international handbooks. London: Routledge.

Hultman, Martin. 2017b. "Natures of masculinities: Conecptualising industrial, ecomodern and ecological masculinities." In Buckingham and Le Masson 2017, 87–103.

Hultman, Martin, and Jonas Anshelm. 2017. "Masculinites of Global Climate Change: Exploring Ecomodern, Industrial and Ecological Masculinity." In *Climate Change and Gender in Rich Countries: Work, Public Policy and Action*, edited by Marjorie G. Cohen, 19–34. Routledge studies in climate, work and society. London: Routledge.

Irving, Simon, and Jenny Helin. 2018. "A World for Sale? An Ecofeminist Reading of Sustainable Development Discourse." *Gender, Work & Organization* 25 (3): 264–78. https://doi.org/10.1111/gwao.12196.

Jayaram, Dhanasree. 2021. "India's Climate Diplomacy Towards the EU: From Copenhagen to Paris and Beyond." In Gieg, Lowinger, Pietzko, Zürn, Bava, and Müller-Brandeck-Bocquet 2021, 201–26.

Jörgensen, Kirsten. 2021. "Low-Carbon Development: An Idea Whose Time Has Come - Unlocking Climate Cooperation Between India and the EU." In Gieg, Lowinger, Pietzko, Zürn, Bava, and Müller-Brandeck-Bocquet 2021, 185–99.

Kangasluoma, Sohvi. 2020. "Drilling for the future: Gendered justifications of the Arctic fossil fuel industry." *Polar Record* 56. https://doi.org/10.1017/S003224742000042X.

Kleimann, David, Niclas Poitiers, André Sapir, Simone Tagliapietra, Nicolas Véron, Reinhilde Veugelers, and Jeromin Zettelmeyer. 2023. "Green tech race? The US Inflation Reduction Act and the EU Net Zero Industry Act." *World Economy* 46 (12): 3420–34. https://doi.org/10.1111/twec.13469.

Leipold, Sina, Peter H. Feindt, Georg Winkel, and Reiner Keller. 2019. "Discourse analysis of environmental policy revisited: traditions, trends, perspectives." *Journal of Environmental Policy & Planning* 21 (5): 445–63. https://doi.org/10.1080/1523908X.2019.1660462.

Lowinger, Timo, Anja Zürn, Philipp Gieg, and Manuel Pietzko. 2021. "Norm Contestation in EU Strategic Partnerships: The Cases of Civil Society Involvement and Climate Justice in EU–India Relations." In Gieg, Lowinger, Pietzko, Zürn, Bava, and Müller-Brandeck-Bocquet 2021, 31–66.

MacGregor, Sherilyn. 2010. "'Gender and climate change': from impacts to discourses." *Journal of the Indian Ocean Region* 6 (2): 223–38.

MacGregor, Sherilyn. 2014. "Only Resist: Feminist Ecological Citizenship and the Post-politics of Climate Change." *Hypatia* 29 (3): 617–33. https://doi.org/10.1111/hypa.12065.

MacGregor, Sherilyn, Seema Arora-Jonsson, and Maeve Cohen. 2022. "Caring in a changing climate: Centering care work in climate action." Research Backrounder. Accessed May 22, 2022. https://oxfamilibrary.openrepository.com/bitstream/handle/10546/621353/rr-caring-in-a-changing-climate-230222-en.pdf;jsessionid=54336C49172B262B67BE866482CEC112?sequence=1.

Manners, Ian. 2002. "Normative Power Europe: A Contradiction in Terms?" *JCMS: Journal of Common Market Studies* 40 (2): 235–58. https://doi.org/10.1111/1468-5965.00353.

McAfee, Kathleen. 2016. "Green economy and carbon markets for conservation and development: a critical view." *Int Environ Agreements* 16 (3): 333–53. https://doi.org/10.1007/s10784-015-9295-4.

Müller-Brandeck-Bocquet, Gisela, Philipp Gieg, Timo Lowinger, Manuel Pietzko, and Anja Zürn. 2021. "India and the European Union: A Growing Responsibility to Cooperate in a Changing World." In Gieg, Lowinger, Pietzko, Zürn, Bava, and Müller-Brandeck-Bocquet 2021, 3–16.

Neumayer, Eric. 2013. *Weak Versus Strong Sustainability: Exploring the Limits of Two Opposing Paradigms.* Fourth edition. Cheltenham: Edward Elgar.

Nonhoff, Martin. 2010. "Hegemonieanalyse: Theorie, Methode und Forschungspraxis." In *Handbuch sozialwissenschaftliche Diskursanalyse*, edited by Reiner Keller. 4. Aufl., 299–331. Wiesbaden: VS Verlag für Sozialwissenschaften.

Paech, Niko. 2012. *Liberation from excess:* oekom verlag.

Peterson, V. S. 2000. "Rereading Public and Private: The Dichotomy that is Not One." *SAIS Review* 20 (2): 11–29. https://doi.org/10.1353/sais.2000.0048.

Scheinert, Christian. 2023. "EU's response to the US Inflation Reduction Act (IRA)." In-Depth Analysis. Accessed January 06, 2024. https://www.europarl.europa.eu/RegData/etudes/IDAN/2023/740087/IPOL_IDA(2023)740087_EN.pdf.

V-Dem Institute. 2021. "Autocratization Turns Viral Democracy Report 2021." Accessed January 04, 2014. https://www.v-dem.net/documents/12/dr_2021.pdf.

Wichterich, Christa. 2012. "The future we want: A feminist perspective." Accessed January 04, 2024. https://www.boell.de/sites/default/files/endf_the_future_we_want.pdf.

SPRINGER NATURE

GPSR Compliance

The European Union's (EU) General Product Safety Regulation (GPSR) is a set of rules that requires consumer products to be safe and our obligations to ensure this.

If you have any concerns about our products, you can contact us on ProductSafety@springernature.com

In case Publisher is established outside the EU, the EU authorized representative is:

Springer Nature Customer Service Center GmbH
Europaplatz 3
69115 Heidelberg, Germany

The manufacturer's authorised representative in the EU is Springer Nature Customer Service Centre GmbH, Europaplatz 3, 69115 Heidelberg, Germany. If you have any concerns regarding our products, please contact ProductSafety@springernature.com

Printed and bound by CPI Group (UK) Ltd, Croydon, CR0 4YY
30/09/2026
02225601-0007